D1292534

Crop Production

Fifth Edition

James J. Vorst
Department of Agronomy
Purdue University

ISBN 0-87563-838-4

Published by

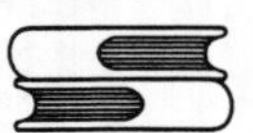

Stipes Publishing L.L.C.
204 W. University Ave.
Champaign, IL 61820

The contributions of Bruce Erickson, Instructor of Agronomy, in developing, revising and updating some of the laboratory exercises are sincerely appreciated.

CONTENTS

TOPIC	PAGES

CROPPING SYSTEMS AND CROP PRODUCTION

I. SCOPE OF U.S. AGRICULTURAL PRODUCTION

Objectives

1. Describe how land in the U.S. is presently used.
2. Recognize U.S. land use changes that have occurred since 1982.
3. Describe national trends in conservation tillage, cropland erosion, ag chemical expenditures, pesticide usage, and fertilizer usage.
4. List the average yield and production for the 4 major U.S. crops.

United States Land Use, and Crop Production Factors

The United States has a land area of 2,265 million acres. Slightly less than one-half is used for agriculture (Fig. 1). Aproximately 382 million acres is used for cropland, whereas permanent pasture and range occupy nearly 520 million acres. Forests cover most of the non-agricultural land.

The use of land is a function of both ownership and land quality. Much of the land is mountainous, desert or swampland, which was never claimed by private owners and remained in the public domain. Nearly all land used for crop production is privately owned.

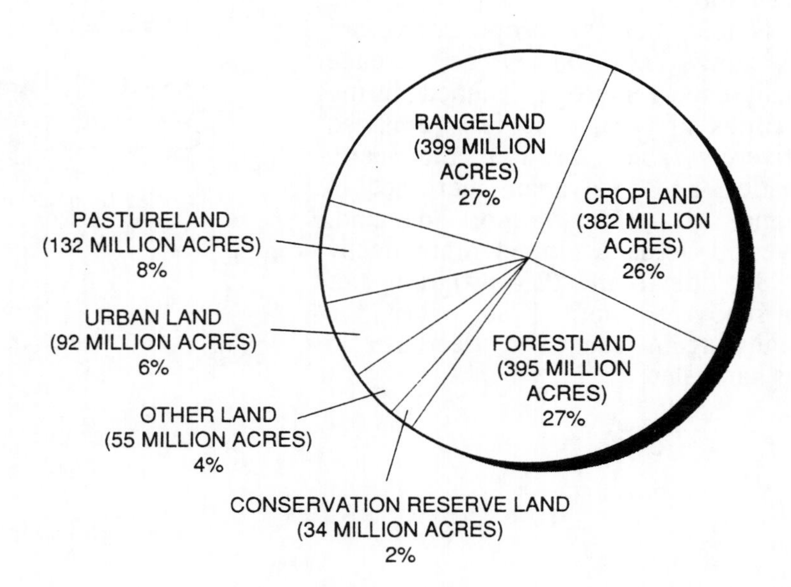

Fig. 1. Major uses of non-federal land in the United States, 1992

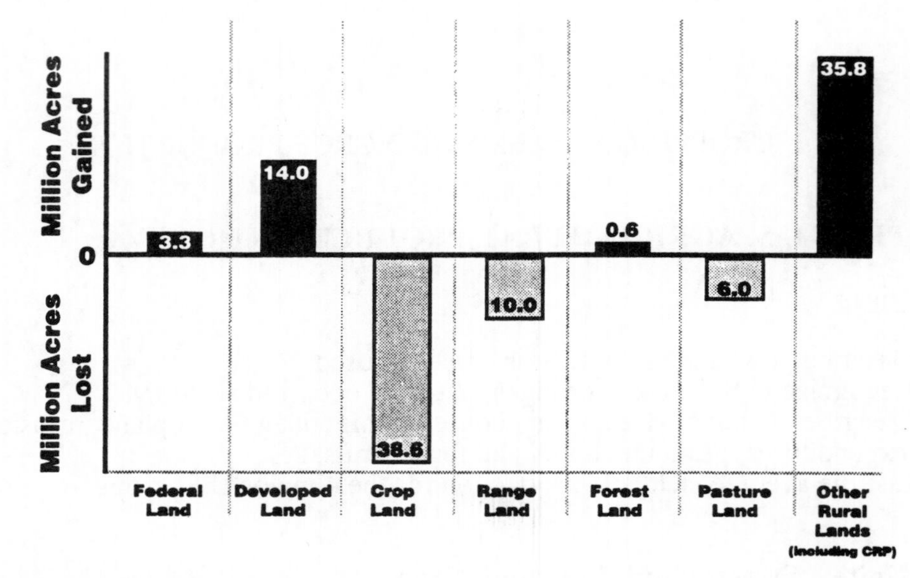

Fig. 2. Land Use Changes 1982-1992.

The amount of land available for agriculture has been decreasing, primarily due to development and the Conservation Reserve Program (CRP).

Table 1. United States Crop Production, 1997

Crop	Acres harvested	Yield per acre
Corn for grain	73.7 million	127 bu.
All wheat	71.0	39.7
Soybeans	70.8	39.,0
All hay	60.8	2.5 tons

Each year, part of the Nation's cropland is used for crops, part for pasture, and part is idle, but the proportions vary significantly (Fig. 3). The acreage actually used for crops declined during the fifties and early sixties, remained relatively stable through the early seventies, and then again increased in response to strong export demand. Harvested acreage dipped dramatically in 1983, due to the PIK program, but then recovered until 1988, when the drought again caused a decrease in acres harvested.

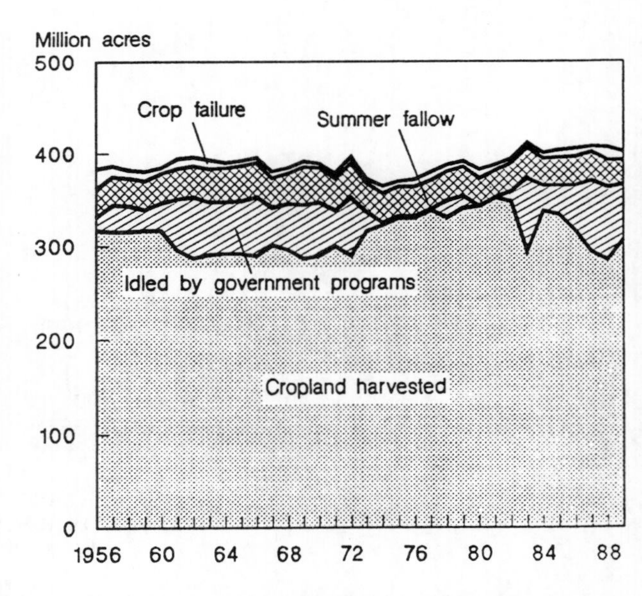

Fig. 3. Major Uses of Cropland

Conservation tillage practices are becoming more prevalent. No-till, a major type of conservation tillage, has increased greatly during the 1980s and 1990s (Fig. 4). The increase in conservation tillage has reduced soil erosion on cropland by about 30% since 1982.

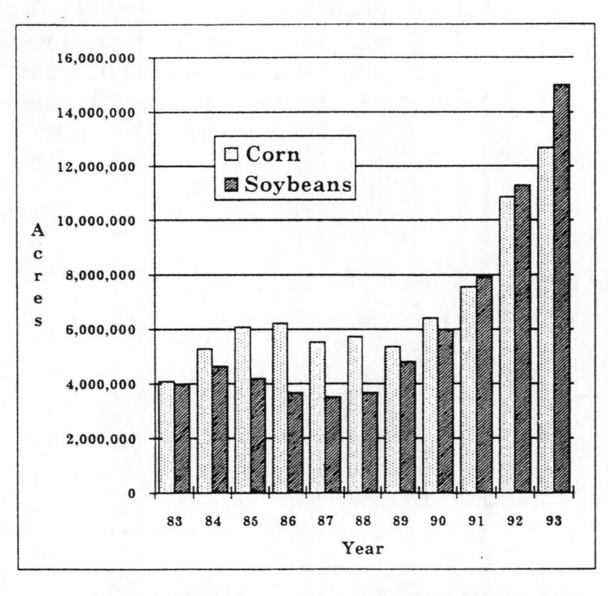

Fig. 4. National no-till adoption trend for corn and soybeans.

Conservation tillage practices reduce input costs by reducing the need for high horsepower tractors and primary tillage equipment. This has led to a reduction in machinery expenditures (Fig. 5).

Total ag chemical expenditures (pesticides and fertilizers) peaked in 1980, and have been declining since then (Fig. 5). However, the rate of pesticides applied per acre has remained steady since 1976, at about 2.2 pounds of active ingredients per acre.

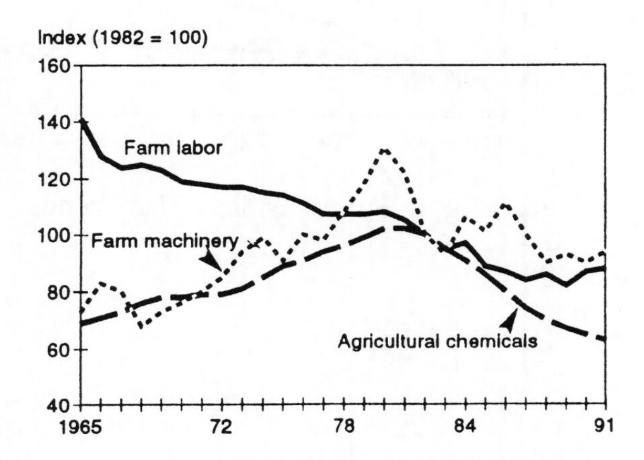

Fig. 5. Expenditures for ag chemicals, machinery, and labor

Fertilizer use is another area where reductions have occurred over the past decade (Fig. 6). Nitrogen use fluctuates widely due to acreage and commodity price fluctuations. The general trend in phosphate and potash use has been steady since the mid 1980s.

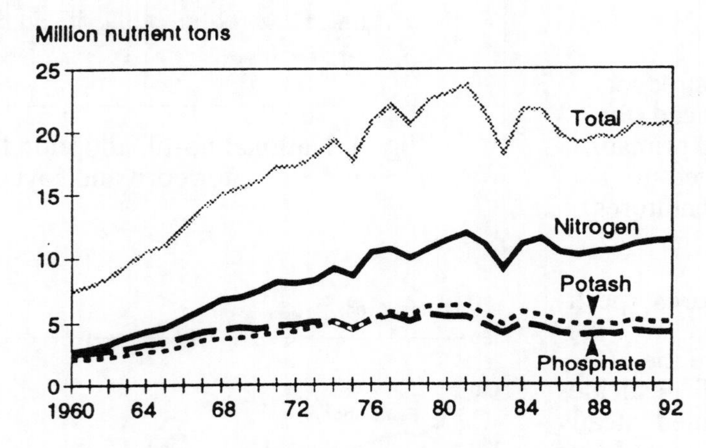

Fig. 6. Farm Fertilizer Use Trends

II. TYPES OF CROPPING SYSTEMS

Objectives

1. Define the following types of cropping systems: rotation, monoculture, relay cropping, intercropping, double cropping, and ratoon.
2. Give one example of each cropping system.

All crops are grown in some type of cropping system. While some cropping systems are fairly simple, others are complex and require high levels of management. Characteristics of six cropping systems are listed below.

A. Rotation

A crop rotation is a systematic plan of producing crops in a field over a period of years. Typical Corn Belt rotations include Corn, Soybeans, and possibly small grains or forage crops. Advantages of rotations include breaking some pest life cycles, better nutrient utilization and better color distribution.

B. Monoculture

Growing the same crop in a field continuously over an extended period of time is a monoculture. Continuous corn is the most common monoculture on the Corn Belt. Farmers may select a monoculture to take advantage of favorable markets, an environment that strongly favors producing a high value crop, or to reduce the need for diverse machinery and facilities.

C. Relay Cropping

Relay cropping (sometimes called relay intercropping) consists of seeding a second crop into an initial crop prior to harvest of the initial crop. This may be done to provide a longer growing season for the second crop. Seeding winter wheat into soybeans prior to soybean harvest in the fall is an example of relay cropping.

D. Intercropping

An intercropping system consists of simultaneously producing two or more crops in the same field. Intercropping is most advantageous when crops with different characteristics are used. Examples of intercropping include growing strips of corn and soybeans, or growing cassava and beans or peas in tropical systems.

E. Double Cropping

The production of two crops, one following another, in the same field during one growing season is called double cropping. The most common double cropping system in the Corn Belt is following winter wheat with soybeans.

F. Ratoon

In a ratoon cropping system, used primarily with sugar cane, after the first harvest is completed the crop is permitted to regrow and a second harvest is obtained from the regrowth prior to destroying it in the fall.

NATIONAL CORN HANDBOOK

CROP MANAGEMENT NCH-50

Corn Cropping Sequences

Joseph P. Zublena, Clemson University

Reviewers

D. R. Christenson, Michigan State University *W. L. Parks, University of Tennessee*
W. C. Morrison, Louisiana State University *R. D. Voss, Iowa State University*

Cropping systems and crop sequences are management tools that can be manipulated to exploit available resources more efficiently. Instead of focusing on one crop, the systems approach looks at two or more crops in different planting sequences and evaluates them for agronomic and economic benefits (Table 1). Designing and evaluating cropping systems are complex tasks because of the vast number of interacting factors. Some of these factors are agronomic, such as the crops involved, hybrids or varieties, soil structure, tilth and fertility, and weed, insect, disease, or nematode pest pressures. Other factors include: climate, managerial capabilities, equipment, cash flow, and market availability.

Research on corn cropping systems usually compare monocropping, the continual planting of corn on the same land year after year, to different rotational sequences. Some of the data supports monocropping, while others support rotational sequences. While different findings are not unusual when working with biological systems, the majority of evidence involving cropping sequences shows an advantage to rotation. The following information will discuss the effect of cropping sequences on soil fertility, soil pH, nitrogen, organic matter, soil tilth, soil erosion, weed control, insect control, disease control, and yields.

SOIL FERTILITY

The effect cropping systems have on soil nutrient balance depends on the initial nutrient status and the quantity of nutrients removed in the harvested portion. Crops harvested as forage remove more nutrients than those harvested for grain. A cropping sequence which includes forages or silages, therefore, will influence the nutrient status more quickly than one including only grain crops. This can be seen in Table 2, which lists nutrients removed by corn silage and grain at two different yield levels.

Rotation also influences soil fertility because crops vary in their nutrient requirements, rooting patterns, and residue compositions. Deep-rooted crops, for example, can absorb nutrients from a lower soil zone and subsequently supply these to a following shallow-rooted crop. Recognizing the nutrient differences in residue composition (Table 3) could be an important step in planning nutrient-efficient crop rotations. If cropping sequences are varied, plants can 1) feed from an entire soil zone and nutrient reservoir; 2) cycle nutrients from subsoil to topsoil; and 3) complement each other in nutrient use and removal.

Planning crop rotations to achieve these benefits is possible, but much research is still needed to determine the practical implications. Until this information becomes available, the ease and convenience of applying commercial fertilizer will overshadow the use of rotations for fertility purposes alone.

Soil pH

Soil pH is primarily affected by soil type, the amount of plant material removed, the type and amount of fertilizer applied, and the level of crop yields attained. Cropping systems where large amounts of plant materials are removed as silage or hay tend to decrease soil pH more readily than systems where only grain is harvested. Likewise, as nitrogen rates and yields increase, soil pH decreases more rapidly. Table 4 illustrates how both crop removal and increased nitrogen rates affected soil pH after 15 years of corn monoculture at the Galva-Primghar Research Center in Iowa.

Table 1. Comparison of rotation and monoculture cropping systems with respect to certain agronomic and economic factors.

Factor	Crop sequence Rotation	Monoculture
Greater feeding range of roots	+	-
Diversity in nutritional requirements	+	-
Climate may favor one crop	-	+
Diversification of markets	+	-
Soil may favor one crop	-	+
Profit may favor one crop	-	+
Legume residues can provide N	+	-
Broader distribution of labor	+	-
Seasonally intensive	-	+
Breaks some pest cycles	+	-
Machinery costs usually lower	-	+
Specialization in production	-	+
Specialization in marketing	-	+
Alters tillage and traffic patterns	+	-

J. Zublena. Nat'l Corn Rotation Survey. 1982. Unpublished.

\+ Advantage   Disadvantage

Table 2. Nutrient removal for 100 and 200 bu./acre of corn.

Corn yield	Harvested portion	N	P_2O_5	K_2O
bu./acre		----------lb.----------		
100	Grain	80	34	24
	Stover	50	11	78
	Total	130	45	102
200	Grain	180	80	54
	Stover	100	35	228
	Total	280	115	282

Data compiled by J. Hanway, Iowa State University

Table 3. Micronutrient composition of certain crop residues.

Residue	B	Mo	Cu	Mn	Zn	Co
	--------parts per million--------					
Bean tops	75	2.2	19	40	551	0.24
Rye	30	1.1	15	32	456	0.20
Rye/vetch	100	11.1	16	80	465	1.32
Rye/grass	30	2.4	20	80	175	0.30

F.E. Bear, SSSA Proc., 13:380 (1948).

Table 4. Effect of crop utilization and nitrogen fertilization on soil pH after 15 years of corn monoculture.

Nitrogen rate	Soil pH Corn silage	Corn grain
lb./acre		
0	5.79	6.09
120	5.29	5.79

Crops & Soils. June/July 1980.

Nitrogen

Before commercial nitrogen fertilizers became readily available, many of the benefits derived from crop rotations were directly linked to nitrogen supplied by legume crops. The quantity and benefits of nitrogen supplied depends on the legume involved and its position in the cropping sequence (Table 5).

Differences between monoculture and rotational corn yields are markedly reduced as nitrogen rates are increased to attain maximum productivity in non-legume cropping sequences. However, a difference still remains in favor of rotations that cannot be duplicated with nitrogen applications (Figure 1).

Factors which cause these benefits are not known, but some speculations are: reduced disease infestations, improved soil physical properties, elimination of phytotoxic substances in corn residues,

Table 5. Effect of crop sequence on the amount of nitrogen required for maximum corn yields.

Corn sequence	Nitrogen required
	lb./acre/yr.
Continuous corn	150-200
CN-Sb	100-150
CN-Sb-C-O-L	20-50
C-Sb-CN-O-L	50-80
CN-C-C-O-L	20-50
C-CN-C-O-L	50-100
C-C-CN-O-L	100-150
CN-O-L-L	0-20

Iowa State Univ. Coop. Ext. Serv. Pm, 905, 1982.

C = Corn, Sb = Soybeans, O = Oats, L = Legume hay, CN=indicates crop specific in nitrogen requirements.

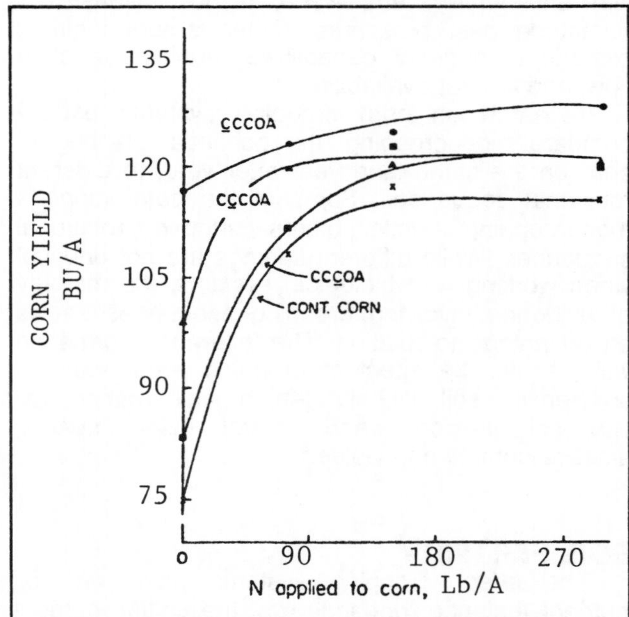

Figure 1. Response of corn to N fertilizer in continuous corn and CCCOA at Lancaster, Wisconsin. Each point is a mean of 20 observations over 10 years. Baldock, Agron. J. 73: (5) 1981.

and addition of growth-promoting substances by legume residues.

An exception to the advantage of using legumes in a rotation occurs in years and regions where perennial or fall-seeded legumes can deplete soil moisture before corn is planted. Under these conditions, corn following any annual crop (including itself) has a greater yield potential.

Organic Matter

Cropping sequences affect soil organic matter by the amounts of residues produced and by the amount of cultivation that stimulates decomposition. In addition to changes in organic matter that occur when land is cultivated, soil N reserves are also changed. Data in Table 6 show that 30 years of monocultured corn reduced soil organic matter and nitrogen by 65 and 62% respectively, while in a five-year rotation with oats, wheat, clover, and timothy, only 24% of the organic matter and 29% of the N reserve was lost. These data also show that by increasing the frequency of meadow or a green manure crop in a rotation, the organic matter (16% decomposed) and N reserve (19%) is more nearly maintained.

SOIL TILTH

Soil tilth refers to the physical condition of soil as related to its ease of tillage and fitness as a seedbed, as well as impedance to seedling emergence and root penetration. Factors influencing tilth include soil texture, organic residue, and tillage practices. In general, as crop residues are increased, tilth increases, and conversely, as tillage is increased, tilth decreases. The effect of 20 years of monocultured corn and rotational corn on two components of soil tilth for a Webster silty clay loam, bulk density and organic matter, are illustrated in Figure 2.

Aggregate development and stability are necessary components of good soil tilth. Clay serves as a major component for maintaining aggregation, but as clay content of a soil decreases, organic matter becomes increasingly more important. Organic matter stabilizes these soils through soil bacteria and fungi which decompose crop residues into polysaccharide gums and by the filamentous hyphae which help to bind soil particles together. Once formed, however, soil aggregates are subject to breakdown from weather and tillage. The replenishment of organic matter will help prevent this.

Table 6. Effect of crop rotation on the organic matter content of arable soils.

Cropping sequence	% reduced after 30 yrs. of cropping: Organic matter	N-reserve
C-C-C	65	62
W-W-W	38	40
C-O-W-CL-T	24	29
C-W-CL	16	19

E. W. Russell. Soil Conditioning and Plant Growth. 10th ed. 1983.

C = Corn, W = Wheat, O=Oats, CL = Clover, T=Timothy.

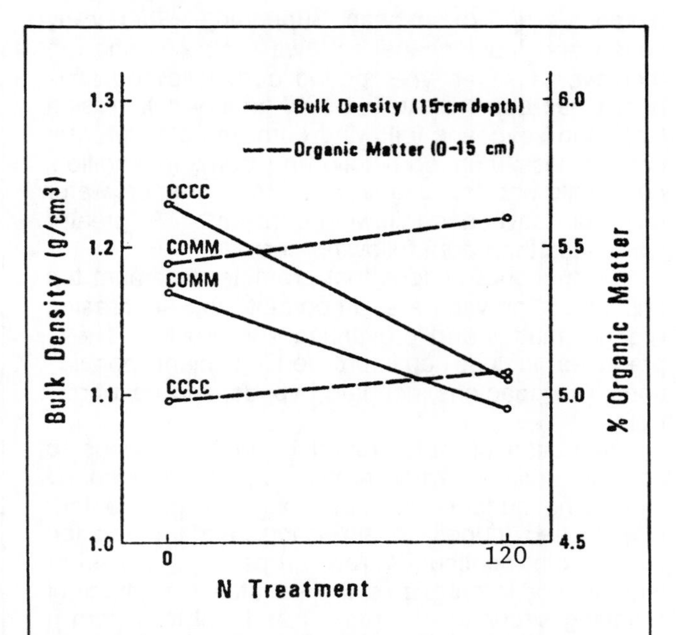

Figure 2. Organic matter content of the 0 to 6 in. layer and soil bulk densisty at the 6 in. depth of the two cropping systems and N fertilizer treatments. Hageman and Shrader, Agron. J. 71: (6) 1979.

SOIL EROSION

Erosion is the symptom and not the cause of soil destruction. The primary cause is use of an unsuitable cropping sequence that fails to adequately protect and maintain the soil. Cropping sequences most subject to erosion utilize tillage operations that expose bare soil to wind and rainfall. As rain droplets impact the soil, unstable aggregates are broken down into sand, silt, clay, and organic matter particles. If rainfall occurs at a rate greater than the soil can absorb, the lighter particles of silt, clay, and organic matter are carried away with the runoff, leaving coarser sands with lower nutrient holding capacities. As slope and rainfall intensity increase, runoff velocities increase and larger particles of sand, stone, and even gravel can be transported. Under these conditions, surface rills and gullies develop and severe soil losses occur.

Erosion can be prevented by selecting cropping sequences that effectively promote soil-water penetration and reduce runoff through the build-up of organic matter and reduction of soil bulk densities. In comparing cropping sequences, this would favor a large biomass crop like corn over a smaller biomass crop like soybean. Supporting evidence by Laflen and Moldenhauer in Iowa (Table 7) showed that over a seven-year period corn following soybeans had approximately a 50% greater soil erosion loss than soybeans following corn and 40% greater erosion loss than corn following corn. In addition, corn following soybeans had 34% greater water loss than soybeans following corn and 13% greater water loss than corn following corn.

Production practices that promote maximum top and root growth are important for increasing organic matter and preventing soil erosion. These practices include optimum fertility, plant populations, row spacings, adapted hybrids, and pest control.

Reduction of water runoff is another means to control erosion. Water runoff can be reduced by increasing surface residues, rough tillage, contour ridges, tied ridges, basin-listing, and/or rotation away from continuous row crops to a grass or legume sod crop. Table 8 illustrates the effect of cropping sequence on run-off and soil loss from a silt loam soil with a 12% slope and an average annual rainfall of 38.5 inches.

Table 7. Effect of cropping sequence on soil loss, water loss and sediment concentration.

Cropping sequence	Soil loss	Water loss	Sediment
	cwt./acre	in./acre	ppm/acre
Sb after C	57.5	5.57	11,800
C after Sb	85.4	7.46	13,000
C-C	62.2	6.59	10,900

J.M. Laflen and W. C. Moldenhauer. Crops & Soils. June/July, 1982.

Silt Loam Soil w/6% slope.

C = Corn, Sb = Soybeans.

Table 8. The effect of crop rotation on the erodability of a soil. Zanesville, Ohio: 9 year mean.

Crop rotation	Avg. runoff of water: Amount	Avg. runoff of water: % rainfall	Soil loss
	in.	%	ton/acre
Continuous corn	15	40.3	99.6
Crops in rotation			
Corn	9	23.7	42.5
Wheat	9	24.8	11.4
1st year grass	7	17.7	0.6
2nd year grass	5	12.8	0.2
Permanent pasture	1.6	4.3	0.2

USDA, Technical Bulletin 888, 1945.

WEED CONTROL

Mechanical cultivation and crop rotation were the primary tools available for general weed control before the introduction of herbicides. Now, crop rotation is only useful in controlling weeds that either escape or are immune to the currently available herbicides. Using rotation to control weeds is unnecessary if all weed species can be controlled with labeled herbicides.

Herbicides have eliminated the need for both cultivation and rotation for weed control in many areas of the country unless "problem" weeds exist. Where they do exist, both cultivation in the corn crop and rotation to an alternate crop with more effective herbicides are necessary for control. Some problem weeds currently requiring both cultivation and rotation for control are listed in Table 9.

Table 9. U.S. weed species requiring rotation for maximum control in corn.

Common name	Scientific name
Bermudagrass	Cynodon dactylon (L.) Pers.
Broadleaf signalgrass	Brachiaria platyphylla (Griseb) Nash.
Johnsongrass	Sorghum halepense (L.) Pers.
Shattercane, wild cane	Sorghum bicolor (L.) Moench
Texas panicum, Texas millet	Panicum texanum Buck L.
Wild proso millet	Panicum miliaceum L.

J. Zublena. Nat'l Corn Rotation Survey. 1982. Unpublished.

INSECT CONTROL

Rotation was used for insect control prior to the development of organic insecticides in the 1950's. As available insecticides improved, the need for and use of rotation decreased. Currently, most major corn insects can be controlled effectively with insecticides in corn monoculture systems. However, many of these insects can also be effectively controlled without insecticides or with reduced rates of insecticides if crop rotation is practiced. The most striking example of rotational control is with the northern corn rootworm. This major corn pest in the north central region of the U.S. can be controlled by a one year rotation, while monocultured corn requires an insecticide. Rotations control this insect because its life cycle consists of one generation per year and it utilizes the corn plant as its primary host. If no corn is present, the larvae starve. Rotation can aid in control of other insects as well, but the degree of success will vary with each species.

DISEASE CONTROL

Disease control through crop rotations is often viewed as a passive process designed to reduce population levels of pathogens in the absence of a suitable host. The key function of a rotation is not to eradicate a pathogen but to reduce it to a low enough level so a profitable yield can be obtained when the susceptible crop is replanted.

The success of rotation as a disease control mechanism in corn is dependent upon the pathogen involved. Pathogens with numerous hosts or pathogens capable of long term survival without a host are less likely to be controlled by rotation. This is also true of pathogens that are either wind-borne or transmitted by insects that travel over long distances. Successful rotations for disease control are generally confined to pathogens in plant residues and soil and where non-host crops are used in the cropping sequence.

SUMMARY

Specific factors such as topography or pest pressures may favor a rotational sequence over monocultured corn, or vice versa, as evidenced in the preceding text. A mystery still remains, however, in that most research reports show higher corn yields using rotation even when all known factors such as fertility, pest pressures, water, etc. are equal. Yield benefits have been obtained by rotating from a non-row crop to corn; from another row crop to corn; and even from one corn hybrid to another hybrid (Figure 3). In a recent (1982) national survey, yield increase due to rotation ranged from 6 to 30 bu./acre. The same survey reported yields to decrease after 2 to 5 years of continuous corn. The specific causes of these yield increases from rotations are still speculative and, as such, warrant further study.

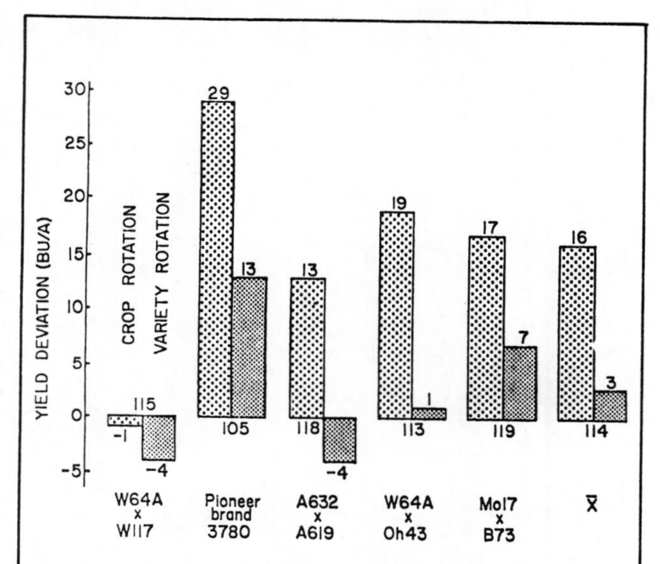

Figure 3. Grain yields of five corn hybrids grown following themselves (yields given horizontally to 0 yield level), following soybeans (crop rotation, left bar for each hybrid), and following the other four corn hybrids (variety rotation, right bar for each hybrid). Yields are averaged from experiments at Lamberton and Waseca during 1979 and 1980. Univ. of Minn. Crop News No. 63 (1981).

A publication of the National Corn Handbook Project

NEW 1/87 (5M)

Cooperative Extension work in Agriculture and Home Economics, state of Indiana, Purdue University and U.S. Department of Agriculture cooperating. H.A. Wadsworth, Director, West Lafayette, IN. Issued in furtherance of the acts of May 8 and June 30, 1914. The Cooperative Extension Service of Purdue University is an affirmative action/equal opportunity institution.

III. CLIMATIC FACTORS THAT AFFECT CROP DISTRIBUTION

Objectives

1. Distinguish between climate and weather.
2. Describe source and distribution of energy in the atmosphere.
3. Describe energy transfer by conduction, convection, radiation, and latent heat.
4. Describe the greenhouse effect.
5. Describe the origin, relative temperature, and relative moisture of air masses.
6. Describe typical weather which occurs during and after the passing of warm and cold fronts.

The climate plays a very important role in determining the adaptation and distribution of crops. Weather and climate differ in that weather refers to the conditions of the atmosphere at a given time, and climate refers to the expected long term atmospheric conditions. While all atmospheric conditions influence the climate, temperature and precipitation have the greatest influence. Climates, therefore, are frequently described on the basis of temperature and moisture.

A. Source and distribution of energy in the atmosphere.

1. The ultimate source of all energy in the atmosphere is the sun, which provides about 2.0 cal/cm^2/min. This value is called the solar constant.

   All points on earth receive an average of 12 hours light and 12 hours darkness each day, but due to the revolving of the earth around the sun and because its axis is 23.5° from perpendicular, we have seasons. The seasons play a very important role in determining crop distribution.

2. Approximate distribution of the sun's energy in Indiana

   14% absorbed by the atmosphere

   43% reflected or re-emitted out of the earth's atmosphere

   43% reaches the earth's surface

B. Energy moves in the atmosphere by 4 methods.

1. Radiation

   The transfer of heat from one source to another
   without physical contact between the two bodies.
   Example: sun heating the earth.

2. Latent heat

   The transfer of heat by evaporation of water requires energy from the earth's surface. The water vapor then condenses, which then releases the energy.

3. Conduction

The transfer of heat by direct contact (Fig. 7)
Example: Transfer of heat to your hand when picking up a hot object.

4. Convection

The transfer of heat by movement of air already heated primarily by conduction (Fig.8).
Example: A forced air heating system in a home.

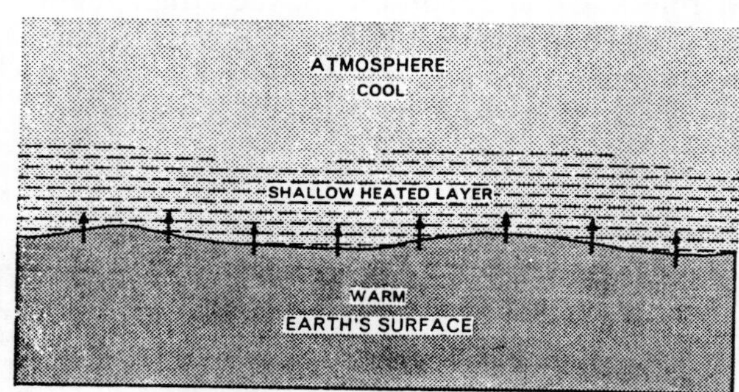

Fig. 7. Conduction

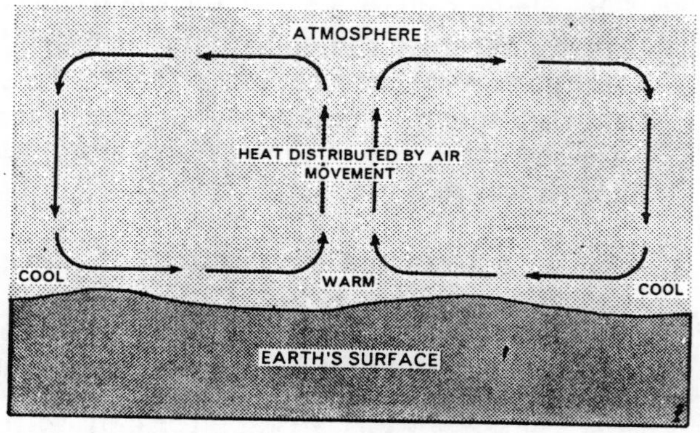

Fig. 8. Convection

C. The Greenhouse Effect

On a sunny winter day the temperature inside a greenhouse may be quite comfortable. This increase in temperature is caused by the conversion of light energy to heat, which is then trapped by the glass (Fig. 9).In the earth's atmosphere, carbon dioxide and water vapor act like the glass of a greenhouse. This atmospheric greenhouse effect modifies diurnal temperature fluctuations and prevents temperature extremes.

Since the early 1800s, atmospheric carbon dioxide levels have risen about 25% and continue to increase at a rapid rate. Scientists speculate this could enhance the greenhouse affect and cause climate changes.

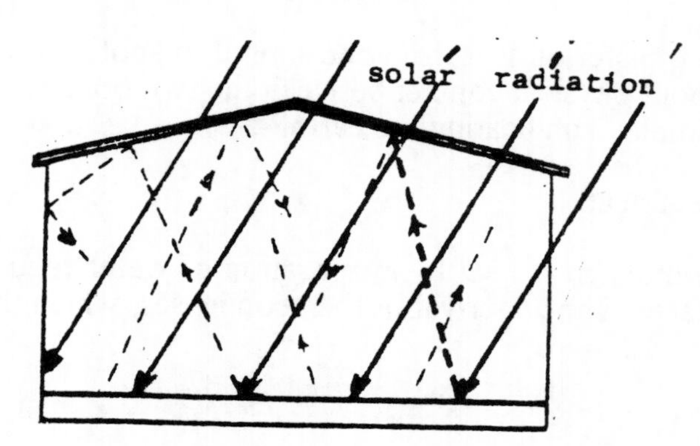

Fig. 9 Incoming solar radiation is converted to heat.

NATIONAL CORN HANDBOOK

CLIMATE & WEATHER NCH-40

Growing Season Characteristics and Requirements in the Corn Belt

Ralph E. Neild, University of Nebraska
James E. Newman, Purdue University

Reviewers

R. F. Dale, Purdue University
D. G. Hanway, Univ. of Nebraska-Lincoln
R. E. Carlson, Iowa State University
G. O. Benson, Iowa State University
R. E. Felch, Control Data Corporation, MN
C. M. Sakamoto, Center for Energy & Ag Assessment, MO
E. A. Runge, Texas A & M University
R. M. Castleberry, Dekalb-Pfizer Genetics, IL
M. W. Seeley, University of Minnesota
R. H. Shaw, Iowa State University
D. R. Hicks, University of Minnesota
R. L. Nielsen, Purdue University

INTRODUCTION

In the simplest terms, a crop growing season refers to that period of the year when seasonal weather is favorable for growth. In the Corn Belt, the "growing season" is often defined as the number of freeze-free days during the year beginning with the last freezing temperature in the spring and ending with the first occurrence of freezing temperatures in the autumn. This definition makes good physical sense because of its relationship to the freezing temperature for water—the most universal compound in living plants. It is also closely related to the normal corn crop growing season across the U.S. Corn Belt.

The average spring planting date within the Corn Belt is rather closely related to the average date of the last spring freeze for most areas. Adapted full-season hybrids at any location must reach maturity under normal seasonal weather conditions, by the average first autumn freeze. In the major portion of Corn Belt, proper full-season hybrids are adapted rather closely to the average frost-free season. For the most southern areas of the Corn Belt and further south, full-season hybrids are adapted to reach maturity in about five months or 150 days. In these geographic areas, the frost-free season is much longer than the corn crop growing season; and because of this, the corn crop growing season is not determined by the freeze-free season.

Across the central Corn Belt, most commercial hybrids are adapted to mature in 130 to 150 days; in the northern portions of the Corn Belt, commercial grain hybrids are adapted to mature in four months, or about 120 days, under normal growing-season weather conditions. In fact, the northern limit of the U.S. Corn Belt, where corn is grown as a grain crop, relates rather closely to the average frost-free growing season of 120 days.

TEMPERATURE REQUIREMENTS

Corn can survive brief exposures to adverse temperatures, such as temperatures ranging from near 32°F (O°C) to over 112°F (45°C). The growth limits are somewhat less, with beginning temperatures of near 41°F (5°C) climbing to near 95°F (35°C). Optimal temperatures for growth vary between day and night, as well as over the entire growing season; for example, during daylight hours, the optimal ranges between 77°F and 91°F (25-33°C) while night temperatures range between 62°F and 74°F (17-23°C). However, the optimal average temperatures for the entire crop growing season ranges between 68°F to 73°F (20-22°C). These relationships are illustrated in Figure 1.

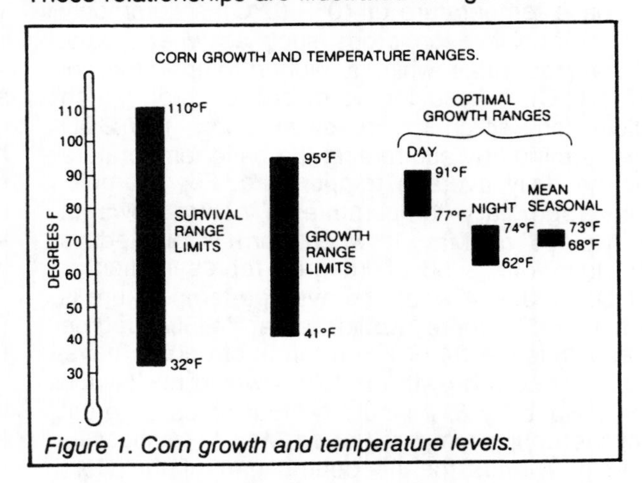

Figure 1. Corn growth and temperature levels.

Corn will germinate and grow slowly at about 50°F (10°C), but the average first spring planting dates across the U.S. usually begin when the average air temperatures reach 55°F (13°C) and soil temperature at seed depth is favorable for seedling growth. Poor germination resulting from below-normal temperatures rather than freezing temperatures is the greatest hazard of planting too early. The most common stress imposed at this time is that of cold soil temperatures. The growing point of germinating seedlings remains below the soil surface and they are not vulnerable until several days after the emergence. By this time, the probability of freezing temperatures has greatly decreased. High-temperature stress during ear formation, reproduction, and grainfill is normally detrimental to yield. Under rain-fed conditions, corn usually begins to stress when air temperatures exceed 90°F (32°C) during the tasseling-silking and grainfill stages. Data from Nebraska showed the yield of dryland corn may be reduced 1½ bushel per acre for each day the temperature reaches 95°F (35°C), or higher, during this critical period.

GROWING DEGREE DAYS

Historically, the number of days from planting to harvest has been used to classify the maturity of corn hybrid; but in recent years, commercial corn hybrid maturity is often determined by growing-degree days (GDD) or heat units (Hu). Growing-degree days are systems used to classify the maturity of corn hybrids. GDD result from summation of mean daily temperature.

The GDD concept assumes that:

- There is a value or base temperature below which plants do not grow or grow very slowly.
- The rate of growth increases as temperature increases above a base temperature.
- Plant growth and development are more closely related to daily temperature mean accumulations above a base value in the absence of other limiting conditions.

A base temperature of 40°F (5°C) is commonly used for cool-season crops such as wheat, oats, and canning peas while a higher temperature of 50°F (10°C) is used for warm-season crops such as field corn, sorghum, and sweet corn. The GDD are determined by subtracting the base temperature from the daily average temperature. For example, the average daily temperature in central Iowa at Des Moines on May 1, when corn is planted, is 56°F; therefore, a 56-50 (degree representation) = 6 GDD. On July 4 when the average temperature is warmer, 74°F, there would be a 74-50 (degree representation) = 24 GDD. In terms of degree days, the rate of corn growth on July 4 would be 24 and when divided by 6, it would be four times as great as at the cooler temperature on May 1. A classification of hybrids for the central part of the Corn Belt representing days from planting to harvest and GDD is presented in Table 1.

Table 1. Field Corn Hybrid Maturity Classification.

Maturity	Days	GDD
Early-season	85-100	2100-2400
Mid-season	101-130	2400-2800
Full-season	131-145	2900-3200

The average mid-season commercial corn hybrid in the central Corn Belt requires about 130 days from planting to maturity, or 2700 GDD. A hybrid of this type requires about 70 days from planting to mid-silk flowering, or 1400 GDD. If planted by mid-May under normal seasonal temperatures, such a hybrid will reach black-lay maturity by mid-September. For the central Corn Belt at Lafayette, Indiana, the relationships for a mid-season hybrid growth and maturity for the central Corn Belt at Lafayette are illustrated in Figure 2.

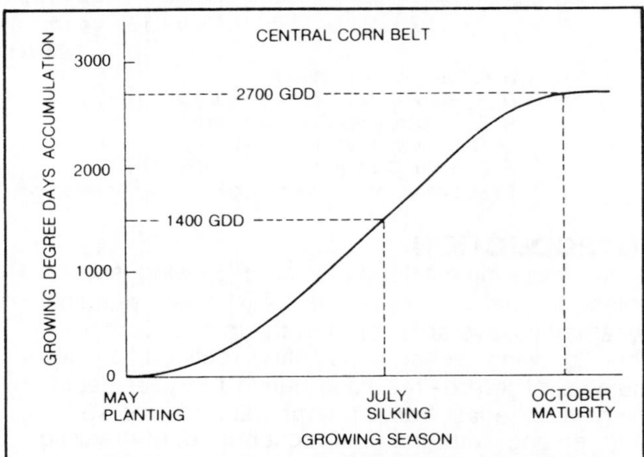

Figure 2. Corn growth and maturity stage predictions based on seasonal accumulation of growing degree days (GDD), Lafayette, Indiana.

The mean GDD values for the Corn Belt and adjacent regions are illustrated in Figure 3. Over a five-month period of May 10 to October 10, the northern boundry of corn grown for grain receives an average of 2400 GDD. The southern boundary receives an average of 3600 GDD. The GDD values in Figure 3 are compiled by the "cut-off" method.

The calculation procedure of the cut-off method assumes that the maximum sustainable corn growth rate is 86°F (30°C). This assumes that higher temperatures do not sustain faster growth rates. So, all daily maximum temperatures above 86°F (30°C) are set to 86°F (30°C) for GDD calculation purposes; and if the daily mean temperature is greater than the "base temperature" for corn of 50°F (10°C) and the daily minimum is below 50°F (10°C), then the daily minimum is set at 50°F (10°C) to calculate the GDD value on such a day.

The mean daily temperature cut-off method assumes that maize begins to grow at 10°C, which is define as the "base" temperature. The maximum growth rate, sustained over a 24-hour period, is

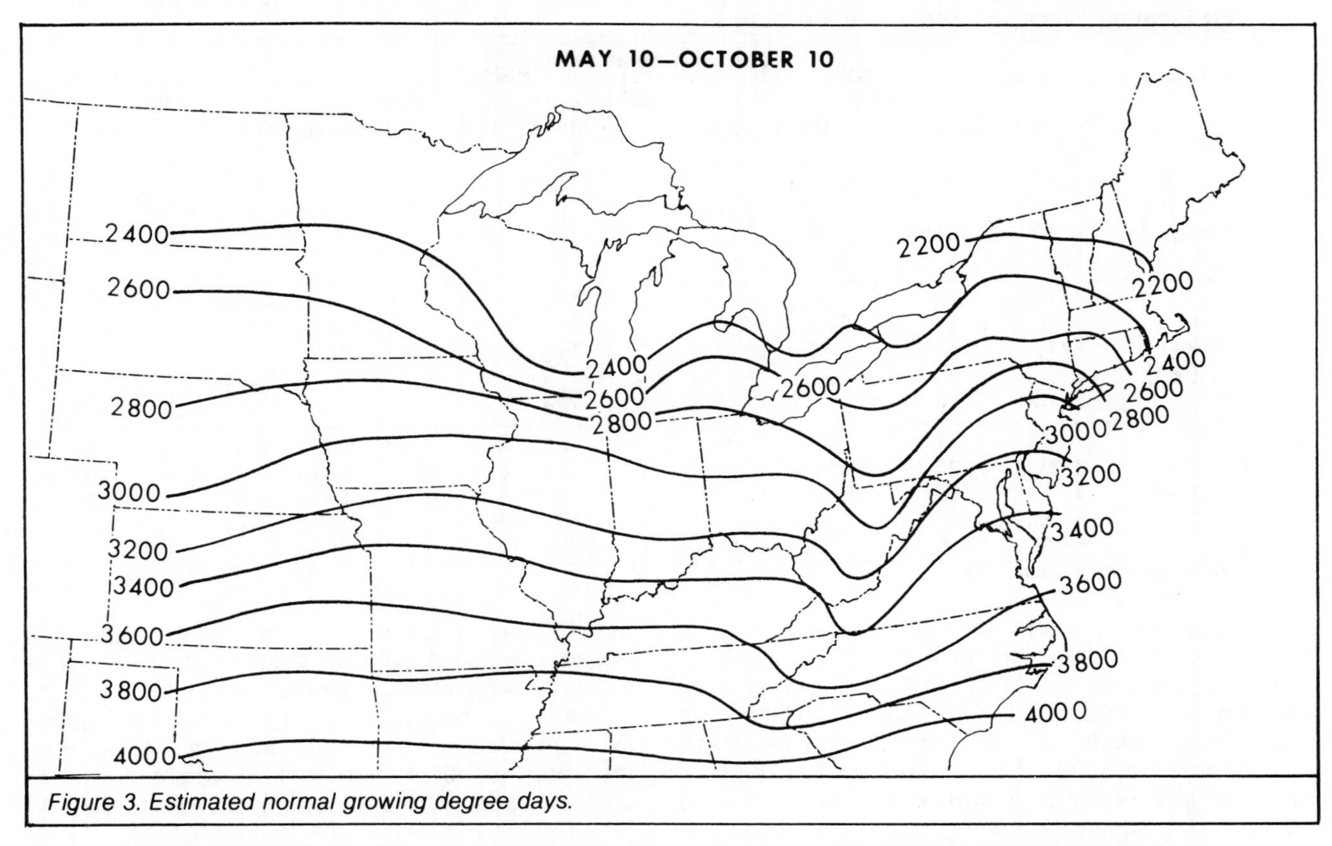

Figure 3. Estimated normal growing degree days.

assumed to be near 30°C. Further, it is assumed that higher daily mean temperatures do not lead to faster growth rates. Therefore, the GDD value for each day is calculated as:

$$GDD = \frac{Tmax \times Tmin}{2} - Tbase$$

where Tmax is set equal to 30°C, (86°F),
when it exceeds 30°C (86°F)
Tmin 10°C is set at 10°C, and
Tbase = 10°C.
In this manner, Tmax will always be entered as 30°C or less and Tmin will always be entered as 10°C (50°F), when less than 10°C (50°F).

By assuming an average of 10°C diurnal change in temperature, this cut-off method produces a stepwise linear simulation of the maize crop temperature dependent response between 5°C and 35°C, as shown in Figure 4, the linear relationships between GDD values and daily mean temperature are as follows: 5 to 15°C, GDD = 0.5T - 2.5; 16 to 25°C, GDD = T - 10; and 26 to 35°C GDD = 0.5T + 2.5.

The effects of seasonal temperatures on the response of corn with different GDD maturity requirements at different regions from north to south through the Corn Belt are presented in Table 2. The average dates and days from planting to accumulate 2200, 2600, 3000, and 3400 GDD in southern Wisconsin, central Illinois, and southern Illinois are shown. The first planting dates for these regions are May 1, April 19, and April 1 respectively, which represents the times when normal spring daily temperatures first rises to 55°F (13°C) and becomes warm enough for planting. These dates are considered "early" planting times for their respective regions. Four additional planting times, each two weeks apart, are shown for these regions.

Comparisons of the response of a particular maturity class at different regions, or of a maturity class at different planting times within a region, can be made from the data. For example, because of cooler temperatures in the North, an early planting of May 1 in southern Wisconsin would need 123 days to accumulate 2200 GDD. Planting dates of May 15, 29, and June 12 would require 116, 113, and 125 days to accumulate 2200 GDD values. In central Illinois a series of five plantings, two weeks

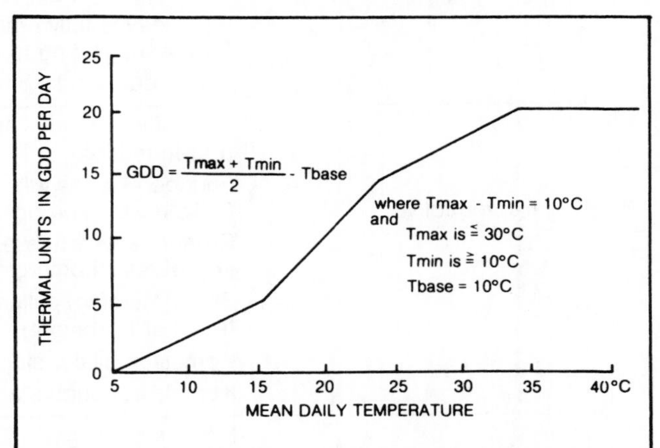

Figure 4. Daily thermal units accumulation in growing degree days (GDD), NOAA-NWS "cut-off" method of calculation.

Table 2. Expected Dates and Number of Days from Planting for Specific GDD Accumulations at Various Planting Times and Regions in the Corn Belt.

		Growing degree days							
		2200		2600		3000		3400	
	Planted	Date	Days	Date	Days	Date	Days	Date	Days
Southern Wisconsin	5/1	9/1	123	10/1*	153	--	--	--	--
	5/15	9/8	116	10/17*	155	--	--	--	--
	5/29	9/19	113	--	--	--	--	--	--
	6/12	10/15*	125	--	--	--	--	--	--
	6/26	--	--	--	--	--	--	--	--
Central Illinois	4/19	8/9	112	8/25	128	9/14	148	10/16*	180
	5/3	8/13	102	8/30	119	9/20	140	--	--
	5/17	8/20	95	9/7	113	10/5*	141	--	--
	5/31	8/30	92	9/20	112	--	--	--	--
	6/14	9/14	92	10/16*	124	--	--	--	--
Southern Illinois	4/1	7/21	111	8/3	124	8/18	139	9/2	154
	4/15	7/24	100	8/7	114	8/21	128	9/6	144
	4/29	7/30	93	8/13	106	8/27	120	9/23	147
	5/13	8/6	85	8/22	101	9/5	115	10/7	147
	5/27	8/16	81	8/31	96	9/18	115	11/8*	--

*Freeze risk greater than 50%

apart, beginning when the average daily spring temperature first rises to 55°F (13°C) on April 19, would need 128, 119, 113, 112, and 124 days to accumulate 2600 GDD. For southern Illinois, the comparison would be 139, 128, 120, 115, and 115 to accumulate 3000 GDD. When the days required for maturity begin to increase for June planting dates, it is rapidly becoming too late to plant the adapted hybrids for any given growing season heat zone of the Corn Belt. Blank cells in this table represent situations where there are not sufficient GDD for the maturity class to be adapted. Maturity dates indicated by an asterisk are situations of seasonal temperature patterns with very high freeze risks. Table 2 shows the effects of temperature changes within a growing season on the development of corn planted at different times.

The growing cycle of corn consists of vegetative, reproductive, and maturation phases, but there are more detailed stages of development within these phases. Different maturity classes require different GDD accumulations to reach these stages. The growing cycle and GDD requirement for different stages of a 2700 GDD hybrid are listed in Table 3.

When seed is planted in moist soil, about 200 GDD above 50°F is required for a seed to germinate and the young seedling to become esta-

Table 3. Growing Degree Day Requirements for Different Phenology Stages of a 2700 GDD Hybrid.

Phase	Development Stage	Number	GDD
Vegetative	Planted	0	0
	Two leaves fully emerged	0.5	200
	Four leaves fully emerged	1.0	345
	Six leaves fully emerged (Growing point above soil)	1.5	475
	Eight leaves fully emerged (Tassel beginning to develop)	2.0	610
	Tenth leaves fully emerged	2.5	740
Reproductive	Twelve leaves fully emerged (Ear formation)	3.0	870
	Fourteen leaves fully emerged (Silks developing on ear)	3.5	1000
	Sixteen leaves fully emerged (Tip of tassel emerging)	4.0	1135
	Silks emerging/pollen shedding (Plant at full height)	5.0	1400
	Kernels in blister stage	6.0	1660
	Kernels in dough stage	7.0	1925
Maturation	Kernels denting	8.0	2190
	Kernels dented	9.0	2450
	Physiological maturity	10.0	2700

blished. By this time, it has two functioning leaves and a root system so it is no longer dependent on the diminishing food supply that was stored in the seed. When corn is planted too early or when a season is cool, it takes too long to accumulate the needed 200 GDD. Consequently, the seedling becomes weak and vulnerable to soil fungus diseases, which then results in a poor stand.

When 475 GDD is accumulated, six leaves have formed and the growing point has risen above the soil surface. At 870 GDD, 12 leaves have formed and a small embryonic ear starts to develop within the plant tissue. It is at this time that the vegetative phase starts to cease and the reproductive stage begins. At 1400 GDD, all the leaves have developed, the tassel has emerged, and the plant has reached its full height; then the silks emerge from the ear and are receptive to the shedding pollen. Grainfill starts to cease, maturation begins, and some kernels become dented at 2190 GDD. At 2700 GDD, a black layer forms near the base of the kernel indicating that dry matter is no longer being translocated to the grain. The corn is at physiological maturity and safe from a freeze; however, the corn is still 35 to 40 percent moisture and will need to dry down before it can be stored.

When used over a large geographic area and over a range of planting times, the GDD system becomes more meaningful in understanding the response of corn to temperature. GDD's are also useful in predicting crop response at different planting dates or when seasonal temperatures deviate from average values.

Figure 5 shows the rates of development as measured by days from planting for different development phases of a 2700 GDD hybrid at eight planting dates, each ten days apart, beginning April 15 in central Illinois. These graphs are based on the GDD requirements in Table 3 and normal daily GDD accumulations for Decatur in central Illinois.

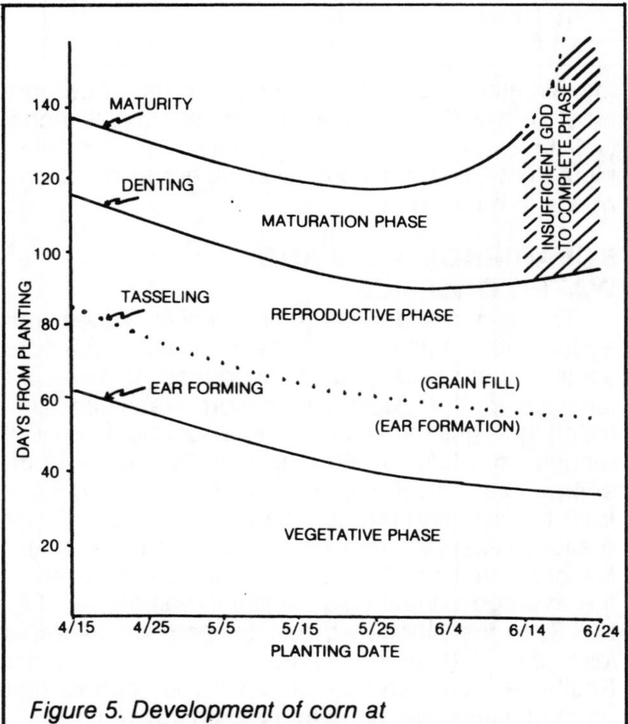

Figure 5. Development of corn at different planting dates in central Illinois.

The temperature available for growth during different development phases varies with time of planting. Because of cooler temperature, early plantings have longer vegetative stages while later plantings require more time to accumulate the GDD necessary to mature grain. The reproductive phase, from ear formation to denting, is less affected by planting time. The variation in days to maturity at different planting times is largely the result of differences in the rates of maturity in the vegetative and maturation phases.

For example, as may be seen in Figure 5, corn planted on April 15 takes 136 days to accumulate 2700 GDD. Almost half of this time, 62 days, is in the vegetative stage. The corn is expected to tassel in 84 days, begin denting in 115 days, and because of warm temperature, require 136-115 = 21 days for grain maturation. A planting in mid-May is expected to accumulate 2700 GDD and be mature in 119 days. Because of warmer temperature at planting, the corn would be in the vegative phase for 44 days. It would begin to dent 96 days after planting and would require 119-96 = 23 days for grain maturation. With still warmer temperatures at planting, corn planted on June 14 would be in the vegative stage only 36 days. The grain would begin denting 91 days after planting. However, delays in planting shift maturation into the cooler part of autumn so that it takes a longer time to accumulate the necessary GDD for this phase of the cycle. The June 14 planting is expected to take 44 days for maturation. This increases the total time to maturity to 135 days. A June 24 planting would have sufficient GDD for the vegetative and reproductive phases, but unless the growing season is warmer than normal, would not accumulate sufficient GDD to complete the maturation phase.

PRECIPITATION REQUIREMENTS

With dryland farming, corn is generally not grown in areas receiving less than 25 inches (60 cm) of annual precipitation. For high yields 18 to 20 inches (45 to 50 cm), or more, moisture should be available during the growing season. This water requirement must be met by natural rainfall, stored soil moisture from precipitation prior to the growing season, or as is the case in the drier western part of the Corn Belt, from supplemental irrigation.

Annual precipitation decreases from over 40 inches (100 cm) in the East to less than 25 inches (60 cm) in the West, across the Corn Belt, see Figure 6. However, as shown in Table 4, the amount of rainfall during the five-month growing season, May through September, is relatively constant. It ranges between 17.51 to 19.39 inches for most of the Corn Belt but drops off rapidly in central

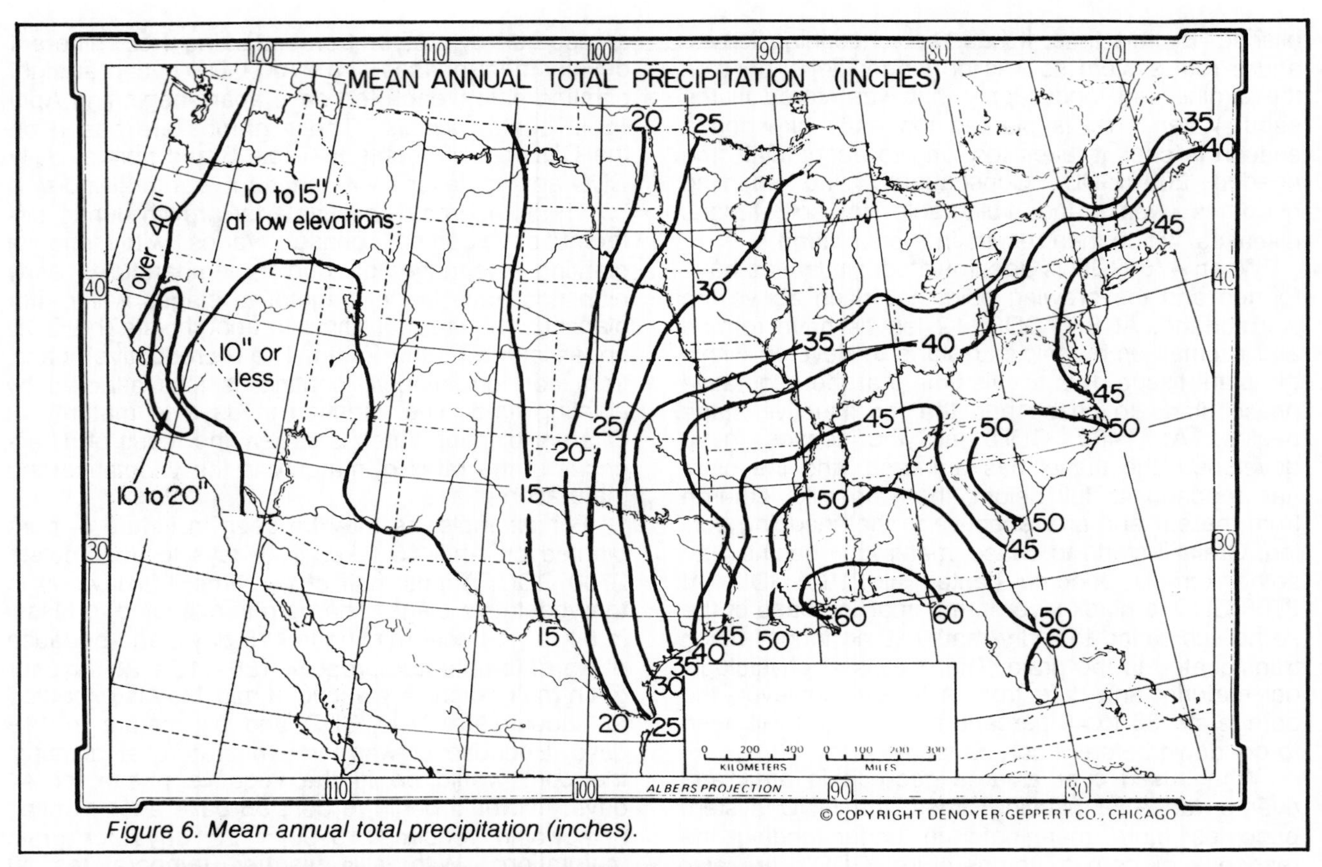

Figure 6. Mean annual total precipitation (inches).

Table 4. Precipitation (inches) Patterns Across the Corn Belt.

	Central Indiana (Kokomo)	Central Illinois (Peoria)	Central Iowa (Des Moines)	Eastern Nebraska (Omaha)	Central Nebraska (North Platte)
May-September	18.41	17.51	19.39	18.38	14.09
October-April	20.32	17.12	12.24	9.67	5.8
Annual	38.73	34.63	31.63	28.05	19.09

Nebraska which is near the western limit for dryland corn. This constancy of rainfall across the Corn Belt is a very significant growing-season characteristic.

Preseason precipitation during October through April is important for recharging soil moisture between seasons. Many Corn Belt soils are able to store about 10 inches of moisture in the five-foot depth to which corn roots may extend. The amount of preseason fall and winter precipitation available for recharge decreases from 20.32 inches at central Indiana in the eastern part of the Corn Belt to 5.8 inches in central Nebraska in the west. Artificial tile drainage to remove surplus soil moisture prior to planting in the East and irrigation to replenish moisture deficit in the West represent major differences in water management practices across the region.

As a corn plant grows, its demand for water increases with increasing leaf area which reaches a maximum near the tasseling stage. The period of time shortly before pollination through grainfill, when the kernels begin to dent, is a critical period during which adequate moisture is important to corn yield. This period begins in mid-July and may extend into mid-August over most of the Corn Belt. Except for the Western part, where irrigation is necessary, most of the Corn Belt averages 3.0 to 3.5 inches (7.5 to 8.8 cm) of rainfall in July. This moisture, plus stored soil moisture, is sufficient to meet crop needs in most years.

SOLAR RADIATION AND WATER BALANCE

The average annual solar radiation varies from about 300 langleys (gram calories per centimeter squared per minute) in the Northeast to over 500 langleys in the Southwest desert areas, as illustrated in Figure 7. The average for the Corn Belt ranges from 330 to 360 langleys. This figure is the energy that evaporates water from ponds and lakes, from the soil, and through plants. The annual evaporation measured from the network of pans, by the National Weather Service, relates rather closely to the average annual solar radiation values, see Figure 8. In the Northeast, pan evaporation averages less than 40 inches (100 cms), while in the Southwest, an average of over 100 inches (250 cms) of water may be evaporated each year.

From an average growing-season water-balance viewpoint, the increased solar energy in

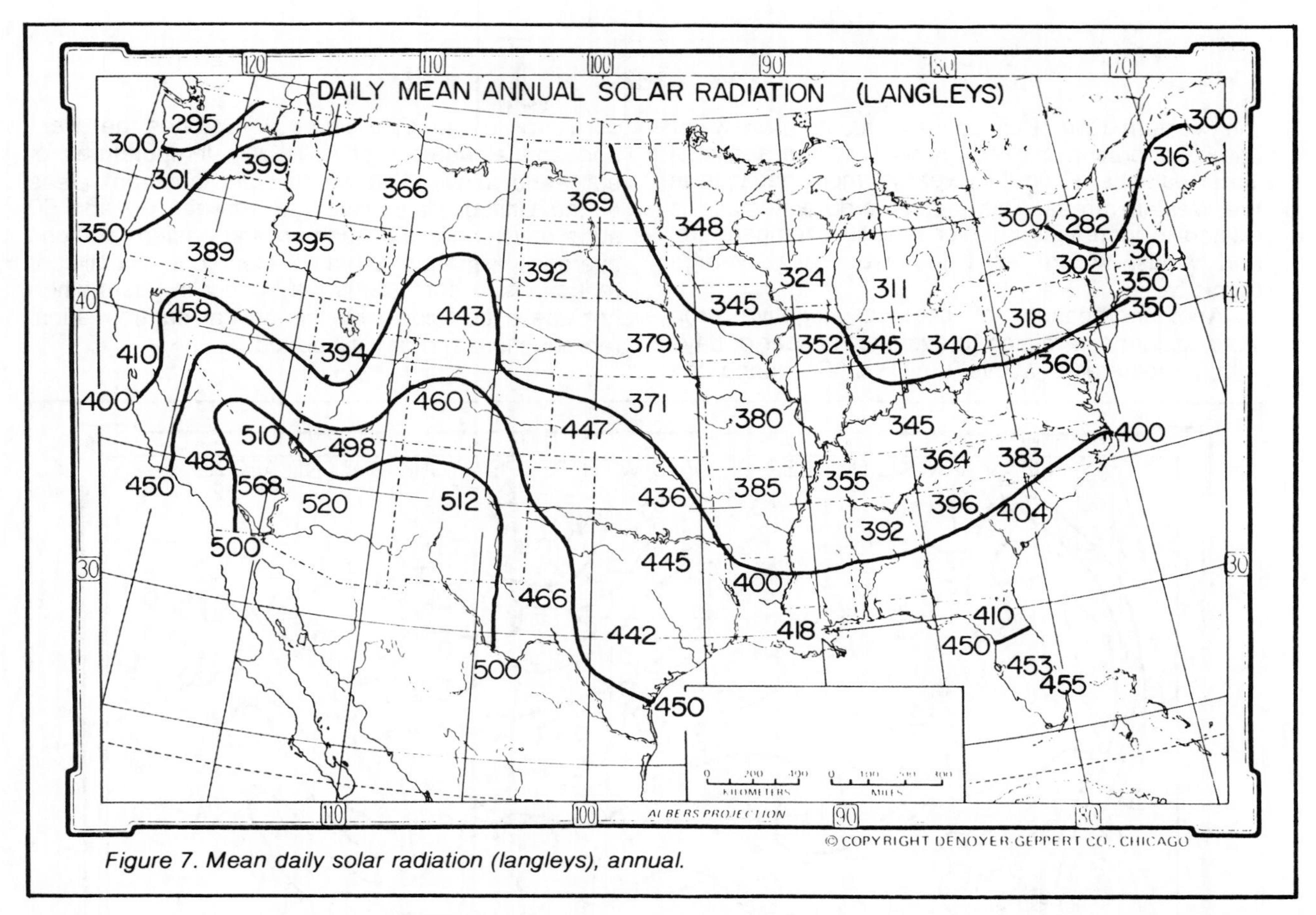

Figure 7. Mean daily solar radiation (langleys), annual.

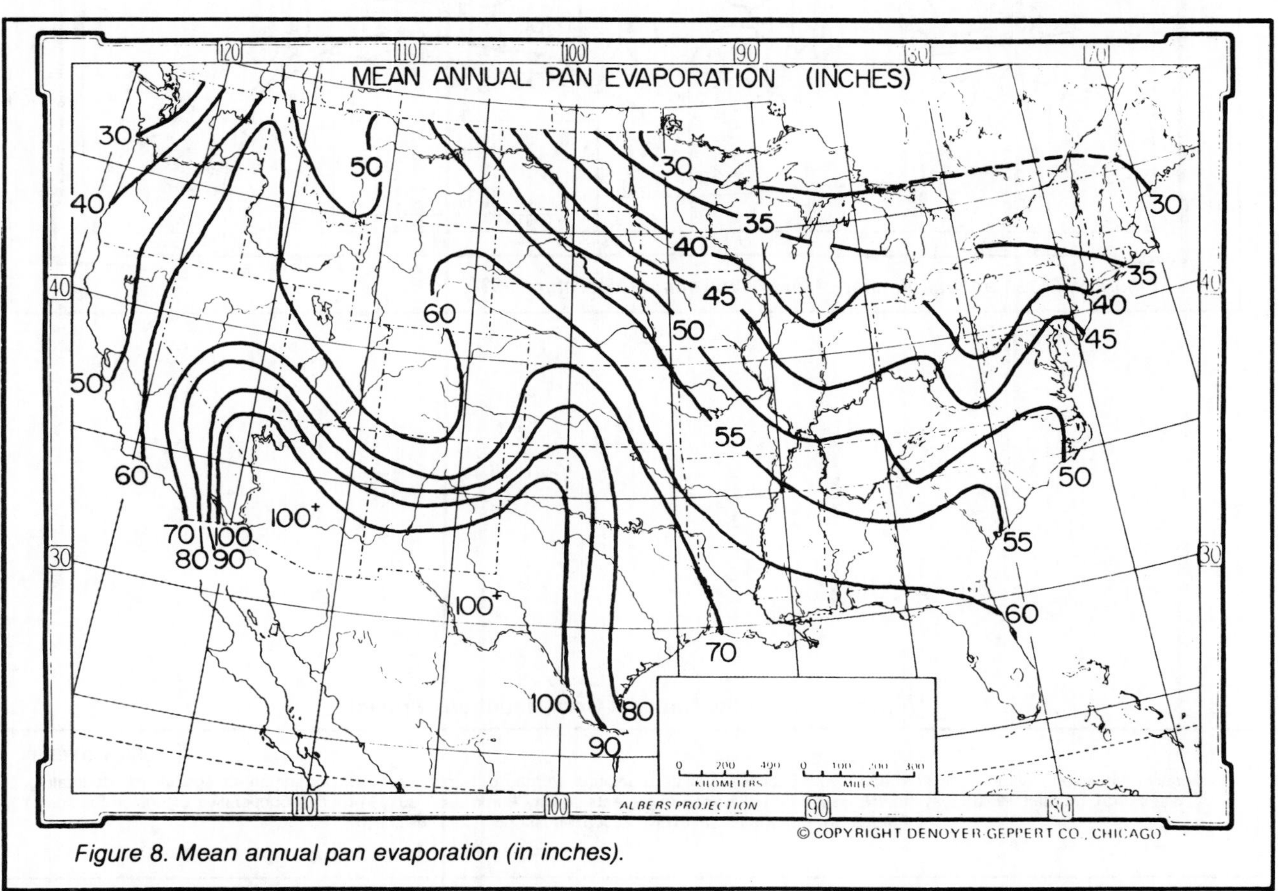

Figure 8. Mean annual pan evaporation (in inches).

the western Corn Belt causes greater crop water use. Evaporation exceeds rainfall, so the seasonal water balance is, on the average, more negative in the Western areas. This greater crop water use is related to increases in solar radiation, temperature, and winds, which result in lower daily average humidity.

Another measure of potential negative crop water balance is related to average number of days with temperatures above 90°F (32°C) during the corn growing season. Figure 9 illustrates the average annual number of days with temperatures of 90°F and above. The western and southern areas of the United States average between 30 and 60 such days, while the northern and eastern portions average less than 20 days per year. Significant reductions in corn yields may occur in years when hot spells coincide with the critical ear formation, reproduction, and grainfill period.

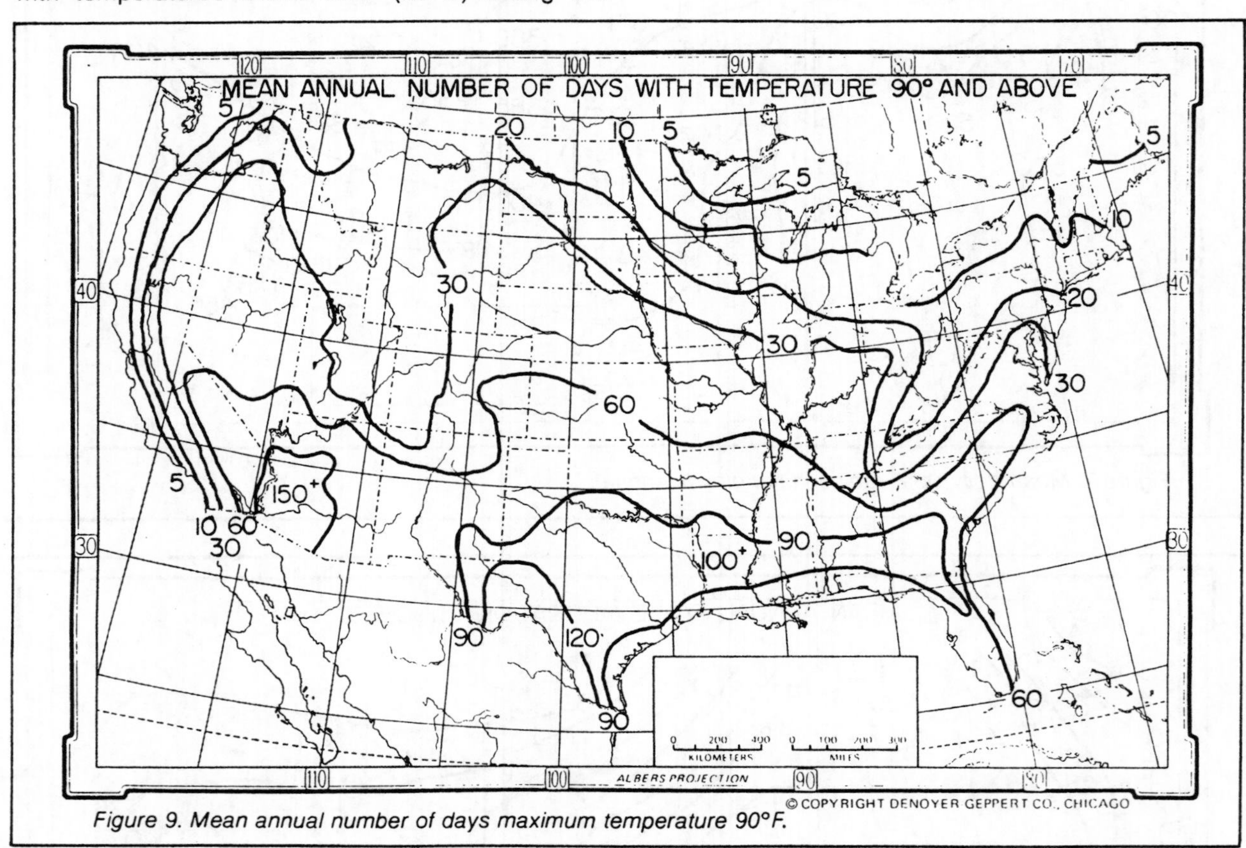

Figure 9. Mean annual number of days maximum temperature 90°F.

A publication of the National Corn Handbook Project

RR 4/90 (5M)

Cooperative Extension work in Agriculture and Home Economics, state of Indiana, Purdue University and U.S. Department of Agriculture cooperating. H. A. Wadsworth, Director, West Lafayette, IN. Issued in furtherance of the acts of May 8 and June 30, 1914. The Cooperative Extension Service of Purdue University is an affirmative action/equal opportunity institution.

D. Air Masses

There are four types of air masses that dominate weather patterns. Three of these influence United States weather patterns.

cP - Continental Polar
Originates over land in cold regions of the world composed of cold, dry air.

mP - Maritime Polar
Originates over water in cold regions of the world composed of cold, moist air.

cT - Continental Tropical
Originates over land in warm regions of the world composed of warm, dry air.

mT - Maritime Tropical
Originates over water in the tropics composed of warm, moist air.

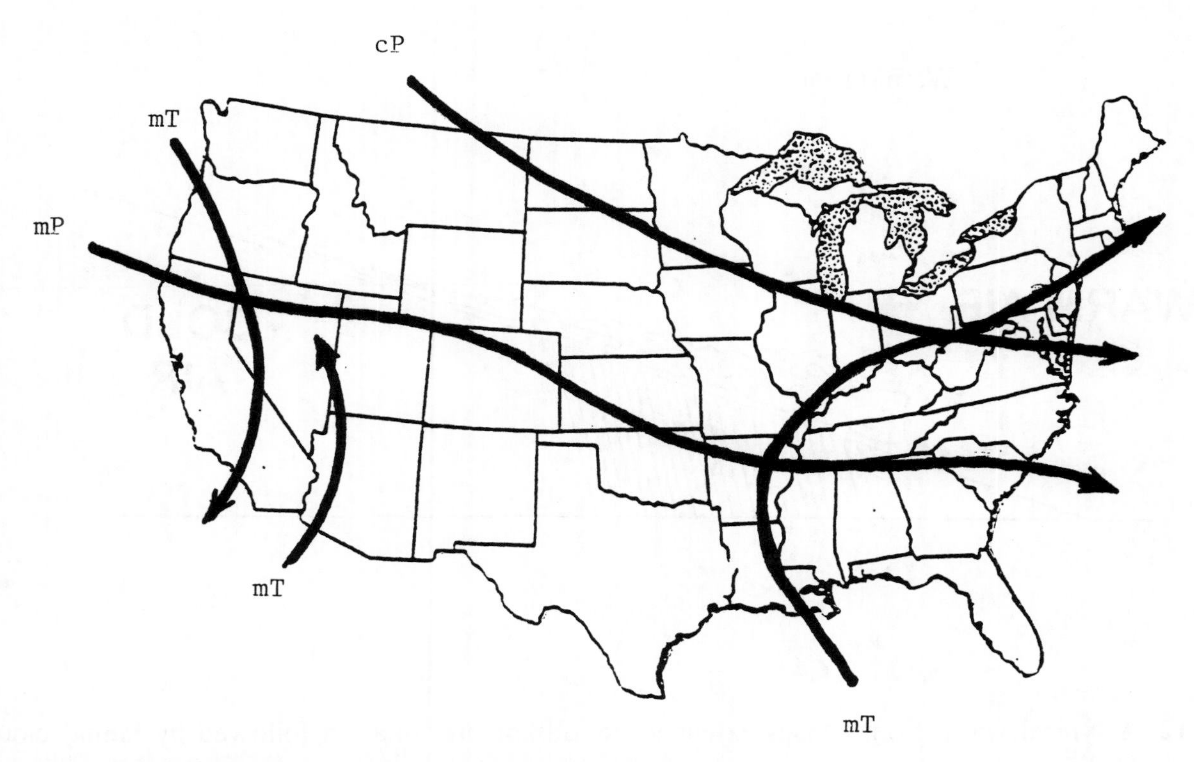

Fig. 10. Typical paths of air masses over the United States

E. Frontal systems

1. Cold front

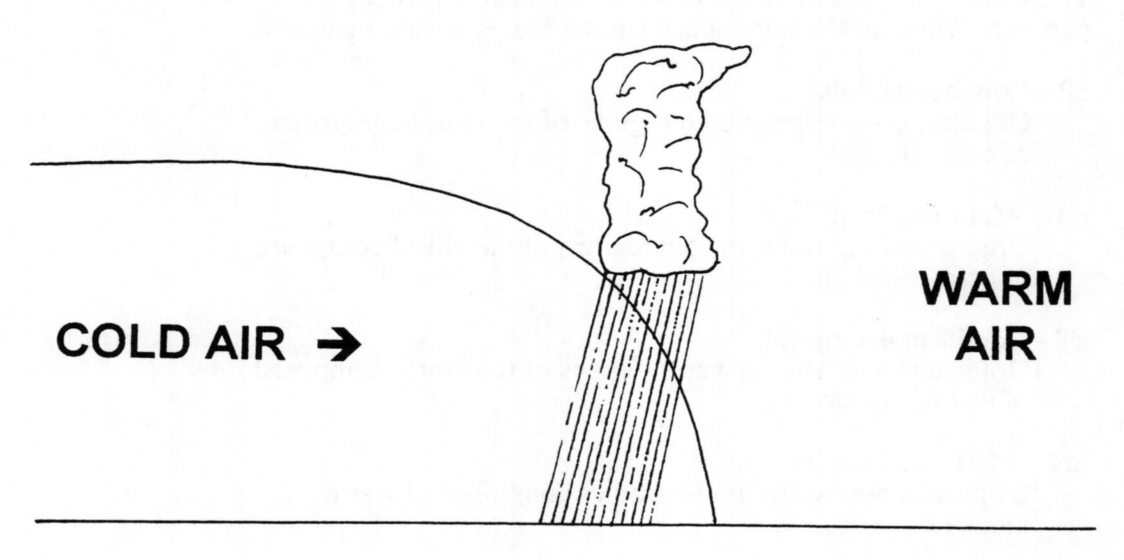

Fig. 11. A typical cold front causes brief intense storms followed by clear skies.

2. Warm Front

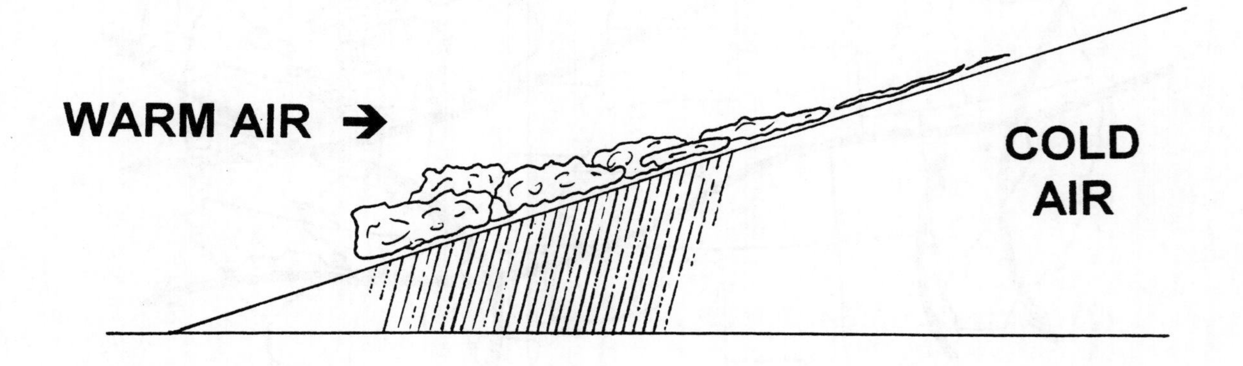

Fig. 12 A typical warm front causes extended periods of precipitation followed by humid cloudy weather.

IV. U.S. CROPPING REGIONS

Objectives

1. Identify the ten major farm production regions of the United States.
2. Describe the major crops, climate, topography, and economic conditions of each region.

Because of its diversity, the continental United States has been divided into Farming Regions, according to the major crop use and land characteristics of each region, as shown in Fig. 1.

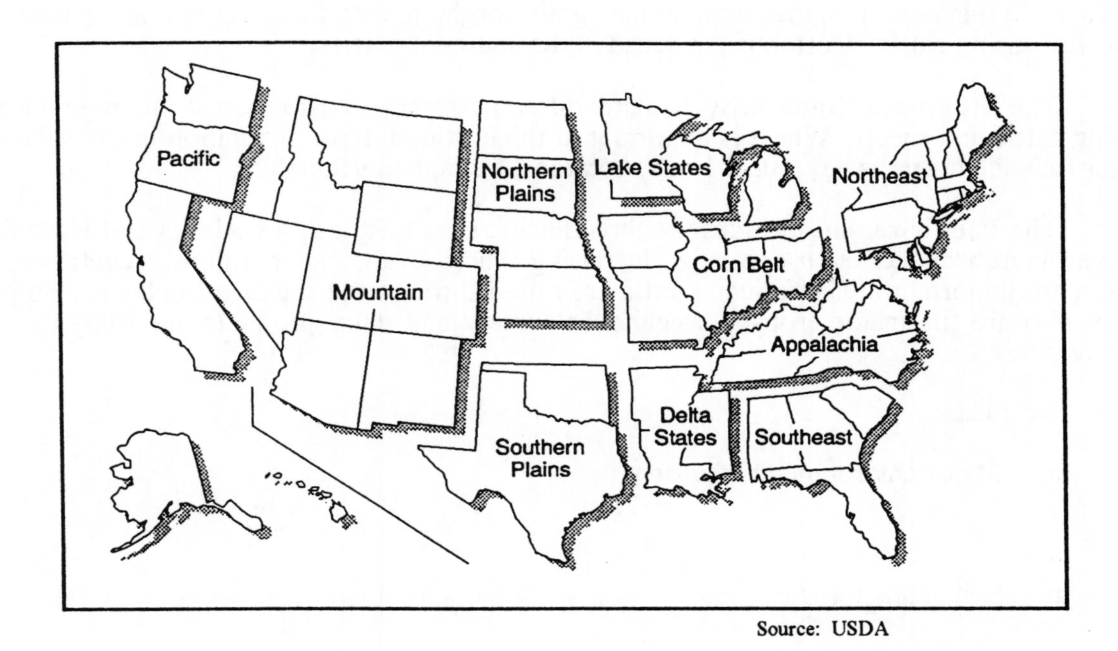

Fig. 1. Farming Regions of the United States

The 10 major farm production regions in the United States differ in soils, slope of land, climate, distance to market, and storage and marketing facilities. Together they comprise the agricultural face of the Nation.

The **Northeastern States** and the **Lake States** are the Nation's principal milk-producing areas. Climate and soil in these States are suited to raising grains and forage for cattle and for providing pastureland for grazing.

Broiler farming is important in Maine, Delaware, and Maryland. Fruit and vegetables are also important to the region.

The **Appalachian** region is the major tobacco-producing region in the Nation. Peanuts, cattle, and dairy production are also important there.

In the **Southeast** region, beef and broilers are important livestock products. Fruits, vegetables, and peanuts are grown in this region. Big citrus groves and winter vegetable production areas in Florida are major suppliers of agricultural goods. Cotton production is making a comeback.

In the **Delta States,** the principal cash crops are soybeans and cotton. Rice and sugarcane are also grown. With improved pastures, livestock production has gained in importance. This is a major broiler-producing region.

The **Corn Belt** has rich soil and good climate for excellent farming. Corn, beef, cattle, hogs, and dairy products are the major outputs of farm in the region. Other feed grains, soybeans, and wheat are also important.

Agriculture in the **Northern** and **Southern Plains,** which extend north and south from Canada to Mexico, is restricted by rainfall in the western portion and by cold winters and short growing seasons in the northern part. About three-fifths of the Nation's winter and spring wheat is produced in this region. Other small grains, grain sorghum, hay, forage crops, and pastures form the basis for raising cattle. Cotton is produced in the southern part.

The **Mountain States** provide a still different terrain. Vast areas of this region are suited to raising cattle and sheep. Wheat is important in the northern parts. Irrigation in the valleys provides water for such crops as hay, sugar beets, potatoes, fruits, and vegetables.

The **Pacific region** includes the three Pacific Coast States plus Alaska and Hawaii. Farmers in Washington and Oregon specialize in raising wheat, fruit, and potatoes; vegetables, fruit, and cotton are important in California. Cattle are raised through the region. In Hawaii, Sugarcane and pineapples are the major crops. Greenhouse/nursery and dairy products are Alaska's top-ranking commodities.

1. Northeast

    a. Major environmental factors

    b. Soil characteristics

    c. Major farming enterprises

2. Lake States

    a. Climate

    b. Major farming enterprises

3. Appalachia
    a. Topography and environmental factors

    b. Major farming enterprises

4. Southeast
    a. Climate

    b. Major farming enterprises

5. Delta States
    a. Major farming enterprises

6. Corn Belt
    a. Environmental factors

    b. Major farming enterprises

7. Northern Plains
    a. Environmental factors

    b. Major farming enterprises

    c. Common production practices

8. Southern Plains

   a. Environmental factors

   b. Major farming enterprises

9. Mountain

   a. Environmental factors

   b. Major farming enterprises

10. Pacific

    This region is very diverse, and can be divided into subregions as follows:

    a. Northwest sub-region

       1. Major Crops

          a.

          b.

          c.

       2. Environmental factors favoring production of these crops.

b. California Valley sub-region

1. Soil characteristics -

2. Climatic characteristics -

3. Major crops produced -

c. Mountain sub-region

1. Climatic characteristics -

2. Major agricultural enterprises -

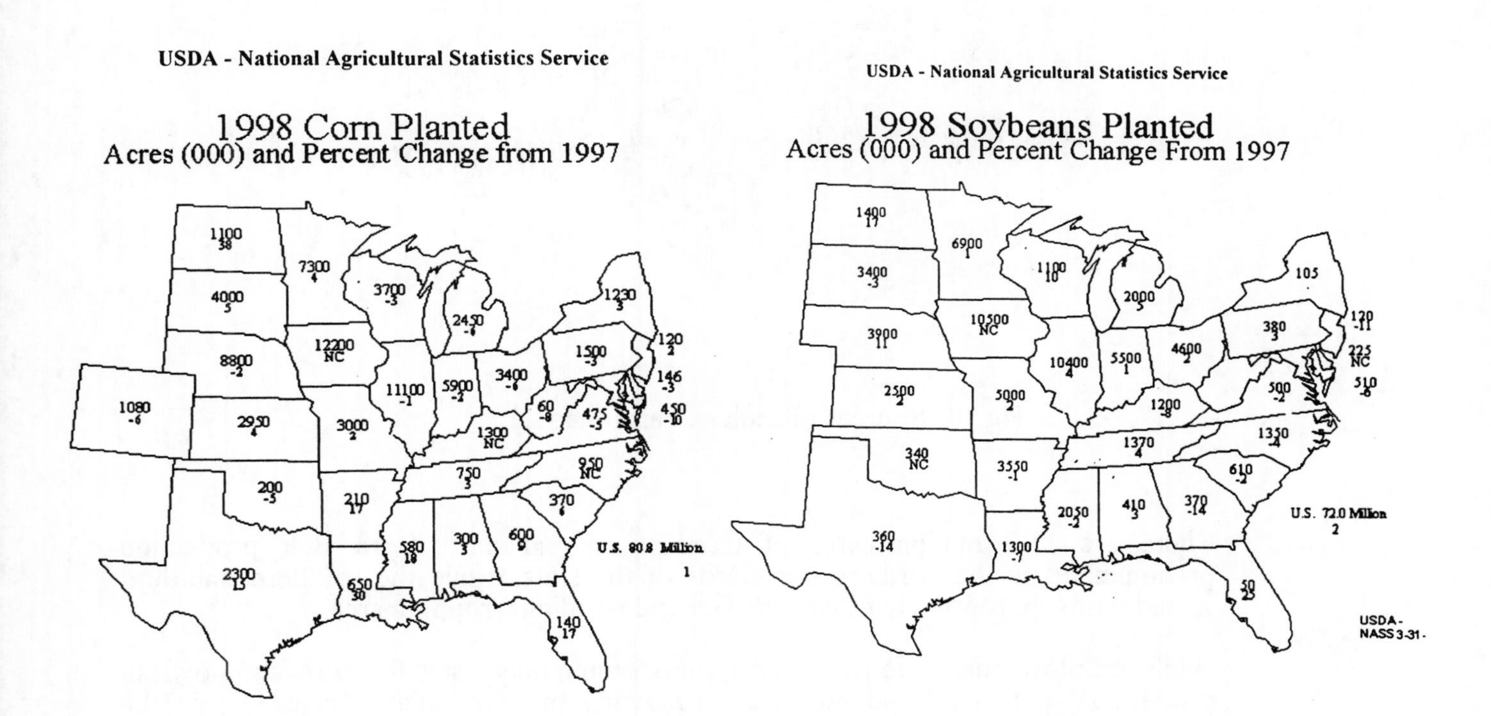

Fig. 2. Corn and Soybean Acres Planted in the Corn Belt.

IV. CROPPING SYSTEM SELECTION

Objectives

1. Describe characteristics of crop production in Indiana.
2. Describe how climate, soil, and economics affect cropping system selection.

A. Indiana Cropping Systems

Agriculture plays an extremely important role to Indiana's economy. In 1994, there were 63,000 farms in Indiana which produced $5.1 billion worth of farm products. Of this total, crop marketings were worth $3.1 billion (Fig. 1). Corn was the leading source of income, meat animals were second, and soybeans were third. The four major crops grown in Indiana are corn, soybeans, wheat and hay.

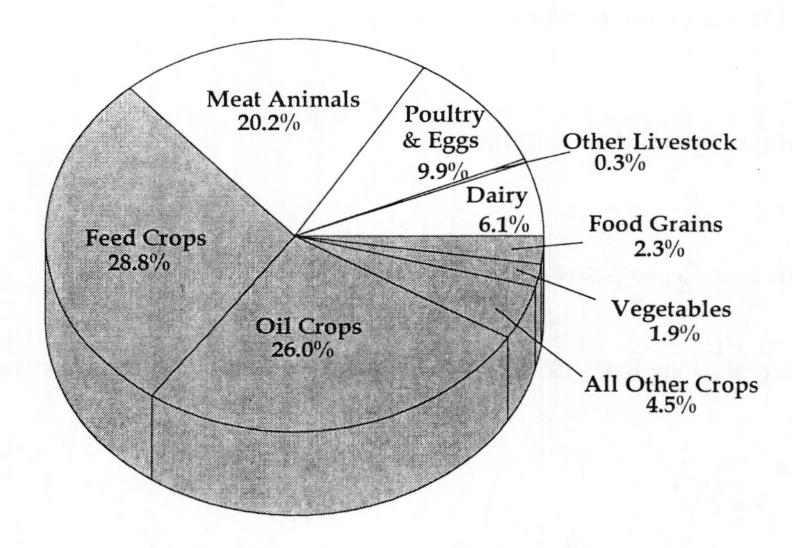

Fig. 1. Sources of Indiana Farm Income.

There are 15.9 million acres of farmland in Indiana. Corn Belt production predominates in the northern two-thirds of the state, while the southern one-third would more appropriately fit into the General Farming Cropping area.

While no-plow conventional tillage is most commonly used for corn and soybean production, mulch till and no-till are becoming more popular (Table 1). In the midwest, 30 inch row spacing is most common for corn, and most soybeans are seeded in 15 inch or narrower rows (Fig. 2).

Table 1. Indiana Tillage Systems - Corn and Soybean Production 1992

Category	Corn	Soybeans
Planted Acres (1,000)	6,100	4,500
	Percent of Acres	
Tillage systems:		
Conventional/w Moldboard Plow	8	15
Conventional/wo Moldboard Plow	54	36
Mulch-till	25	25
No-till	13	24
	Percent of Soil Surface Covered	
Residue Remaining After Planting		
Conventional/w Moldboard Plow	2	2
Conventional/wo Moldboard Plow	16	19
Mulch-till	38	41
No-till	64	72
Average	26	35

Source: Economic Research Service

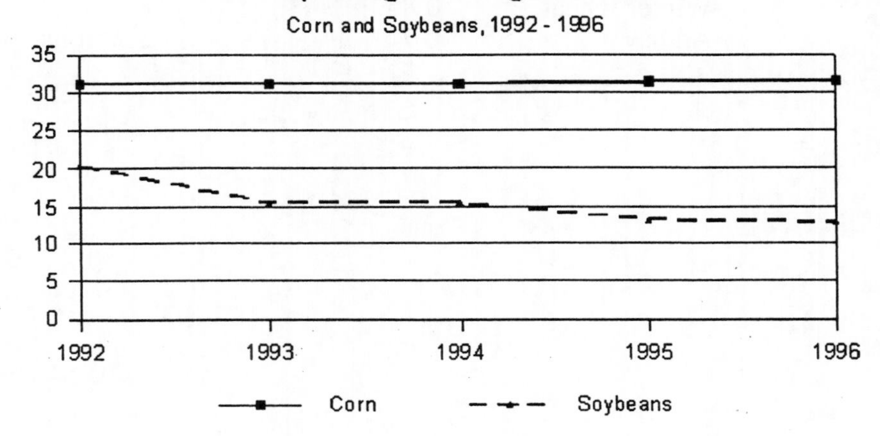

Fig. 2. Indiana Corn and Soybean Row Spacing.

B. Cropping System Selection

In order for a cropping system to be successful on a farm it must meet 3 criteria:

1. The crop or crops must be environmentally adapted.

   The relatively level, fertile soils of much of Indiana, and the favorable sunlight, temperature, precipitation patterns combine to make the production of row crops and small grains extremely attractive.

2. There must be a market or use for the crops produced.

   Crops such as sugar beets are adapted to Indiana, but are not grown because of the lack of market.

3. Costs of production and expected returns must allow for a reasonable profit.

C. Major Crops Grown in Indiana

The four major crops grown in Indiana are corn, soybeans, wheat, and hay (table 2). Other crops grown in Indiana include oats, peppermint, tomatoes, melons, turf, and a wide variety of orchard and horticultural crops. However, because of Indiana's unique location and combination of soil and climatic factors which make row crop production profitable, corn and soybeans will likely continue to dominate the crop production picture for some time to come.

Table 2. Indiana Crop Production, 1997.

	Indiana	
Crop	Acres harvested	Yield per acre
Corn for grain	6.0 million	116 bu.
Soybeans	5.5	44
Winter wheat	0.85	58
All hay	0.7	3.2 tons

INFLUENCE OF SOILS ON CROP PRODUCTION

I. SOIL FORMATION

Objectives

1. Define "soil".
2. Describe the 5 soil-forming factors.
3. Describe the following types of parent material:
   a. residual
   b. glacial
   c. loess
   d. alluvial
   e. colluvial
   f. lacustrine
4. Distinguish a mineral soil from an organic soil.

A. What is Soil?

Soil is defined differently by different people. To a mining engineer it might be defined as the bothersome material covering the valuable mineral reserves. A contractor might define soil as the structural fill on which to build houses or roads. The agronomist considers the soil as the thin, naturally occurring dynamic material covering the terrestrial land surface in which plant may grow. You should be able to derive your own definition of "soil".

B. Soil Forming Factors

The soils on a farm result from the interaction of five factors. These factors interact through the process of weathering to result in the material we call "soil." Soil formation may occur through chemical weathering, the chemical breakdown of parent material; physical weathering, the mechanical breakdown of parent material; or biological weathering, the breakdown of parent material due to microorganism activity.

The five factors affecting soil formation are:

1. Parent Material

   a. Residual materials

   b. Transported materials

      glacial
      loess
      alluvial
      colluvial
      lacustrine

2. Organisms

3. Climate

4. Topography

5. Time

C. Types of Agricultural Soils

1. Organic soils:

   a. Muck

   b. Peat

2. Inorganic, or mineral soils (most agricultural soils are mineral soils). The composition by volume of a typical mineral soil in good productive condition consists of soil solids (50%) and pore space (50%). The soil solid fraction contains the mineral and organic portions which remain fairly constant. The amount of water and air in the pore space of soil changes rapidly (Fig. 1).

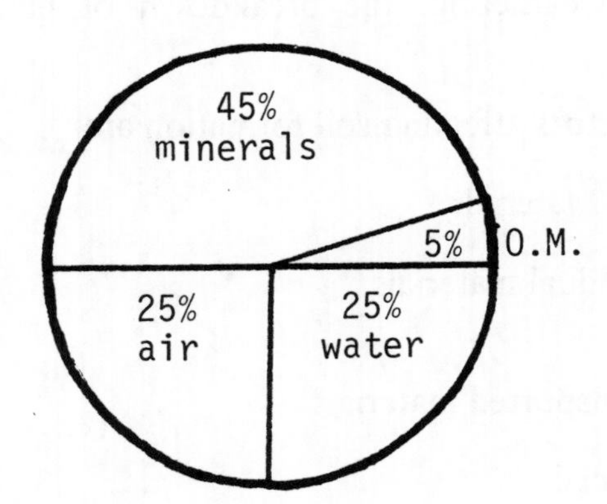

Fig. 1. Volume composition of a typical mineral soil.

II. SOIL PROFILE CHARACTERISTICS

Objectives

1. Define soil horizon and soil profile.
2. Name the four major horizons.
3. Use color, texture, and structure to distinguish soil horizons.

A section of soil which exposes its various layers of development is called a soil profile. A soil horizon is a more or less horizontal layer of development with fairly uniform chemical and physical characteristics.

Characteristics of each horizon:

Oi:

Oa:

A:

B:

C:

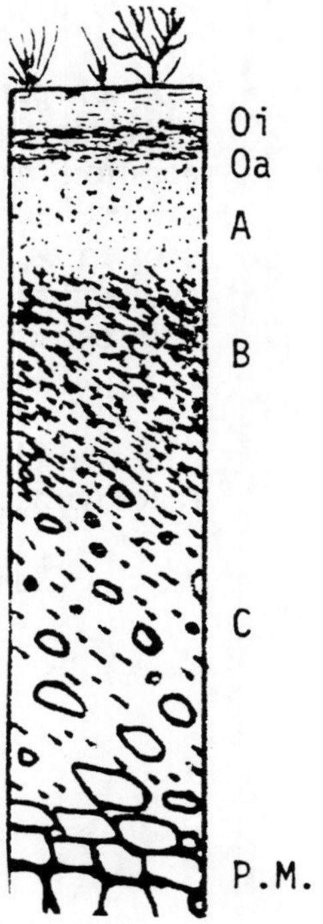

Oi
Oa
A
B
C
P.M.

Profile of a forest soil.

A prairie soil may be similar, except it has no O horizons.

III. SOIL PROPERTIES

Objectives

1. Describe basic biological, chemical and physical characteristics of soil.
2. List the two chemically reactive portions of the soil.
3. Compare and contrast 1:1 and 2:1 clays.
4. Recognize the role of cation exchange in plant nutrition.
5. List characteristics of soil organic matter.
6. Define humus.
7. Describe the relationship between soil organic matter and:
   a. soil color
   b. soil structure
   c. cation exchange capacity

Soil properties result from the interaction of physical, chemical, and biological factors. Physical properties, such as texture and structure affect conditions such as soil moisture, aeration, root growth, and soil tilth. Chemical properties affect management decisions such as fertilizer recommendations, liming needs, and herbicide application rates. Biological properties may affect herbicide carryover, organic matter decomposition, and nutrient availability. All of these factors interact. For example, a soil that is physically compacted will restrict biological activity, thus reducing the rate of organic matter decomposition.

The two chemically reactive portions of the soil are clay and organic matter. These portions react with, and hold nutrients, soil amendments, and pesticides applied to the soil:

A. Clays

1. Clay micelle structure

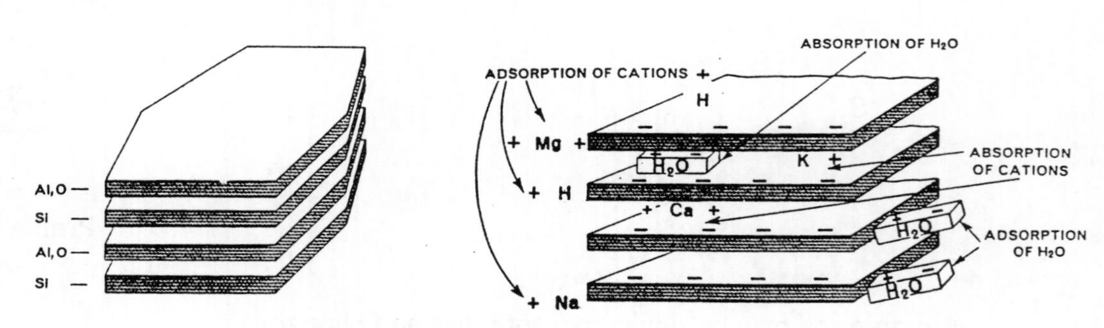

2. Cation exchange capacity (CEC)

Read AY 238, Fundamentals of Soil Cation Exchange Capacity, for a discussion of cation exchange capacity and its importance in soil management.

B. Soil organic matter

There are two types of organic matter in the soil. The **original material** consists of the plant and animal residue on or in the soil that has not yet had time to decompose. The **humus** is the dark colored more resistant products of decomposition that remains after the easily decomposed material has disappeared. The CEC of organic matter is largely a function of the humus in the soil.

Agronomy Guide

Purdue University Cooperative Extension Service

Soils (Fertility) AY-238

Fundamentals of Soil Cation Exchange Capacity (CEC)

David B. Mengel, Department of Agronomy, Purdue University

Soils can be thought of as storehouses for plant nutrients. Many nutrients, such as calcium and magnesium, may be supplied to plants solely from reserves held in the soil. Others like potassium are added regularly to soils as fertilizer for the purpose of being withdrawn as needed by crops. The relative ability of soils to store one particular group of nutrients, the cations, is referred to as *cation exchange capacity* or CEC.

Soils are composed of a mixture of sand, silt, clay and organic matter. Both the clay and organic matter particles have a net negative charge. Thus, these negatively-charged soil particles will attract and hold positively-charged particles, much like the opposite poles of a magnet attract each other. By the same token, they will repel other negatively-charged particles, as like poles of a magnet repel each other.

Forms of Nutrient Elements in Soils

Elements having an electrical charge are called ions. Positively-charged ions are *cations*; negatively-charged ones are *anions*.

The most common soil cations (including their chemical symbol and charge) are: calcium (Ca^{++}), magnesium (Mg^{++}), potassium (K^{+}), ammonium (NH_4^{+}), hydrogen (H^{+}) and sodium (Na^{+}). Notice that some cations have more than one positive charge.

Common soil anions (with their symbol and charge) include: chlorine (Cl^{-}), nitrate (NO_3^{-}), sulfate ($SO_4^{=}$) and phosphate ($PO_4^{\equiv}$). Note also that anions can have more than one negative charge and may be combinations of elements with oxygen.

Defining Cation Exchange Capacity

Cations held on the clay and organic matter particles in soils can be replaced by other cations; thus, they are *exchangeable*. For instance, potassium can be replaced by cations such as calcium or hydrogen, and vice versa.

The total number of cations a soil can hold—or its total negative charge—is the soil's cation exchange capacity. The higher the CEC, the higher the negative charge and the more cations that can be held.

CEC is measured in millequivalents per 100 grams of soil (meq/100g). A meq is the number of ions which total a specific quantity of electrical charges. In the case of potassium (K^{+}), for example, a meq of K ions is approximately 6×10^{20} ions or 6×10^{20} positive charges. With calcium, on the other hand, a meq of Ca^{++} is also 6×10^{20} positive charges, but only 3×10^{20} ions because each Ca ion has two positive charges.

Following are the common soil nutrient cations and the amounts in pounds per acre that equal 1 meq/100g:

Calcium (Ca^{++}) — 400 lbs./acre
Magnesium (Mg^{++}) — 240 lbs./acre
Potassium (K^{+}) — 780 lbs./acre
Ammonium (NH_4^{+}) — 360 lbs./acre

Measuring Cation Exchange Capacity

Since a soil's CEC comes from the clay and organic matter present, it can be estimated from soil texture and color. Table 1 lists some soil groups based on color and texture, representative soil series in each group, and common CEC value measures on these soils.

Table 1. Normal Range of CEC Values for Common Color/Texture Soil Groups.

Soil groups	Examples	CEC in meg/100g
Light colored sands	Plainfield Bloomfield	3-5
Dark colored sands	Maumee Gilford	10-20
Light colored loams and silt loams	Clermont-Miami Miami	10-20
Dark colored loams and silt loams	Sidell Gennesee	15-25
Dark colored silty clay loams and silty clays	Pewamo Hoytville	30-40
Organic soils	Carlisle muck	50-100

Cation exchange capacity is usually measured in soil testing labs by one of two methods. The direct method is to replace the normal mixture of cations on the exchange sites with a single cation such as ammonium (NH_4^+), to replace that exchangeable NH_4^+ with another cation, and then to measure the amount of NH_4^+ exchanged (which was how much the soil had held).

More commonly, the soil testing labs estimate CEC by summing the calcium, magnesium and potassium measured in the soil testing procedure with an estimate of exchangeable hydrogen obtained from the buffer pH. Generally, CEC values arrived at by this summation method will be slightly lower than those obtained by direct measures.

Buffer Capacity and Percent Base Saturation

Cations on the soil's exchange sites serve as a source of resupply for those in soil water which were removed by plant roots or lost through leaching. The higher the CEC, the more cations which can be supplied. This is called the soil's *buffer capacity.*

Cations can be classified as either acidic (acid-forming) or basic. The common acidic cations are hydrogen and aluminum; common basic ones are calcium, magnesium, potassium and sodium. The proportion of acids and bases on the CEC is called the *percent base saturation* and can be calculated as follows:

$$\text{Pct. base saturation} = \frac{\text{Total meq of bases on exchange sites (i.e., meq } Ca^{++} + \text{meq } Mg^{++} + \text{meq } K^{+})}{\text{Cation exchange capacity}} \times 100$$

The concept of base saturation is important, because the relative proportion of acids and bases on the exchange sites determines a soil's pH. As the number of Ca^{++} and Mg^{++} ions decreases and the number of H^+ and Al^{+++} ions increases, the pH drops. Adding limestone replaces acidic hydrogen and aluminum cations with basic calcium and magnesium cations, which increases the base saturation and raises the pH.

In the case of Midwestern soils, the actual mix of cations found on the exchange sites can vary markedly. On most, however, Ca^{++} and Mg^{++} are the dominant basic cations and are in greater concentrations than K^+. Normally, very little sodium is found in Midwestern soils.

Relationship Between CEC and Fertilization Practices

Recommended liming and fertilization practices will vary for soils with widely differing cation exchange capacities. For instance, soils having a high CEC and high buffer capacity change pH much more slowly under normal management than low-CEC soils. Therefore, high-CEC soils generally do not need to be limed as frequently as low-CEC soils; but when they do become acid and require liming, higher lime rates are needed to reach optimum pH.

CEC can also influence when and how often nitrogen and potassium fertilizers can be applied. On low-CEC soils (less than 5 meq/200g), for example, some leaching of cations can occur. Fall applications of ammonium N and potassium on these soils could result in some leaching below the root zone, particularly in the case of sandy soils with low-CEC subsoils. Thus, spring fertilizer application may mean improved production efficiency. Also, multi-year potash applications are not recommended on low-CEC soils.

Higher-CEC soils (greater than 10 meq/100 g), on the other hand, experience little cation leaching, thus making fall application of N and K a realistic alternative. Applying potassium for two crops can also be done effectively on these soils. Thus, other factors such as drainage will have a greater effect on the fertility management practices used on high-CEC soils.

Summary

The cation exchange capacity of a soil determines the number of positively-charged ions—cations—that the soil can hold. This, in turn, can have a significant effect on the fertility management of the soil.

RR 3/93 (2.5M)

Cooperative Extension Work in Agriculture and Home Economics, State of Indiana, Purdue University and U.S. Department of Agriculture Cooperating. H. G. Diesslin, Director, West Lafayette, IN. Issued in furtherance of the Acts of May 8 and June 30, 1914. It is the policy of the Cooperative Extension Service of Purdue University that all persons shall have equal opportunity and access to its programs and facilities without regard to race, religion, color, sex or national origin.

IV. MAINTAINING A FERTILE SOIL

Objectives

1. List the 16 essential nutrients.
2. Distinguish between macronutrients and micronutrients.
3. Describe 4 sources of nutrients.
4. Explain the information presented on a fertilizer tag or bag.
5. List characteristics of dry, liquid, gas, and slurry fertilizers.

A. Elemental Composition of Agronomic Crops

Carbon, hydrogen, and oxygen make up the bulk of an actively growing crop in the field.

B. The Essential Elements:

1. Non-soil supplied -

2. Macronutrients -

3. Micronutrients -

C. Sources of Nutrients

The source of a plant nutrient is immaterial, as long as the nutrient is in an available form, and is not applied with other harmful materials. For example, corn is able to utilize nitrogen from anhydrous ammonia fertilizer equally as well as from manure. Four sources of nutrients for the crop are:

1. The earth's atmosphere.

   Each time a lightning storm occurs, a small amount of nitrogen is converted to plant available nitrogen. The amount has been estimated at about 5 lbs. per acre per year.

   Precipitation may also indirectly supply plant nutrients by washing pollutants from the atmosphere. The kind and amount of nutrients supplied in this manner vary with location and amount of annual precipitation. In addition, atmospheric nitrogen can be fixed by bacteria living in nodules on roots of legumes.

2. Organic sources.

Common organic sources of plant nutrients are crop residues, barnyard manure, and biosolids, or municipal sludge. When residue from vegetation is incorporated into the soil, the nutrients remaining in the residue may be once again available for crop use. However, the amounts of nutrients available from crop residues are inadequate to meet the needs of high yielding crops.

The nutrient content of barnyard manure is quite variable. While an excellent source of nutrients, manure may not have the right ratio of nutrients to meet, but not exceed, fertilizer recommendations. Therefore, manure is often supplemented with commercial fertilizers. Read, "Crediting Manure in Soil Fertility Programs" for a discussion of manure management.

3. Soil weathering

During natural weathering of soils, nutrients are made available. For many of the micronutrients this process may supply adequate amounts for most agronomic crops. However, crop needs for other nutrients, such as N, P, and K, usually cannot be met through release by soil weathering.

CREDITING MANURE IN SOIL FERTILITY PROGRAMS

By Dr. Randy Killorn

Many crop producers routinely apply manure to their fields. Failure to account for the nutrients in manure can lead to uneconomical application of fertilizer and increased potential for environmental contamination.

The amount of nutrients in a ton or 1,000 gallons of manure that will be available for crop production is affected by many factors. Such things as the type and age of animal, how the manure is collected and stored, and the method of land application all affect the nutrient credit that should be taken.

The average nutrient content of manure from different species of animals and various handling systems is listed in Table 1. However, the best way to determine the nutrient content of manure is through chemical analyses since it will vary from farm to farm.

Samples can be collected from the spreader as the manure is applied. The results from analyses of these samples should be valid for several years because the nutrient concentration is stable as long as the animal management remains the same. By basing credits on a recent chemical analyses you can avoid having to estimate the effect of animal type, age, and storage system.

It is important to remember that animal manure contains plant nutrients in both inorganic and organic compounds. Only the inorganic forms are available to plants.

NITROGEN

About half the total nitrogen in manure is ammonium. The rest is in various organic compounds. The ammonium is immediately available to plants. Organic N is not available to plants until it is released through decomposition of organic matter by soil microorganisms. The rate that this occurs depends on soil temperature and moisture conditions. However, not all of the organic N will become available the year the manure is applied. Nitrogen available the year of application ranges from 25% to 75% depending on the type of manure and how it is handled (Table 2). The remaining organic N will become available at a rate of about 5% per year.

The amount of credit given to N in animal manure should reflect the method of application. If the manure is broadcast and not incorporated the ammonium component can be lost through volatilization, and the credit for N should be reduced as shown in Table 2.

PHOSPHORUS AND POTASSIUM

Estimates of P and K availability the year of application range from 50% to 100% of the total. A conservative estimate is that about 70% of both will be available. The effect of application of P and K in manure can be monitored using a soil testing program.

TABLE 1. NUTRIENT CONTENT OF MANURE.

Type of Manure	Total N	Total P_2O_5	Total K_2O
Liquid		**lb/1,000 gallons**	
Swine			
Finisher/Grower	49	35	25
Nursery	37	28	22
Farrowing	13	10	8
Mixed	29	19	15
Dairy			
W/out parlor waste	45	21	31
W/parlor waste	23	11	23
Beef			
W/out parlor waste	45	21	31
W/parlor waste	2	1	3
Poultry			
Pit	80	36	96
Solid Manure		**lb/ton**	
Swine Scrape & Haul			
Summer	33	34	15
Winter	21	17	9
Dairy Feedlot			
Summer	17	9	16
Winter	12	7	7
Dairy Bedded Pack	15	6	21
Poultry			
W/out litter	33	48	34
W/litter	56	45	34

These estimates were extracted from extension pamphlets from several states.

POULTRY LITTER GOOD NPK SOURCE

Poultry litter is a good source of nitrogen for growing corn, according to Monroe Rasnake, University of Kentucky extension agronomist. He points this fact out because the development of an intensive poultry industry in Kentucky will make poultry litter available to many farmers.

Rasnake provided this evaluation of litter:

Poultry litter consists of a bedding material such as saw dust, wood shavings and rice hulls mixed with manure. It contains an average of about 20% moisture and 55 pounds of nitrogen, 50 pounds of phosphate and 40 pounds of potash per ton. These figures vary considerably; therefore, litter should be tested to determine the actual nutrient content.

NUTRIENT VALUE

The value of nutrients contained in poultry litter can be estimated by comparing with the cost of commercial fertilizer. For example, using the average figures mentioned previously, and nutrient costs per pound of $.20 for nitrogen, $.20 for phosphate and $.12 for potash, litter would be worth $25.80 per ton.

Not all of these nutrients will be available to the crop the first year, however. Because most of the nitrogen is in the organic form, only about 60% will be available the first year. About 20% will be lost to leaching, denitrification, etc. Very little of the phosphorus and potassium will be lost, but only about 75% will be available the first year. The fertilizer value during the application year could be calculated at about $18 per ton.

Surface applications are less efficient than mixing with the soil, especially in dry years, Rasnake says. In situations where large applications of P and K are not needed, it may be better to use lower rates of litter and use fertilizer to supply some of the nitrogen. This would better utilize nutrients in the litter and reduce the risk of water pollution.

SELECTING THE RATE OF APPLICATION

Applying animal manure for crop production presents at least one significant problem: We cannot alter the amount of N, P and K in the manure so as to apply the nutrients in the exact amounts called for in the fertilizer recommendation. Figure 1 illustrates that on Iowa soils testing low in P and K, applying liquid swine manure to supply the N required for a 140 bushel corn yield works well. The P and K needs are also met.

If the soil tests high in P and K, applying the manure at the same rate will result in an over-application of P and K, exceeding the rates needed to maintain the soil test. In this case the manure should be applied to meet the phosphorus need. Any additional N and K that is needed can be supplied by fertilizer.

TABLE 2. PERCENT OF TOTAL N AVAILABLE.

Type of Manure	Year of application 1	3	5
Swine or beef, liquid			
Incorporated	75 %	85 %	95 %
Not incorporated	50	65	75
Swine or beef, feedlot			
Stockpiled, incorporated	35	50	55
Dairy, fresh			
Incorporated	50	70	78
Dairy, liquid			
Incorporated	40	60	68
Dairy, fresh or liquid			
Not incorporated	25	35	45

MANURE MANAGEMENT PLAN

The best way to help crop producers manage the nutrients in manure is to base the applications on the soil tests from their fields and on the nutrient content of the manure. This information should be coupled with crop plans and manure applied to those fields that will benefit most.

For example, priority fields for manure application should be those with low soil tests for P and K intended for corn production. Lower priority should be given to fields that will be planted to legumes such as soybeans.

SUMMARY

The nutrients contained in manure can and should be used for crop production. The best way to do so is to determine the nutrient content of the manure and apply it to fields that require the nutrients to support profitable crop production. Take care to ensure that the availability of the nutrients is realistically assessed. Under-application reduces yields and profitability while over-application reduces profitability and increases the potential for contamination of the environment. ■

Dr. Randy Killorn is Extension Agronomist and Associate Professor of Soil Fertility at Iowa State University.

4. Commercial fertilizers

Because high yielding crops remove considerable quantities of nutrients, these crops usually receive applications of commercial fertilizers.

The fertilizer label contains information that you will need to insure proper fertilizer application. Study the information on the label below. You should be familiar with the following terms:

a. Analysis

b. Ratio

c. Carrier

d. Micronutrient guarantee

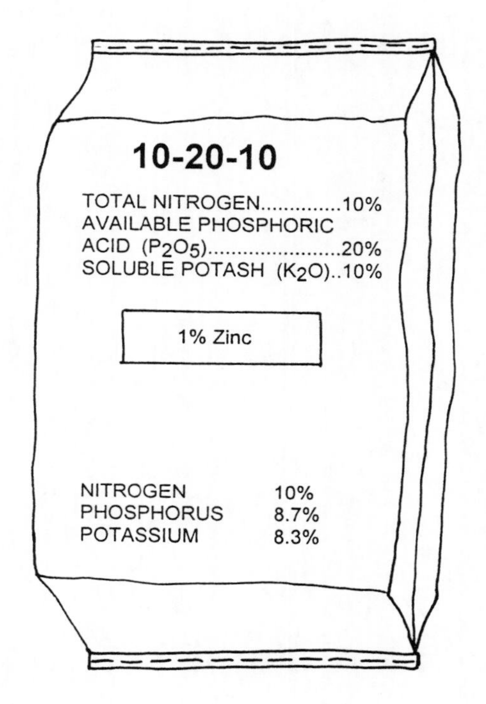

Commercial fertilizers are produced in the following formulations:

Gas -

Liquid -

Slurry -

Solid -

Agronomy Guide

Purdue University Cooperative Extension Service

SOILS/FERTILITY AY-268

Fertilizing Corn Grown Using Conservation Tillage

David B. Mengel, Department of Agronomy, Purdue University

The use of conservation tillage in corn production is increasing rapidly in Indiana. There are a number of reasons for this. One is the belief shared by many Hoosier farmers that the land must be protected and preserved. But other reasons also are important. Reduced tillage techniques can be used to reduce costs of labor and machinery and to save fuel and energy. Yield increases are also possible on droughty soils, and reduced tillage can allow a person to farm more land with the same resources. Conservation compliance which ties soil conservation efforts to the federal farm program will also force some farmers growing crops on highly erodible land to begin to adopt conservation tillage in the near future.

The adoption of conservation tillage systems requires some change in the management practices used for corn production, particularly in the areas of weed control, pest management, and fertilization. This publication describes the adjustments in fertilizer management programs which have been shown to be important when conservation tillage is adopted in Indiana.

Soil Testing

The first step in developing a fertilizer program, whether using conventional plowing or conservation tillage, is a good job of soil testing. Soil tests are primarily used in Indiana to make routine recommendations for phosphorus (P), potassium (K), and agricultural limestone. Indiana soils tend to be quite variable. It is common to find soils of widely differing physical and chemical properties in the same field. Thus, developing a soil sampling program that addresses this variability is important.

Two basic approaches are used when measuring soil spatial variability in sampling to make fertilizer recommendations. The University of Illinois has developed a system of grid sampling where samples are taken using a systematic sampling pattern with a density of about one composite sample per 2½ acres. The farmer or consultant sets up sampling lines or rows 300 feet apart and simply takes a composite sample (8 to 10 cores) every 100 yards on each line. Thus, a farmer using this system would have a map of fertility patterns based on sampling points on a 100-yard by 100-yard grid. In Indiana, modern soil surveys, with soil maps superimposed on aerial photographs, are commonly used to define sampling areas. This generally results in a single composite soil sample representing a 10- to 25-acre area of similar soils. Both systems provide insight into the variability of fertility levels in a field and allow a grower to apply fertilizers only where needed. This in turn should lead to more efficient and economical use of fertilizer.

Most cornbelt P and K fertilizer recommendations are made based on soil tests calibrated to a 6- to 8-inch sampling depth. Research to date has not shown any reason to deviate from that sampling depth when sampling with the intent of making routine P and K recommendations for conservation tillage.

Stratification of P and K in the surface few inches of soil in fields where reduced tillage has been used for some time has been routinely observed. However, in most situations this has not led to any particular problems. One exception to this has been the interaction of drought stress and nutrient stratification. Under dry conditions, when root activity in the surface of the soil has been reduced, availability of P and K can be reduced. This can enhance the normal reductions in P and K uptake found in dry years and reduce yields further in reduced tillage situations.

One sampling method proposed for use in tillage systems such as no-till, where nutrient stratification can be a problem, is to split the normal 8-inch core into two sub-samples representing the top 4 inches of the root-zone and the bottom 4 inches. Routine P and K recommendations are made from the results from a normal 8-inch sample. However, if the test results from the bottom 4 inches are very low in available P or K, then the use of deeper fertilizer placement techniques, or tillage, to mix fertility into that zone may be called for.

A second deviation from routine 8-inch sampling depth that should be used, is to sample the surface 1 to 2 inches for soil pH only. When nitrogen fertilizers such as urea or urea-ammonium nitrate (UAN) solutions are applied directly to the surface in no-till systems, the surface few inches of soil can be acidified very quickly. Thus, sampling the surface on a regular basis can detect these changes and prevent this low pH from adversely effecting plant growth or herbicide activity.

Liming and pH Control

Most of the soils of the eastern corn belt of the U.S. have naturally acidic surface horizons. The use of nitrogen fertilizers further compounds the acidity problem. In Indiana, optimum pH for corn and soybeans is thought to be in the range of 6.0 to 6.5 on most soils. Below this level, both plant growth and triazine herbicide activity can be reduced. Agricultural limestone is normally used to increase the pH of acid soils.

Current limestone recommendations are based on the use of the Shoemaker-McLean-Pratt (SMP) buffer to estimate exchangeable acidity. These recommendations assume the use of conventional tillage. Changing to reduced tillage will normally reduce the volume of soil liming materials react with. Thus, the amount of lime used at each application must be reduced in very reduced tillage systems to prevent over-liming of the surface and micronutrient deficiencies such as zinc in corn or manganese in soybeans. The frequency of liming is usually increased in reduced tillage since the volume of soil that acidifying fertilizers react with is also reduced. In continuous corn, no-till systems this normally means soil sampling and liming every 2 to 3 years, but with lime rates of about half those normally used in conventional plowing systems.

Phosphorus and Potassium Fertilization

One of the factors of reduced tillage that concerned both farmers and agronomists is the accumulation of phosphorus and potassium in the surface few inchs of soil when the moldboard plow is no longer used. The data in Table 1 clearly illustrate the problem.

This type of data has created a major concern for many people. However, root distribution data from these tillage systems (Table 2) show that roots tend to accumulate in the areas where the fertility is highest. Thus, in normal conditions, the stratification may be less of a problem than first thought and may actually lead to increased nutrient availability.

Weather, however, is one of a number of factors that can influence the availability of the nutrients concentrated near the surface. The data in Table 3, collected from a long-term study in Indiana, clearly illustrate this point. Under the drier than normal conditions found in Indiana in 1980, P and K levels in the leaf of corn were reduced, as compared to the "normal" season of 1981. This was particularly true in the case of K in the chisel and no-till plots, where nutrients and roots were concentrated near the soil surface. Yields followed similar trends, with reductions in yields in reduced tillage in the drier year.

This type of data has led to the conclusion that with the weather and soil fertility levels common in the eastern cornbelt (where the subsoil contributes little or no fertility), P and K fertilization should present few problems when reduced tillage systems are adopted if a few simple precautions are taken.

First, farmers should make an effort to ensure that soil pH and P and K levels are properly adjusted before going to a continuous reduced tillage program. This means that the surface 6 to 8 inches of soil would have a pH of 6.0 or greater, Bray P1-P level of 20 ppm (40 lb./acre) or greater, and an exchangeable K level of 125 to 150 ppm (250 to 300 lb./acre) or greater. If these conditions are met, then routine applications of P and K can be made based on soil tests from the surface 6 to 8 inches of soil, and the fertilizer can be applied directly to the soil surface.

If either P or K levels are not built up prior to adoption of a reduced tillage system, then there are some indications that the use of alternate placement methods such as deep banding (6 to 8 inches deep) or surface stripping may prove to be of some advantage. These same placement systems would also be considered in conventional tillage under similar circumstances.

Over long periods of time surface applications of P and K, together with the natural cycling of nutrients due to plant uptake which will occur in continuous no-till systems, can lead to marked stratification of P and K and to low levels of these nutrients in the lower portions of the old tillage zone. This will occur even when soil test levels were high when the reduced tillage system was established. For this reason many agronomists suggest using a split soil sample to monitor nutrient level of the lower portion of the old plow layer. Farmers can then switch to an alternative fertilizer placement system or use tillage to increase the fertility level of the lower portions of the plow layer to prevent nutrient uptake problems in dry years.

Starter Fertilizer

Recent research in Indiana has shown that the probability of response to banded starter fertilizer increases markedly as reduced tillage systems are used. The data in Table 4 illustrate this point. In 11 experiments where starter fertilizer treatments were used in both no-till and conventional tillage for corn, starter fertilizer responses were obtained only in one case in conventional tillage, but in eight of the 11 experiments with no-till.

Traditionally, American farmers think of P as the nutrient responsible for starter fertilizer response. Current work in Indiana has shown, however, that nitrogen (N) is responsible for most of the starter fertilizer responses observed. The data in Table 5 illustrate this point and again emphasize the tillage effect on starter response.

All of the studies in Table 5 were conducted at soil test levels above which responses to applied P and K would normally be expected. At low soil test levels (Bray P1-P <10 ppm, exchangeable K <75 ppm) starter applications of P and/or K would also be recommended regardless of tillage system.

Work in the northern cornbelt (Wisconsin and Minnesota) has also shown excellent responses to starter fertilizer when reduced tillage systems are used. The work indicates a large response to K in starter fertilizers, especially in no-till. Thus, while the response to starter appears to be consistent across the eastern corn belt, there seem to be some geographic differences in the nutrients responsible for these responses. In Indiana and southern states, N is a principle contributor to the response, while in Wisconsin and Minnesota, K appears to be more important.

Nitrogen Fertilization

The key changes in a fertilizer program that should occur when reduced tillage systems are adopted relate to nitrogen fertilization. Reduced tillage means more crop residue on the soil surface, which can affect N use efficiency, particularly when the N fertilizer is applied directly on the residue.

Table 1. P and K soil test levels found at various depths in the soil. Cruz, 1982.

Depth	Plow	Chisel	No-till	Plow	Chisel	No-till
inches	----------Bray P_1, ppm----------			----------Exch. K, ppm----------		
0-3	37	85	90	150	230	285
3-6	47	35	27	165	105	100
6-9	30	15	18	140	100	100
9-12	8	8	8	100	100	100

Table 2. Influence of tillage systems on corn root density at different depths in the soil. Cruz, 1982.

Soil depth	Root density		
	Plow	Chisel	No-till
inches	-----mg roots/cm^3 soil-----		
0-3	0.75	0.83	1.88
3-6	0.98	0.98	0.75
6-9	0.51	0.50	0.50
9-12	0.23	0.22	0.23

Table 3. Nitrogen, phosphorus, and potassium content of corn leaves as affected by tillage system. Cruz, 1982.

Tillage system	Nutrient concentration in corn leaf							
	Percent N		Percent P		Percent K		Yield	
	1980	1981	1980	1981	1980	1981	1980	1981
	------------------------%------------------------						---bu./a.---	
Plow	2.82	2.79	0.24	0.31	1.77a	2.27	150	169
Chisel	2.73	2.89	0.23	0.32	1.56b	2.21	136	171
No-till	2.77	2.84	0.23	0.33	1.49b	2.27	135	166
NS	NS	NS	NS	NS	**	NS	*	NS

* Statistically significant at the 5% level.
** Statistically significant at the 1% level.
NS - No significant difference.

Table 4. Effect of tillage on starter response, Indiana.

	Number of responses over 300 bu./a.	Average yield increase when response	Overall yield response
		--------bu./a.--------	
Conventional	1/11	12	0.9
No-till	8/11	10	7.8

Mean of 11 studies using starter fertilizer placed 2 inches to the side and 2 inches below the seed and at least 100 lbs of fertilizer material and with direct tillage comparisons conducted from 1982 to 1987.

Table 5. Nutrients responsible for starter response, Indiana.

Starter fertilizer	Tillage system		
	Chisel	Ridge	No-till
lb. N-P-K/a.	------Corn yield bu./a.------		
0-0-0	150	141	130
25-0-0	148	144	137
25-17-0	148	146	137
25-17-7	152	145	137

Mean of 6 studies conducted using 2 x 2 placement from 1985-1987.
Soil P level H to VH.

A study conducted in Indiana using N solutions clearly illustrates this point(Table 6). In continuous corn grown using conventional tillage, no significant differences in grain yield or in leaf or grain N content were noted when N solutions were incorporated into the soil immediately before planting, broadcast on the soil surface immediately after planting, or knifed into the row middles shortly after planting. However, in no-till, both yield and leaf N content were reduced with surface application of N directly on the residue, as compared to knifing the fertilizer below the residue.

This observation has been made many times, and it is now well established that for optimum efficiency, every effort should be made to minimize the contact between surface residue and N fertilizers. Unfortunately, it is not always possible to knife N below the residue. The high power requirement of this application method, together with the erosion potential on sloping lands from the disturbed fertilizer slot, make this an unacceptable system in some situations. In these situations alternative approaches can be used.

In Maryland and Pennsylvania research has focused on selecting more efficient N sources for surface application in high residue situations. Both studies report that ammonium nitrate is the preferred N source for no-till corn as compared to urea or UAN solutions. Other work has focused on alternate methods of placement of N fertilizers. Researchers in Alabama have shown that surface banding N solutions or using a shallow band placement at planting time can substantially enhance N use efficiency as compared to broadcasting N fertilizers on top of the residue.

There are a number of mechanisms responsible for the reduced efficiency from surface applied N in high-residue situations. These include enhanced leaching and denitrification due to increased rates of water infiltration and reduced evaporation, immobilization of N in the residue decomposition, and volatilization of ammonia from urea-based fertilizers. The key to an efficient N fertilizer program in high residue situations is to minimize fertilizer-residue contact.

Summary

Soil testing is the key to developing a good fertilizer program regardless of the tillage system used. Sampling patterns should take into account spatial soil variability and provide adequate information to make needed adjustments in application rates.

Liming to neutralize soil acidity will still be important in reduced tillage systems. However, rates may need to be adjusted, and applications may need to be more frequent in very reduced systems such as no-till where nitrogen is surface applied.

Phosphorus and potassium rates should be based on soil tests, and rates should probably be based on a 6- to 8-inch sampling depth. If soil test levels are built up to high levels prior to switching to reduced tillage, P and K problems should be minimal. Over time, however, it may be advisable to monitor the levels of phosphorus and especially potassium in the lower portions of the old tillage zone. Gradual depletion of nutrients in the lower portions of the plow layer may make the crop more susceptible to moisture stress. Tillage or deep banding can both be used to correct the stratification problem.

Starter fertilizer responses have been shown to be more common in reduced tillage. While the nutrient responsible may vary, the trend towards more frequent response is very consistent throughout the U.S.

Nitrogen management is probably the key to a successful fertilizer program for reduced or conservation tillage. Developing systems which will minimize fertilizer-residue contact has been shown to greatly enhance efficiency.

In conclusion, changing to reduced or conservation tillage requires some changes in how fertilizer and plant nutrients are managed. However, research has identified ways that ensure that plant nutrition will not become a limiting factor.

Table 6. Effect of nitrogen source and placement on corn yield, leaf N concentration, and grain N concentration in both plow and no-till production systems. Southeast Purdue Agricultural Center, 1982.

Nitrogen source and placement	Yield bu./a.	Ear leaf %N	Grain %N	Yield bu./a.	Ear leaf %N	Grain %N
		Plowed			No-till	
UAN Broadcast-not incorp.	145	2.48	1.21	128	1.63	1.08
UAN Broadcast-incorp.	153	2.34	1.23	--	--	--
UAN injected	149	2.44	1.29	156	2.13	1.04
LSD.05	NS	NS	NS	17	0.27	NS

NS - No significant difference.

Mengel, 1986.

RR 6/90 (3.5M)

Cooperative Extension work in Agriculture and Home Economics, state of Indiana, Purdue University, and U.S. Department of Agriculture cooperating; H.A. Wadsworth, Director, West Lafayette, IN. Issued in furtherance of the acts of May 8 and June 30, 1914. The Cooperative Extension Service of Purdue University is an affirmative action/equal opportunity institution.

V. SOIL ACIDITY AND ALKALINITY

Objectives

1. Define soil pH and the pH scale.
2. List the most favorable pH range for agronomic crops.
3. List 3 factors that lower soil pH.
4. List 2 factors that raise soil pH.
5. Describe the mechanisms by which soil pH affects plant growth.
6. Recognize cropping system, soil pH, and soil type affecting lime requirement.
7. Describe how neutralizing value influences lime rate.
8. Describe how lime fineness influences rate of pH change.
9. List 5 types of liming material and their neutralizing values.

A. Soil pH

Soil pH refers to the concentration of hydrogen (H+) ions when the soil is put into solution. As the concentration of H+ ions in a soil solution increases the pH decreases. Conversely, the pH increases as the H+ ion concentration decreases. A soil with a pH of 7.0 is neutral, above pH 7.0 is alkaline (basic) and below pH 7.0 is acidic. Since pH is a logarithmic scale, for each unit change in pH, there is a 10 fold change in the concentration of H+ ions.

When correcting a nutrient problem in the soil, pH should be corrected first, because this affects availability of the other nutrients.

B. Suitable pH ranges for several crops are as follows:

CROP	pH RANGE
Alfalfa	6.5-7.5
Red clover	6.0-7.0
Wheat	5.5-7.0
Corn	5.7-7.0
Oats	5.5-7.0
Tobacco	5.2-5.8
Soybeans	5.8-6.8

Soil pH can be modified by both naturally occurring processes and management practices. In humid areas, such as the eastern U.S., soils are usually too acidic and pH must be raised.

C. Factors which decrease pH are:

1. Leaching of ions by water -

2. Removal of ions by crops -

3. Application of acid-farming fertilizers and soil amendments

D. Factors which increase soil pH are:

1. Addition of liming materials -

2. Accumulation of basic ions due to poor drainage or low rainfall-

3. Irrigation with high pH water -

E. Liming and plant growth

Addition of lime to an acid soil is the first aspect of crop nutrition that should be corrected. Maintaining proper soil pH is important because it affects crop growth and nutrition through its effect on:

1. nutrient availability -

2. solubility of toxic substances -

3. activity of soil microorganisms -

4. plant cell growth and development -

F. Determining rate of lime application

The amount of lime needed to increase pH of an acid soil to a desirable level will vary, depending on the crops grown and characteristics of both the soil and the liming material.

Two important soil factors are:

1. Soil pH -

2. CEC -

Two important lime characteristics:

1. fineness -

2. neutralizing value -

G. Types of liming material

The type of lime you select will depend upon its neutralizing value, cost which includes transport and application, and availability. Also, if your soil needs Mg, you should use a dolomitic limestone, which contains Mg. The most commonly used liming materials include:

1. $CaCO_3$ -

2. $Ca(OH)_2$ -

3. CaO -

4. Marl -

VI. NUTRIENT MANAGEMENT

Objectives

1. Describe the traditional and grid methods of soil sampling.
2. Identify information required to make fertilizer recommendations.
3. List types of tests conducted at a soil testing laboratory.
4. Describe site-specific nutrient management.

A. Nutrient management recommendations

Producers are faced with many nutrient management decisions, including how much fertilizer to apply, when to apply it, and what kind to use.

Most producers rely on experts to some degree to help them make the best decisions. Soil testing labs often make fertilizer recommendations using soil test results along with information supplied by the grower. This information includes: previous crop, crop to be planted, yield level desired, and fertilizer, lime, or manure recently applied.

B. Soil testing

The most critical step in the soil testing procedure is obtaining representative samples. Suggestions for proper sampling procedures are given on the sheet entitled "How to Take a Soil Sample".

Chemical tests are used to determine P and K levels and soil pH. In the humid portion of the Corn Belt, nitrogen is not determined chemically because it is very closely related to organic matter and can be estimated with reasonable accuracy. The soil test report tells the level of nutrients available, and may also include any recommendations the grower requested.

C. Site-specific nutrient management

Profitable and environmentally sound nutrient management may be enhanced by managing within-field variability. Site-specific nutrient management involves recording yield, soil test, and soil properties with a precise description of the location within the field where the data was collected (geo-reference). Geo-referencing is made possible by the Global Positioning system (GPS) of earth orbiting satellites. Computer software called Geographic Information Systems (GIS) allows the geo-referenced records to be analyzed and displayed as management maps. Computers use the maps to automatically change fertilizer rates and blends during application (Variable Rate Technology).

D. Soil Sampling

1. Traditional soil sampling techniques

The traditional method of soil sampling is illustrated in "How To Take a Soil Sample" on the next page.

2. Grid soil sampling

Grid soil sampling is used in site-specific nutrient management. A field is divided up into a grid of 2-5 acre rectangular sections and a soil sample is submitted from each section. A sample is composed of 5 or more subsamples pulled from the same locations within each grid section.

Compared to traditional sampling methods, more samples are submitted for analysis. Unlike traditional methods, the number of samples from a field do not correspond to the number of soil types in that field.

How To Take A Soil Sample — And Prepare It For Testing

1 2 3

4

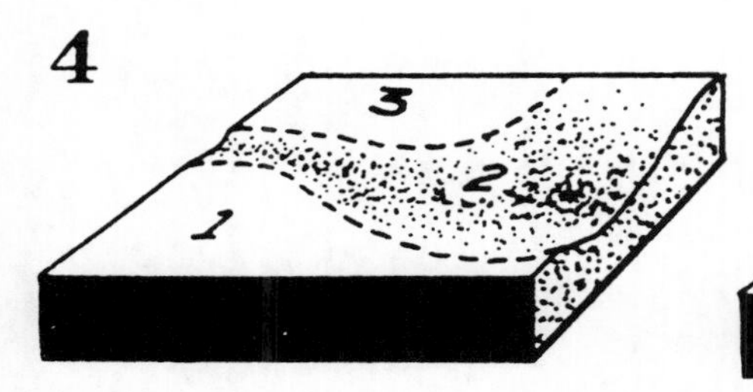

Divide the field into uniform *areas.* Each area should have the same soil color and texture and the same past cropping, fertilizing and liming treatment.

5

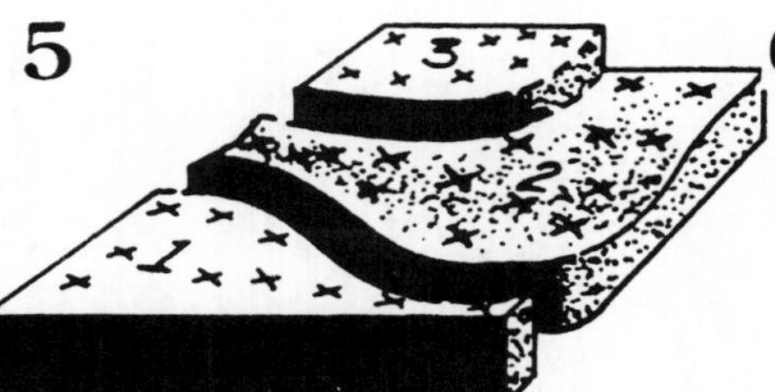

Sample each area separately. Get equal-sized cores or slices from 15 or more places using probe, auger, or spade. Do not mix light and dark colored soils together.

6

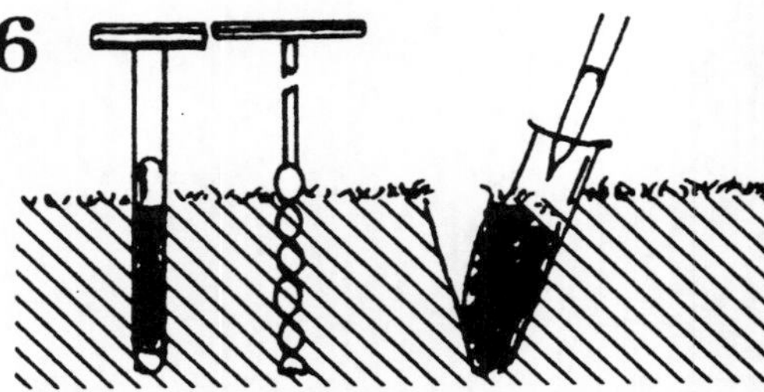

Take soil from surface to plow depth (6-8 inches) except in permanent pasture (2-3 inches). If a spade is used, dig out a V-shaped hole, then remove a slice from the side.

7

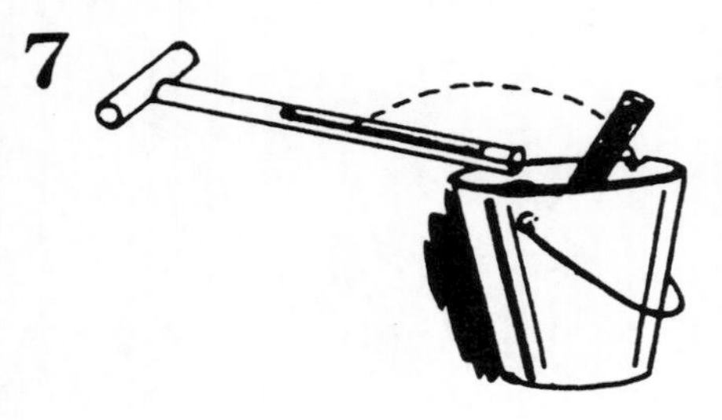

8

Spread mixture out on clean paper to dry. Do not heat in oven or on stove. Do not dry in places where fertilizer or manure may get in sample.

9

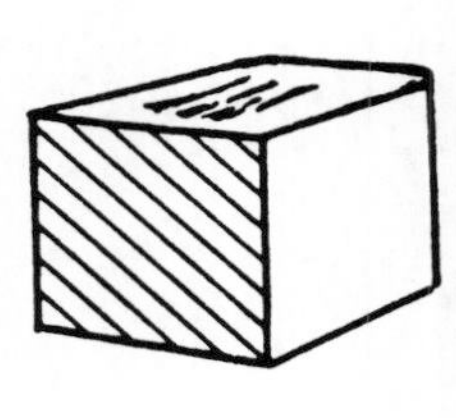

Caution! *In sampling, stay away from:*

1. Farm lanes and field borders.
2. Fertilizer bands in row crops and small grains.
3. Areas within 8-10 rods of gravel roads.
4. Any other areas which are distinctly different, such as potholes, sandy ridges and eroded spots.

Take separate sample if interested.

Important! Your recommendations will be no better than the information that you send in with the samples. Be sure to . . .

FOR LAWNS:

Take the soil to a depth of 2-3 inches (See Fig. 6).

FOR GARDENS AND FLOWER BEDS:

Sample the soil to plow depth of 6-8 inches.

The *field record sheet* is available from your county agent's office.

Proper Liming for Optimum Corn Production

J. V. Baird, North Carolina State University, and L. W. Murdock, University of Kentucky

Reviewers

J. R. Anderson, North Carolina State University — *K. D. Frank, University of Nebraska*
H. F. Reetz, Potash & Phosphate Institute, IL — *R. L. Nielsen, Purdue University*
K. L. Anderson, Mississippi State University — *J. W. Johnson, Ohio State University*
C. J. Overdahl, University of Minnesota — *B. W. Remick, National Crushed Stone Assn., DC*
W. I. Segars, University of Georgia

Throughout the United States where corn is grown on "acid soils," lime should be applied periodically both to prevent soil acidity from limiting yields and to ensure efficient fertilizer use. A lime application will influence two to four (perhaps even more) subsequent crops, depending on such factors as amount applied, rainfall, soil type, nitrogen use, and legume cover crop growth.

This publication deals with the why, what, and how of liming for optimum corn production. Discussed first is the matter of soil acidity and how it develops. Next, we will look at soil pH requirements for corn, how well those requirements are being met in various corn-growing regions through the application of lime, and the many benefits to be realized by proper liming. The rest of the publication then focuses on selecting and applying liming materials, including what determines quality, how it's measured, types of materials that can be used, and how and when to apply them.

SOIL ACIDITY AND HOW IT DEVELOPS

Degree of soil acidity or pH is determined by the amount of exchangeable hydrogen and aluminum ions in the soil. Hydrogen (H^+) is formed through decay of organic matter and plant residues, the metabolic activity of plant roots, and certain reactions of fertilizer materials.

This H^+ replaces the soil's basic cations, especially calcium (Ca^{++}) and magnesium (Mg^{++}), and also causes partial decomposition of clays around the exchange sites. The altered clay then releases aluminum (Al^{+++}), which becomes the predominant exchangeable cation at each site. Thus, very acid mineral soils will have a high content of exchangeable aluminum moving from the exchange sites into soil solution and producing hydroxy aluminum plus free hydrogen—

$$[Al^{+++} + HOH \text{ (water)} \longrightarrow Al(OH)^{++} + H^+].$$

Many factors can contribute to an increase in soil acidity (i.e., decline in pH). Major ones are climatic influence, nitrogen fertilization, plant root respiration, and soil erosion.

Acid soils are common in regions where precipitation is high enough to leach nutrient cations from the surface layers, and in regions where climate is warm and humid enough to accelerate organic matter and plant material decay.

Acidity can also be caused by use of nitrogen fertilizers, including urea, anhydrous ammonia, nitrogen solution, manure, and other ammonium sources. The large amounts of ammonium nitrogen (NH_4^+) traditionally used on corn, once nitrified, produce large amounts of hydrogen, which will drastically lower the soil pH. This is especially true in sandy soils and in the top 2 inches of soil for no-till crops, especially corn.

Plant root respiration contributes to soil acidity by producing hydrogen (H^+) and biocarbonate (HCO_3^-) ions as by-products of the reaction. These H^+ ions displace basic cations, which then combine with the HCO_3^- ions to form highly leachable calcium, magnesium, and potassium bicarbonates.

And finally, undissolved limestone on or near the soil surface can be removed by erosion processes.

SOIL pH REQUIREMENTS FOR CORN

Every soil series has a high and low critical pH beyond which corn growth will be substantially reduced. Growth is affected at the low end by toxic levels of aluminum and manganese and at the high end by micronutrient deficiencies such as zinc, iron, and possibly manganese.

For the most efficient use of fertilizers, research has shown that soil pH should range between 6.0 and 7.0. However, research also indicates that, with good fertility, maximum corn yields are realized when pH is 5.5 to 6.5. In fact, only where alfalfa is grown in rotation with corn does pH need to be above 6.0. In calcareous soils (those where free calcium carbonate occurs), a pH below 6.0 may be desirable to increase the availability of iron and zinc.

For many corn producers, especially in the eastern part of the U.S., low soil pH (i.e., less than 5.5 on mineral soils) is a serious but sometimes unrecognized or ignored problem. For instance, recent surveys of soil testing laboratories in 12 northeastern states indicate that only one-third of the agricultural lime tonnage recommended according to soil tests was being applied. Similarly, laboratories in 10 southeastern states report that agricultural lime usage was less than half the recommended tonnage.

Soil test results from numerous Corn Belt laboratories illustrate a progressively greater need for lime as one moves eastward through the region. For example, Iowa summaries show that state's present lime application rates as being adequate to maintain a desirable pH for corn; whereas in Wisconsin, soil pH is not being maintained at the rates currently applied.

BENEFITS OF PROPER LIMING

Proper liming means applying the proper amount of a suitable material in the right way. Guided by soil tests and field or greenhouse "lime rate" research, proper liming will provide the following benefits to ensure maximum corn root development, plant growth, and yield:

- Reduces solubility of manganese and aluminum, which are always present in acid soils (below pH 5.5) in large amounts to hinder growth.
- Increases efficiency of fertilizer phosphorus by reducing amount of exchangeable aluminum that can react with it to form insoluble compounds.
- In acid soil, increases plant uptake of potassium, calcium, and magnesium by reducing amount of exchangeable aluminum.
- Reduces leaching of potassium and magnesium in sandy soils.
- Provides an economical source of both magnesium and calcium for plants if dolomitic limestone is used, or of calcium if calcitic limestone is used.
- Improves nitrogen fixation by nodule-forming bacteria in legumes grown as a cover crop or in rotation with corn.
- Increases availability of some micronutrients, including molybdenum, which is important to legumes in rotation with corn.
- Maintains activity of pH-sensitive herbicides, such as atrazine, simazine, cyanazine, or metribuzen, whose effectiveness decreases as soil pH decreases.

QUALITY OF LIMING MATERIALS

Quality or effectiveness of liming materials differs not only between different products, but also between different batches of the same product. Thus the need for some measure of equivalency that is understood and accepted by both suppliers and farmers. Soil testing laboratories have established such a measure by basing recommendations on the amount of lime that will effectively react with soil acidity rather than on a lime-per-acre application rate.

Factors Determining Lime Quality

The things that determine the effectiveness of a particular material are its purity, its particle size or fineness, and its calcium and magnesium content.

Purity is measured by a material's calcium carbonate equivalent (i.e., a weight percentage of the pure calcium and magnesium present) and is expressed as a relative "neutralizing value" (i.e., ability to neutralize soil acidity compared to pure calcium carbonate). Table 1 shows the neutralizing values of some agricultural liming materials relative to pure calcitic limestone with a value of 100.

Table 1. Relative Neutralizing Values of Common Agricultural Liming Materials (Expressed as Percent of Calcium Carbonate Equivalent).

Materials	Composition	Neutralizing value
Calcitic limestone	$CaCO_3$	100
Dolomitic limstone	$CaMg(CO_3)_2$	109
Burned lime	CaO	150-175
Hydrated lime	$Ca(OH)_2$	120-135
Calcium carbonate sludge	$CaCO_3$	100
Slags	$CaSiO_3$	60-90
Coal ash	$CaSiO_3$*	0-40
Cement dust	$CaSiO_3$*	80-85
Sea and egg shells	CaO	90-100
Marl	$CaCO_3$	70-90

* The principal component that increases soil pH.

The presence of clay, sand, organic matter, or other "impurities" will lower a product's calcium carbonate equivalent (CCE). Although there is no official minimum neutralizing value standard in the U.S., a material will usually not be used for agricultural purposes when impurities decrease its CCE below about 80%. It is possible, though, that limestone deposits having a CCE below 80% will be mined and sold locally if no other nearby satisfactory liming sources exist.

Particle size is another major factor affecting quality. As particle size decreases, the amount of reactive surface area of a given quantity of lime increases, resulting in more rapid dissolving in the soil solution. For example, a 1 cubic inch piece of limestone has a surface area of 6 square inches. If that piece was ground up enough to pass through a 100-mesh screen, it would have 60,000 square inches of surface area—10,000 times as much! (See Table 2.)

Table 2. Approximate Weight Percent of Dissolved Limestone of Various Fineness Ranges at 1, 4, and 8 Years after Application to the Soil.

	Years after application		
Size fraction	**1**	**4**	**8**
Passing 60-mesh	100	100	100
30- to 60-mesh	50	100	100
8- to 30-mesh	20	45	75
Retained on 8-mesh	5	15	25

Source: NCSA AgLime Fact Book.

Calcium and magnesium content, as the third determinant of quality, is related to the fineness factor. Calcitic limestone ($CaCO_3$) dissolves faster than the magnesium-containing dolomitic limestone [$CaMg(CO_3)_2$]. When particle size is 60- to 70-mesh or finer, these two liming sources are equally effective. If coarser than 60-mesh, however, dolomitic lime will be only half as effective as calcitic lime of equal reactive surface area.

These findings of E. J. Kamprath at North Carolina State University and other soil scientists have led to the adoption of the following differential size standard—*35% of a dolomitic source vs. 25% of a calcitic source must pass a 100-mesh screen to be acceptable agricultural limestone.*

Measurement of Lime Quality

Since crushed limestone contains a range of particle sizes, it is important that a lime sample be fractionated (separated out) to obtain a meaningful measure of lime quality. Among the terms being used to define quality are: effective calcium carbonate equivalent (ECCE), total neutralizing power (TNP), effective neutralizing value (ENV), and effective neutralizing power (ENP).

Basically, they are all determined the same way—i.e., multiplying the material's percent calcium carbonate equivalent (CCE) by the percent availability of that CCE based on fineness of grind ("fineness efficiency").* To illustrate:

* State regulatory agencies differ in how they arrive at fineness efficiency. For instance, the equation used in Oklahoma is: (0.5 x % passing 8-mesh) + (0.5 x % passing 60-mesh); the one used in Iowa is: (0.1 x % passing 4-mesh) + (0.3 x % passing 8-mesh) + (0.6 x % passing 60-mesh); and Wisconsin's is: (0.0 x % larger than 8-mesh) + (0.2 x % between 8- and 20-mesh) + (0.6 x % between 20- and 60-mesh) + (1.0 x % passing 60-mesh). Check with the regulatory agency of the state where the lime is purchased for their definition of fineness efficiency.

Assuming liming material A has a 90% CCE and a 75% fineness efficiency, and material B has a 90% CCE but only a 68% fineness efficiency, their ECCE or neutralizing ability would be—

A = 0.90 x 0.75 = 0.675 x 100 = 67.5%
B = 0.90 x 0.68 = 0.612 x 100 = 61.2%

Therefore, on a per-ton basis, material A would contain 1350 pounds of ECCE (2000 lb. x 0.675) and material B 1224 pounds of ECCE (2000 lb. x 0.612). As you can see, small variations in a material's calcium carbonate equivalent and/or fineness efficiency will markedly affect its neutralizing value.

Most agronomists and crop consultants agree that agronomic and economic considerations are by far the main ones when selecting between several lime sources. For instance, if test results show that a soil's magnesium supply is low as well as its pH, a farmer might choose a dolomitic limestone, which contains magnesium. Or if transportation of lime happens to be a major expense, he should probably buy high CCE products because they contain less inert material, which costs the same to transport but does nothing to reduce soil acidity.

Definitions of quality, such as ECCE, are attempts to characterize lime so that these and similar economic and agronomic decisions can be better made.

TYPES OF LIMING MATERIALS

Sources of lime that is applied to cropland soils include limestone, limestone derivatives, certain manufacturing process by-products, marl, and shells. Following is a brief discussion of the more common specific liming materials from these sources, including major advantages and/or disadvantages. Their composition and neutralizing values are summarized in Table 1.

Agricultural Limestone

Commonly called ag lime, agricultural limestone is made by crushing or grinding mined limestone rock to a selected degree of fineness (or mesh size). This degree of fineness determines the rate at which a particular material will react with the soil acids.

Ag lime may be calcitic or dolomitic. Calcitic limestone contains calcium carbonate and very little or no magnesium carbonate; whereas dolomitic limestone contains calcium carbonate and approximately 4.4 to 22.6% magnesium carbonate. Pure dolomite has nearly equal amounts of calcium and magnesium carbonate.

Quality of an ag lime material depends on the type and amount of impurities it contains, such as clay and sand. Its neutralizing values, based on calcium carbonate equivalence, may range from about 60% to slightly more than 100%.

Water content of quarried agricultural limestone varies from 0 to 10%, which is an important consideration from the user's standpoint for two reasons. First, since the material is sold by weight, the more water it contains, the less lime is being purchased. Also, at higher moisture levels, lime particles tend to conglomerate or clump, which will affect lime distribution over the soil surface.

Fluid Lime

Fluid lime or lime suspensions are terms used to describe the mixing and applying of liming materials in water or fertilizer solutions. Advantages of fluid lime are its almost-immediate availability to the soil and its compatibility with present liquid fertilizer equipment. Drawbacks are its low concentration (about 1/2 water) and thus the need to apply great volumes, and its relatively high cost per unit of lime applied.

One particular fluid suspension perhaps to avoid is calcium oxide or by-products containing calcium oxide added to nitrogen fertilizer solutions containing urea and ammonium nitrate. Studies show that this combination can result in appreciable loss of ammonia.

Burned Lime

This is a limestone derivative made by heating crushed limestone in an oven or furnace, which drives off carbon dioxide to form calcium oxide. Also known as unslaked or quick lime, burned lime is sold in bags because of its powdery, unpleasant handling properties and its reaction to humidity.

Because of its minute particle size, quick lime dissolves quickly in the soil. However, if not immediately incorporated, it will form conglomerates on the surface, thereby reducing its "fineness" advantage.

Quality of burned lime depends on the purity of the limestone from which it is made. Its neutralizing value is extremely high—ranging from 150 to 175 compared to pure calcitic limestone, which has a neutralizing value of 100.

Hydrated Lime

Another limestone derivative, hydrated lime is made by reacting burned lime with water to form calcium hydroxide. Also known as slaked or builders' lime, its characteristics are similar to burned lime—i.e., powdery, quick-acting and unpleasant to handle. Its neutralizing value usually ranges from 120 to 135, depending on the purity of the burned lime from which it is made.

Sludges and Slags

These are by-products of manufacturing processes. Those used for liming purposes include calcium carbonate sludge and blast-furnace, basic and electric- furnace slags.

Calcium carbonate sludge is a by-product of the paper pulp wood industry. It contains from 20 to 50% water, depending on how much dewatering takes place in the holding pit. Unlike most of the other materials, its application rate should be based on volume rather than weight.

Blast-furnace slag, a by-product in the manufacture of pig iron, is primarily a calcium silicate but contains an appreciable amount of magnesium. If finely ground and applied at rates equivalent to ground limestone, it is nearly as effective, having a neutralizing value of 75 to 90.

Basic slag is a by-product from pig iron in the open-hearth steelmaking process. It contains 5-10% phosphorous and, if finely ground, has a neutralizing value of 60-70.

Electric-furnace slag comes from the electric furnace process for extracting elemental phosphorus from rock phosphate. The product is primarily calcium silicate, has a neutralizing value of 60-80 and contains approximately 6% P_2O_5.

Coal Ash

Coal ash, sometimes called fly or bottom ash, is a by-product of coal-fired electricity-generating or heating plants. It is collected during removal of noxious gases and particulate matter from smokestacks.

Coal ash can be basic or acidic in reaction, depending on the type of coal burned or the type and amount of materials added to improve combustion. Its acid-neutralizing potential ranges from 0 to 40 but can be greater than predicted by its calcium carbonate equivalence.

Cement Dust

Cement plant flue or stack dust, which is collected similarly to coal ash, usually contains calcium oxide as the primary acid neutralizer. Its value is proprotional to its calcium carbonate equivalence, which may vary from 60 to 80%.

As the name implies, cement dust is extremely fine. While this causes some handling problems, it does make the material convenient to apply as a lime suspension. But as previously noted, it should not be applied with urea or ammonium nitrate fertilizer solutions because of high ammonia loss.

Shells

Sea and egg shells are primarily calcium carbonate and provide a neutralizing value ranging from 90 to 100+. If finely ground, they make a satisfactory liming material, particularly for land located near the source of supply.

Marl

Although not considered a mineral, marl is soft calcium carbonate occurring as natural deposits in poorly drained, low-lying areas. Its usefulness as a liming material depends on clay content, which affects neutralizing value.

Because of its generally wet state, marl is handled by volume rather than weight. For high quality marl, 2 cubic yards is considered the equivalent of 1 ton of ground limestone.

WHEN AND HOW TO APPLY LIME

How often to lime and how much to apply should be determined by soil test results, not by habit or guess. Over-application of lime is costly, wasteful of material and energy resources, and can cause manganese, zinc, and iron deficiencies. Under-application, on the other hand, will hinder the achievement of high yields.

Periodic soil testing for nutrient availability and pH level is essential to realizing optimum corn production. In the case of liming needs, normally the soil should be analyzed every 2-4 years.

Limestone particles are not mobile in the soil. So for maximum effectiveness, lime should be incorporated into the plow-depth layer where most root growth takes place.

If corn is grown under a no-till system, the normally high rates of broadcasted nitrogen may lower pH within the 0- to 2-inch soil layer. Studies show that topdressing the lime will correct or prevent acidification in no-tilled fields. However, rate of application should be based on soil samples taken 0-4 inches deep.

Whether incorporated or not, it's important that lime be applied as uniformly as possible to avoid "streaking"—i.e., irregular plant growth in streaks parallel to the direction that the spreader truck traveled. This irregular growth occurs both where swaths didn't overlap (resulting in inadequate pH) and where they did overlap (resulting in higher-than-anticipated pH that may cause a manganese deficiency).

The key to uniform spreading is properly maintained and adjusted spreading equipment. When having lime custom-applied, the crop producer has every right to observe the operation and question suspected irregularities. A good vendor will take it upon himself to conduct "spread pattern" tests.

SOURCES OF ADDITIONAL INFORMATION

Goodwin, J. H., 1979. "A guide to selecting agricultural limestone products," Illinois Mineral Note, Illinois State Geological Survey, Urbana, IL.

Proceedings, National Conference on Agricultural Limestone, 1980. National Fertilizer Development Center, TVA, Muscle Shoals, AL.

Handbook: AgLime Fact Book, 1981. National Crushed Stone Associates, Washington, DC.

Pearson, R. W. and Adams, F. (eds.), 1967. "Soil acidity and liming," Agronomy Monograph 12, American Society of Agronomy, Madison, WI.

Research and extension agronomists in many states have written publications about lime that address their states' specific conditions.

A publication of the National Corn Handbook Project

NEW 1/87 (5M)

Cooperative Extension work in Agriculture and Home Economics, state of Indiana, Purdue University and U.S. Department of Agriculture cooperating. H. A. Wadsworth, Director, West Lafayette, IN. Issued in furtherance of the acts of May 8 and June 30, 1914. The Cooperative Extension Service of Purdue University is an affirmative action/equal opportunity institution.

HYBRID AND VARIETY SELECTION

I. PRINCIPLES OF CROP IMPROVEMENT

Objectives

1. Distinguish between a hybrid and a variety
2. Define heterosis
3. Describe how hybridization is used in plant breeding
4 Define true-breeding line
5. Differentiate among natural, mass, and pure line selection techniques
6. Describe how to use inbreeding to develop crop hybrids
7. Describe techniques used to produce corn hybrids
8. Define transgenic plant

Through natural processes, genetic changes to plants have occurred since plants have existed. The plant breeder intervenes in these natural processes by using artificial selection to obtain plants that have the characteristics desired for modern crop production systems. Plant breeders have improved crop characteristics such as yield, maturity, pest resistance, herbicide resistance, grain quality, and harvestability.

A. Hybrids and varieties

All crops are classified as either a **hybrid** or a **variety (cultivar)**.

A hybrid is:

A variety is:

Crops that farmers grow as hybrids exhibit increased vigor, growth, and yield over their parents. This **hybrid vigor**, or **heterosis**, is not found to the same degree in all crops. Crops that have little heterosis are usually grown as varieties.

B. Selection techniques

One way a crop may change or improve is through selection. **Natural selection** occurs without direct human intervention. **Artificial selection** includes techniques practiced by plant breeders.

1. Natural selection

Through natural selection, plants that are well adapted to their environment thrive and flourish, while plants that are poorly adapted fail to reproduce and are eliminated. Natural selection is usually a very slow process, and will not result in all of the improvements desired by humans.

2. Artificial selection

Artificial selection includes techniques used by plant breeders to quickly improve a crop. The two most common types of artificial selection are **mass** and **pure line selection**, which are commonly used to develop new crop varieties. The major difference between mass and pure line selection is the manner in which the seed is grouped and grown out.

a. Mass selection

Mass selection is not a modern invention. Native North Americans selected the most desirable corn plants and plants with the best endosperm for making corn meal and flour. They also selected plants that would mature at a desirable time for harvesting.

Mass selection is a relatively quick, simple, and effective method of improving plant varieties. It is often used to purify mixed varieties. Or, mass selection may be used after a plant breeder makes a hybrid to combine desirable characteristics of two or more plants. With mass selection, plant breeders select hundreds of the best plants and harvest their seed. The seed is mixed together and planted the second year, and the best of these plants are tagged and harvested. This process is repeated for 7-10 generations. After this time, the seed **breeds true**, which means the plants all have identical characteristics. If the plants prove to be superior, the supply of seed is increased and released to farmers as a variety.

Mass selection works best for self-pollinated crops. Improvement is slow with cross-pollinated crops such as corn because when the breeders select the ears they are taking advantage of only the female half the gene pool. They do not know the characteristics of the male parent. In self pollinated crops, the same plant serves as both the male an female parent. Therefore, the genetic characteristics of the resulting plant are more completely known.

b. Pure line selection

Pure line selection is similar to mass selection except in the the way the seeds are grouped. In pure line selection, seeds from a desirable plant are kept together to become an experimental line. With pure line selection, the breeder can identify the original parents of a line. The lines are grown in plots, and the best plants from the best lines are saved. This process is repeated for about five generations, after which the seeds of each line breed true. The best lines are tested, and the ones that show superior performance are increased and sold to farmers.

Pure line selection is used in wheat breeding programs to develop resistance to the Hessian fly, a serious insect pest. It is also used to select for resistance to the cereal leaf beetle.

Pure line selection is also commonly used in corn to develop inbred lines that are used as the parents of hybrids.

C. Inbred development and hybridization

Certain crops, such as corn, sorghum, and sugarbeets, are grown by farmers in the United States as hybrids. Hybrids grown by farmers are the result of a cross between two or more **inbred lines**. An inbred is formed when a plant self-pollinates for 5-7 generations. Inbred lines are carefully selected and crossed to produce hybrid seed. An inbred line often exhibits a mixture of desirable and undesirable traits. For example, plant breeders might develop an inbred line of corn that is true-breeding for resistance to to northern corn leaf blight, but the line might also be low yielding. The breeder may cross this line with a genetically different, high-yielding inbred line. The plant breeder hopes the result of the cross will be a hybrid that has the desirable characteristics of both inbred lines.

D. Developing corn hybrids

Corn is particularly well-suited to be grown by farmers as a hybrid. It has a high degree of heterosis, and its separate male and female flowers make it easy for a plant breeder to manipulate.

1. Development of inbred lines

The first step in developing a corn hybrid is to develop inbred lines. Breeders identify the characteristics they desire in an inbred line and select the most desirable plants using pure line selection. After 5 generations, the inbred is genetically pure and will breed true.

After inbred lines are developed, they are crossed to create hybrid seed. If the inbred lines were selected carefully, the resulting hybrid plants will exhibit a high degree of heterosis and will have the desired traits. After performance trials, the hybrid may be released for sale to farmers if it meets the standards of the seed company.

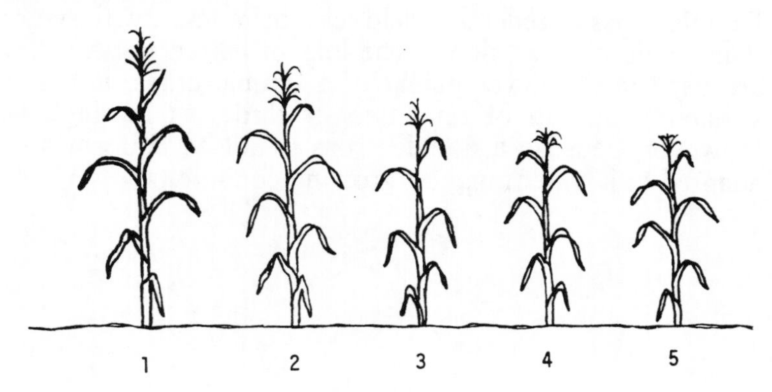

Generations of Inbreeding
The vigor of corn is reduced with each generation of inbreeding.
After 5 generations the inbred line is uniform and will breed true.

2. Single crosses

The single cross is the most popular type of hybrid seed corn used today because it has the highest yield potential. This is because a single cross has the maximum amount of hybrid vigor. Created from high-quality inbred lines, the single-cross population is very uniform in appearance and in its response to the environment and management. Early inbred lines gave poor yield and had low pollen production. Trying to create hybrids with them was difficult and expensive because to required large numbers of plants and highly trained workers. Today's inbred lines generally give higher yields and produce ample pollen.

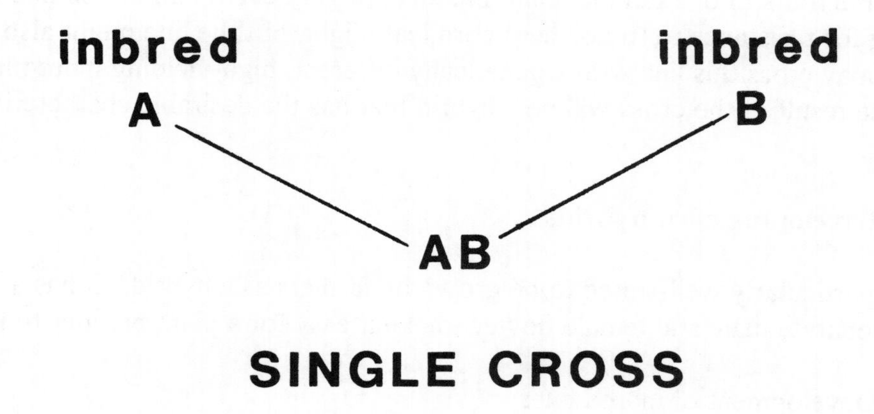

3. Double crosses

A double cross is a cross of two single-cross hybrids. Double-crossed seed was the first type of hybrid corn sold to farmers. The original single crosses were produced from low yielding (5 to 20 bushels per acre) inbred plants and were prohibitively expensive to sell to farmers. Double-cross seed, on the other hand, was produced from high-yielding single-cross plants, which greatly reduced seed costs.

Double crosses generally yield less, have less hybrid vigor, and are less uniform than single crosses, due to the loss of inbred purity. Because four inbred lines are used in the development of a double cross, it has four sources of genetic variability instead of only two or three, as in single and three-way crosses. However, because a double cross is not as uniform as the other crosses, it is adapted to a wider range of growing conditions.

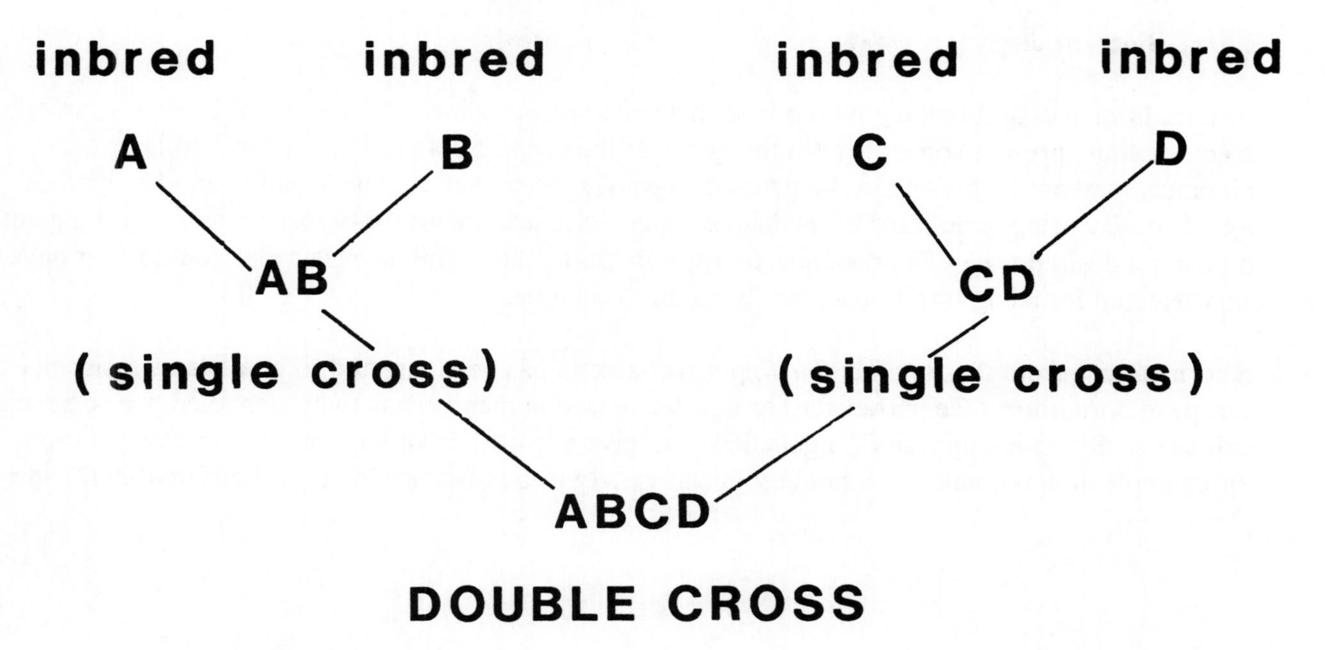

DOUBLE CROSS

4. Three way crosses

A three-way cross is formed by crossing a single cross with an inbred line. It is usually attempted when there is an exceptionally good pollen parent (C) available. This hybrid type has the genetic advantage of the more vigorous single-cross ear parent (AB). A three-way cross is more uniform than a double cross because one parent is inbred. Its potential yield is usually lower than that of single crosses but higher than that of double crosses. The single cross is always used as the female parent because it produces more seed for sale, thus reducing farmers' seed costs.

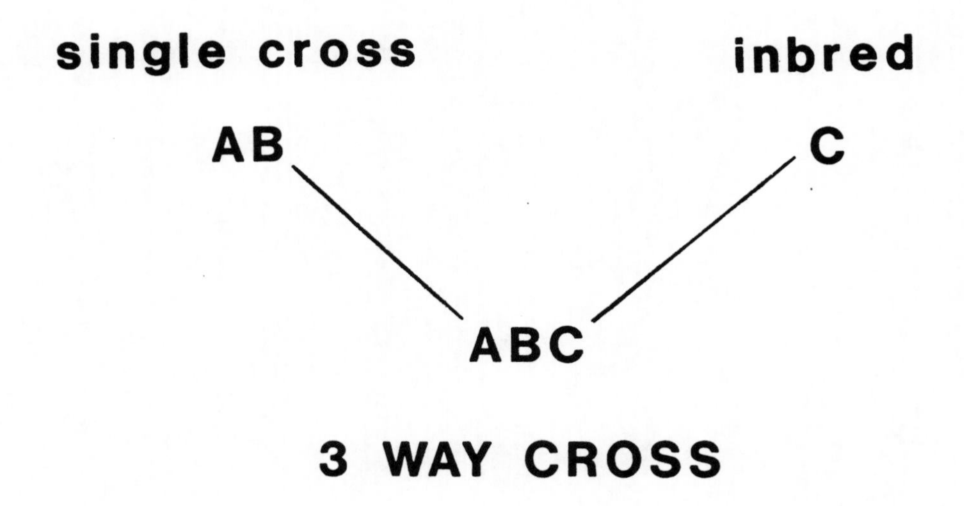

3 WAY CROSS

E. Biotechnology

With traditional plant breeding methods such as mass or pure line selection, plant breeders can only select for traits present somewhere in the genes of the crop. Some traits, such as herbicide tolerance, may not be present in the genes of a plant species, but may be found in another kind of organism. By using gene transfer techniques, geneticists can incorporate foreign genes into a plant to give it a desired trait. The resulting **transgenic plant** thus exhibits a trait that could never have been selected for using traditional plant breeding techniques.

In corn, a gene from the *Bacillus thuringiensis* bacteria has been inserted to give plants resistance to European corn borer. Gene transfer has also been used in many crops to confer tolerance to specific herbicides. By "stacking traits", a geneticist can give a hybrid or variety multiple transgenic traits. For example, it is possible for a single soybean variety to be resistant to several different herbicides.

D. Terms you should know for a complete understanding of Section I.

For each of the following terms, write a brief definition. Refer back to the study guide if necessary.

1. Mass selection -

2. Natural selection -

3. Pure line selection -

4. Inbred line -

5. Single cross -

6. Double cross -

7. 3-way cross -

8. Fertilization -

9. Hybrid -

10. Hybrid vigor -

11. Variety -

12. True-breeding line

13. Transgenic plant

NATIONAL CORN HANDBOOK

Origin, Adaptation, and Types of Corn

W. L. Brown, Pioneer Hi-Bred International, Inc., IA; M. S. Zuber, University of Missouri;
L. L. Darrah, USDA-ARS, University of Missouri; and D. V. Glover, Purdue University

Reviewers

R. G. Creech, Mississippi State University
K. F. Schertz, USDA-ARS, Texas A&M University
A. A. Fleming, University of Georgia
A. F. Troyer, DeKalb-Pfizer Genetics, IL

ORIGIN OF CORN

Corn (*Zea mays* L.) is the only important cereal indigenous to the Western Hemisphere. Apparently originating in Mexico, it spread northward to Canada and southward to Argentina. While the possibility of secondary centers of origin in South America cannot be completely ruled out, the oldest (7000 years) archaeological corn was found in Mexico's Valley of Tehuacan.

The earliest "corn" of which there is record is unmistakably corn. The female inflorescence of this 5000 B.C. corn had reached a degree of specialization that precluded the possibility of natural seed dissemination. Thus, the oldest corn of record was dependent upon man for its survival.

Numerous theories of origin have been offered over the years, only two of which receive serious consideration today. One is that teosinte (*Zea mexicana*) is the wild progenitor of corn; the other is that a wild pod corn, now extinct, was the ancestor of domesticated corn. While perhaps more students of corn seem to accept the first theory, others are equally convinced of the second.

Aside from its possible role in the origin of corn, teosinte has had major impact on its evolution. In Mexico particularly, introgression between corn and teosinte has likely occurred for centuries and continues to this day. The effects are apparent in the morphology and cytology of both species. There is also reason to believe that genes for resistance to certain viruses have reached corn through its introgression with teosinte.

The origin of corn may never be known with certainty. One reason is that the hypotheses purporting to explain origin cannot be tested experimentally. Therefore, science would perhaps be better served if less attention were given to determining corn's origin and more to understanding the remarkable variability found within the species.

Variability and Races

Regardless of origin, corn has proven to be one of the most adaptable and variable members of the grass family. Its evolution, a large part of which apparently occurred under domestication, has resulted in biotypes with adaptation ranging from the tropics to the north temperate zone, from sea level to 12,000 feet altitude and growing periods (planting to maturity) extending from 6 weeks to 13 months.

Almost 300 races of corn have been described from Mexico, Central and South America, and the Caribbean. Although many appear synonymous, at least 150 distinct entities have been collected in these areas. It was from certain of these races that most of the corns of North America were ultimately derived.

Spread from Center of Origin

Following discovery, corn moved quickly to Europe, Africa and Asia. From Spain, it spread northward to the short-growing-season areas of France, Germany, Austria and eastern Europe, where selection for early maturity has produced some of the earliest commercial varieties of corn now available. In Italy and Spain, early counterparts of many South American races are evident even today.

Although introduced into Africa soon after discovery, much of the corn now found in that continent is derived from later introductions from the southern U.S., Mexico and parts of eastern South America. Most of southern Africa's corn traces back to varieties grown in the southern U.S. in colonial and post-colonial times. Corn of tropical middle (lowland) Africa is similar to the lowland and tropical corns of Central and South America. Africa has always preferred white corns; and until recently, the

The most widely used and productive corns of Asia are derived from Caribbean-type flints introduced in relatively recent times. However, older and quite distinct types of corn can also be found, for example, among the hill people of Mindanao in southern Philippines. Included are some small-eared, early-maturing flints or pops that either have the capacity to grow and mature quickly before being devastated by downy mildew or carry some genetic resistance to downy mildew.

In the eastern Himalayas (Sikkim and Bhutan), a distinctive type of popcorn is found whose Western Hemisphere progenitors seem to have disappeared. When and how the ancestor of this corn reached Asia is not known. However, since it is not present on the Indian subcontinent to the south, it may have reached the Himalayas by way of China and Tibet. In any event, a search for similar corns in south China and Tibet would seem justified.

Corn of the U.S. Corn Belt

Whereas most of the modern races of corn are derived from prototypes developed by early native agriculturists of Mexico, Central and South America, one outstanding exception is solely the product of post-colonial North America—the yellow dent corns that dominate the U.S. Corn Belt, Canada and much of Europe today. The origin and evolution of this remarkable race of corn have been clearly documented and confirmed.

In the early 1800's, two predominant races of corn of North America's eastern seaboard—the late-maturing Virginia Gourdseed and the early-maturing Northeastern Flints—were first crossed, and the superiority of the hybrid recognized and described. The cross was repeated many times during the western migration of settlers; and out of these mixtures eventually emerged the Corn Belt dents, the most productive race of corn found anywhere in the world.

It was the highly selected varieties of Corn Belt dents that formed the basis of hybrid corn and the source of the first inbred lines used to produce hybrids. Germplasm from some of these varieties (Reid, Lancaster, Krug, etc.) still figure prominently in the ancestry of hybrids used in the Corn Belt even to this day.

ADAPTATION OF CORN

Definition and Adaptation Worldwide

Adaptation in corn means good performance with respect to yield and other agronomic characteristics in a given environment. The environment includes all conditions to which the plant is subjected during the growing season (from pre-seedling emergence to harvest maturity).

The major environmental factors are: (1) daily maximum and minimum temperatures, (2) soil atmosphere and moisture levels, (3) humidity of the atmosphere immediately surrounding the plant, (4) wind movement, (5) day length, (6) light intensity, (7) air pollution, (8) soil type, (9) soil fertility, (10) number of days exceeding 50°F, (11) competition from other plants including its intended neighbors as well as weeds, and (12) the disease-insect complex.

Many of these factors interact in a complex manner to produce stress on the plant. The plant's reaction to stress is under genetic control, and differences among hybrids exist. Corn breeders are continuously developing and testing new genetic combinations in differing environments to find types that give the best agronomic performance over a range of geographical locations and years.

Corn is grown from sea-level to altitudes of more than 12,000 feet and from the equator to about 50° north and south latitudes. Compared to environmental conditions of the U.S. Corn Belt, many producing areas would be considered very much substandard. In harsh-environment areas, the varieties grown would be considered adapted because they responded the best under the unfavorable growing conditions. The original open-pollinated varieties have become adapted through selection over time by both man and the environment.

Adaptation Within the Corn Belt

World production of corn in the early 1980's approached 450 million metric tons annually, with the U.S. contributing over 48 percent of the total. Of U.S. annual production, the 13 Corn Belt states account for about 82 percent.

No area of the world equals the Corn Belt for high yields. This is due to a combination of ideal soils and climate, advanced farmer know-how, and the success of corn breeders in developing hybrids with high genetic potential.

Climate. Corn is considered a warm-weather crop. In the Corn Belt, average summer temperatures range from 70° to 80°F daytime and exceed 58°F at night. The average frost-free growing period is over 140 days. Greatest yields are obtained where 30 or more inches of rain occur during the growing season. In areas where rainfall is less than 20 inches, yields are much reduced unless irrigation is used.

Rainfall distribution greatly influences maximum yields, especially for the 3-week period centered around tasseling. In the southern part of the Corn Belt, high-temperature stress and rainfall deficiencies often occur in late July and August. Therefore, farmers of this region try to avoid having corn tassel during this dry, hot period by planting earlier and using earlier-flowering hybrids.

Maturity. Maturity of corn hybrids is a genetic characteristic and is generally defined as the period from germination to when the kernel ceases to increase in weight. In the northern Corn Belt, early hybrids often reach physiological maturity in less than 100 days; whereas in the extreme south, 150 or more days may be needed. These day periods are measures of relative maturity.

Hybrids have traditionally been classified into 15 maturity groups ranging from Agricultural Experiment Station (AES) 100 to AES 1500 (earliest to latest). However, maturity classification can be made more precise by determining the total *heat units* required from emergence to physiological maturity. A heat unit measure commonly used accumulates the daily excess of average temperature over 50°F, where:

average temperature = (maximum + minimum) ÷2.

Early hybrids grown in the northern Corn Belt classified as 100-day maturity may require 90 or less days to reach maturity when grown further south where heat unit accumulation is more rapid.

Adapted hybrid development. Hybrids have been developed that are adapted from Nebraska to Ohio. However, the environmental conditions differ widely from west to east especially rainfall patterns, daily minimum and maximum temperatures, and the disease complex. For example, the hotter, drier conditions of the western Corn Belt are less favorable for leaf diseases caused by *Helminthosporium* spp. and anthracnose, but more conducive to viral infection, bacterial wilt and smut. For this reason, development of hybrids adapted from west to east has been relatively difficult and requires large-scale testing over a number of locations and years.

Number of frost-free days decreases from the southern to northern Corn Belt; however, hours of daylight on June 22nd are much longer in the north than in the south. Sunlight intensity is greater in the western and northern Corn Belt because these sections have less cloudy weather.

The result of intensive corn breeding efforts over the past four decades has been better adaptation to the many environments under which corn is grown.

TYPES OF CORN

Corn variation may be artificially defined according to kernel type as follows: dent, flint, flour, sweet, pop and pod corn. Except for pod corn, these divisions are based on the quality, quantity and pattern of endosperm composition in the kernel and are not indicative of natural relationships.

Endosperm composition may be changed by a single gene difference, as in the case of floury (*fl*) versus flint (*Fl*), sugary (*su*) versus starchy (*Su*), waxy (*wx*) versus non-waxy (*Wx*), and other single recessive gene modifiers that have been used in breeding special-purpose types of corn. The quantity or volume of endosperm conditioning the size of the kernel (e.g., the difference between dent and flint corns or flint corn and popcorn) is polygenic and, in the latter example, is of some taxonomic significance.

The pod corn trait is monogenic and more of an ornamental type. The major gene involved (*Tu*) produces long glumes enclosing each kernel individually, such as occurs in many other grasses.

Dent Corn

The U.S. Corn Belt dents originated from the hybridization of the Southern Dent or late-flowering maize race called Gourdseed, and the early-flowering Northern Flints. Dent corn is characterized by the presence of corneous, horny endosperm at the sides and back of the kernels, while the central core is a soft, floury endosperm extending to the crown of the endosperm where, upon drying, it collapses to produce a distinct indentation. Degree of denting varies with the genetic background. Nearly all varieties grown in the U.S. are yellow, with only a few white endosperm types grown.

Dent corn is used primarily as animal food, but also serves as a raw material for industry and as a staple food. Upwards of 93 percent of dent corn produced (including the corn equivalent of by-product feeds from corn processing) is used as animal feeds. However, it is still an important human food and industrial material, entering into many specialized products via the dry- or wet-milling industry in the U.S.

Yellow dent corn sells at market price as it enters the normal feed grain or milling channels. However, white dent often receives a premium price in the dry-milling industry, where it is utilized for certain human food products because of its whiter starch.

Flint Corn

The flint corns mostly have a thick, hard, vitreous (glassy) or corneous endosperm layer surrounding a small, soft granular center. The relative amounts of soft and corneous starch, however, vary in different varieties. Generally, the kernels are smooth and rounded, and the ears long and slender with a comparatively small number of rows or kernels. In temperate zones, flint corn often matures earlier, germinates better, has more spring vigor, more tillers and fewer prop roots than dent strains.

Very little flint corn is produced and utilized in the U.S. today, although it was undoubtedly grown extensively up through colonial times. Generally, yields are lower than our Corn Belt dents, in part because of relatively little breeding work done. Flints are more extensively grown in Argentina and other areas of South America, Latin America and southern Europe where they are used for feed and food.

Flour Corn

This is one of the oldest types of corn, tracing back to the ancient Aztecs and Incas. American Indians ground the soft kernels for flour. Floury maize types have soft starch throughout, with practically no hard, vitreous endosperm and thus are opaque in kernel phenotype. Kernels tend to shrink uniformly upon drying, so usually have little or no denting. When dry, they are easy to grind, but may mold on the mature ear in wet areas.

In the U.S., flour corn has limited production and

is restricted to the drier sections. It is grown widely in the Andean region of South America.

Sweet Corn

The following genetic model featuring primary isolation groups for naming "vegetable corns" has been suggested by the industry:

I. Sugary mutants
  A. Standard sugary (*su*)
  B. Augmented sugary
    1. Partial modification
      a. Heterozygous shrunken-2 (*sh*2)
      b. Heterozygous sugary enhancer (*se*)
      c. Heterozygous shrunken-2 and sugary enhancer (*sh*2 and *se*)
    2. Complete (100%) modification
      a. Homozygous sugary enhancer (*se*)

II. Shrunken-2 (*sh*2)
III. Brittle (*bt*)
IV. Brittle-2 (*bt*2)
V. Amylose-extender (*ae*) Dull (*du*) Waxy (*wx*)
VI Dent (vegetable)
VII. Additional classes as new genes are used

Isolation will be required between major groups identified by a Roman numeral. Isolation is suggested but not required between subgroups within a major group. No isolation is needed for cultivars within the same classification.

Standard sugary kernel types. Sweet corn, commonly referred to as the standard sugary (*su*) corn, is thought to have originated from a mutation in the Peruvian race Chullpi. Most certainly it was grown and used by native American Indians in pre-Columbian times.

In sweet corn, the sugary gene prevents or retards the normal conversion of sugar into starch during endosperm development, and the kernel accumulates a water-soluble polysaccharide called "phytoglycogen." As a result, the dry, sugary kernels are wrinkled and glassy. The higher content of water-soluble polysaccharide adds a texture quality factor in addition to sweetness. In the U.S., sweet corn is eaten in the immature milk stage and is one of the most popular vegetables.

Sweet corn in the U.S. is more important economically than its limited commercial production would indicate, because it is consumed directly as human food (fresh market or canned and frozen products) rather than indirectly as livestock feed. The bulk of sweet corn production is confined to the northern tier of states and to southern Florida as a winter crop.

In the broader sense, vegetable corns include all corn harvested and eaten while the kernels are still tender and before all of the sugars are converted to starch. This definition includes "roasting ears" of selected field corns.

Today, the standard sugary corns are being modified with other endosperm genes and gene combinations that control sweetness to develop new cultivars. As a result, growers must consider genetic type when making selections for planting. The genetic type is not readily identifiable by cultivar name alone. At least 13 endosperm mutants, in combination with sugary, have been studied for improving sweet corn. Except for sugary, the genes used in breeding act differently to produce the taste and texture deemed desirable for sweet corn.

Augmented sugary kernel types. In these sweet corns, the sugars are modified (increased) by the action of other genes, either partially or completely.

Major modifier genes of kernel sweetness are shrunken-2 (*sh*2) and sugary enhancer (*se*). In partial modifications, the sugary (*su*) kernels are modified by the segregation of major modifier genes such that about 25 percent of the kernels are double-mutant endosperm types possessing the enhanced benefits of the modifier. The addition of the sugary enhancer (*se*) gene along with one of the major modifier genes (e.g., *sh*2) will further modify some of the sugary kernels to about 44 percent double-mutant endosperm types rather than 25 percent.

In complete (100 percent) modification, the sugary (*su*) kernels are all modified with the sugary enhancer (*se*) gene to produce the double combination (*su se*) for obtaining maximum benefit from the *se* gene. Other major modifier genes of kernel sweetness are: brittle (*bt*), brittle-2 (*bt*2), shrunken (*sh*) and shrunken-4 (*Sh*4). Other genes with minor modifying effects of kernel sweetness are: dull (*du*), floury (*fl*), floury-2 (*fl*2) opaque (*o*), opaque-2 (*o*2), sugary-2 (*su*2), and waxy (*wx*). Some are known to be present in sweet corn backgrounds either in the segregating or homozygous state.

Other mutants producing sugary kernels include the single-mutant endosperm genes shrunken-2 (*sh*2), brittle (*bt*) and brittle-2 (*bt*2), and the multiple-mutant endosperm genes amylose-extender, dull, waxy (*ae du wx*).

Precautions with modified endosperm sweet corn to avoid xenia. Isolation of "sweet corn" cultivar plantings of different genetic types is necessary to prevent cross-pollination. Xenia is the immediate effect of foreign pollen on a variety; on sweet corn (*su*), it will produce a starchy kernel. Isolation can be obtained by planting at a different time, planting cultivars of different maturities, planting "upwind" of prevailing wind direction, or providing barriers and border rows. All of these methods will reduce the isolation distances necessary. On a practical basis, commercial growers should provide at least 50 feet separation, plant upwind of normal field corn, and use four or more border rows.

Popcorn

Popcorns are perhaps the most primitive of the surviving races of maize. This corn type is characterized by a very hard, corneous endosperm containing only a small portion of soft starch. Popcorns are essentially small-kerneled flint types. The kernels may be either pointed (rice-like) or round (pearl-like). Some of the more recently developed popcorns have thick pericarps (seed coats), while some primitive semi-popcorns, such as the Argentine popcorns, have thin pericarps.

Popcorn is a relatively minor crop compared to dent corn. It is used primarily for human consumption as freshly popped corn or as the basis of popcorn confections. Isolated planting is not necessary, since there are no major xenia effects on popping expansion and many popcorns are cross-sterile with field corn.

Most popcorn acreage is grown under contract. Although conditions for growing popcorn are the same as for dent corn, special harvesting, drying and storage practices are necessary to maintain popping quality (see NCH-5, "Popcorn Production and Marketing").

Pod Corn

Pod corn (tunicate maize) is more of an ornamental type. The major gene involved (*Tu*) produces long glumes enclosing each kernel individually, such as occurs in many other grasses. The ear is also enclosed in husks, as with other types of corn.

Homozygous pod corn usually is highly self-sterile, and the ordinary type of pod corn is heterozygous. Pod corn may be dent, sweet, waxy, pop, flint or floury in endosperm characteristics. It is merely a curiosity and is not grown commercially.

Special-Purpose Corns

Corn may be altered by genetic means to produce modifications in starch, protein, oil and other properties. As a result of modifications of ordinary dent types, new corn specialties have been created. Among them are waxy-maize, amylomaize, and high-lysine or modified-protein corn.

Waxy corn. This special-purpose type was introduced to the U.S. from China in 1908. Although China was the original source, waxy (*wx*) mutations have since been found in American dent strains. Its name derives from the waxy appearance of the endosperm exposed in a cleanly cut cross-section. Common corn starch is approximately 73 percent amylopectin and 27 percent amylose, whereas waxy starch is composed entirely of amylopectin, which is the branched molecular form. Ordinary cornstarch stains blue with 2 percent potassium iodide solution, whereas waxy cornstarch stains a reddish brown. The waxy gene also expresses itself in the pollen with this staining reaction, which is an aid in breeding.

Significant advances in yields have been made with the newer waxy hybrids. While the overall average may run somewhat less than dent corn hybrids, the newer waxy hybrids are more comparable to the better dents in yields.

Waxy corn has carved out a formidable position as the raw material of waxy cornstarch produced by certain wet-corn millers in the U.S., Canada, Europe, etc., for industry and food uses. Waxy starch and modified waxy starches are sold extensively worldwide because of their stability and other properties of their solutions.

Products made from waxy corn are used by the food industry as stabilizers and thickeners for puddings, pie fillings, sauces, gravies, retorted foods, salad dressings, etc. Other waxy products are used as remoistening adhesives in the manufacture of gummed tape, in adhesives and in the paper industry. Waxy grain is also grown as a feed for dairy cattle and livestock.

Waxy corn is usually grown under contract for the major wet millers and exporters. Premiums are paid to the growers of waxy corn for wet milling because it must be isolated during production, harvesting, transporting and storing. Since waxy is a recessive characteristic, isolation from dent corn is necessary to prevent loss of its peculiar starch properties.

High-amylose corn. Amylomaize is the generic name for corn that has an amylose content higher than 50 percent. The endosperm mutant amylose-extender (*ae*) found by R. P. Bear in 1950, increases the amylose content of the endosperm to about 60 percent in many dent backgrounds. Modifying factors alter the amylose contents as well as desirable agronomic characteristics of the grain. The amylose-extender gene expression is characterized by a tarnished, translucent, sometimes semi-full kernel appearance.

High-amylose grain is grown exclusively for wet milling. The two types produced commercially are Class V (amylose content, 50-60 percent) and Class VII (amylose content, 70-80 percent). The starch from high-amylose corn is used in the textile industry, in gum candies (where its tendency to form a gel aids production), and as an adhesive in the manufacture of corrugated cardboard.

High-amylose corn yields vary depending upon location, but average only 65-75 percent of that of ordinary dents. Present production acreage is limited to that grown under contract arrangements for wet millers. Premiums are paid to growers because of decreased yields and the necessity to isolate high-amylose corn during production, harvesting, transporting and storing. The premium depends upon class, year and desired acreage.

High-lysine corn. This is the generic name for corn having an improved amino acid balance, thus a better protein quality for feeding and food use compared to ordinary dent types. E. T. Mertz in 1964 discovered that the single recessive gene, opaque-2 (*o2*), reduced zein in the endosperm and increased the percent of lysine to improve nutritional quality. Other genes with similar gross effects on protein quality exist in corn, but attempts to improve corn protein quality have been primarily based on use of the opaque-2 gene and modified opaque-2 germplasm.

The opaque-2 gene is characterized by a soft, chalky, non-transparent kernel appearance, having practically no hard vitreous or horny endosperm. Undesirable kernel characteristics (e.g., kernel and ear rots) and insect and rodent damage can be a problem with the soft opaque-2 chalky phenotypes. Improvements in resistance to ear and kernel rots have been substantial with selection, and a number of good hybrids exist. On the average, the opaque-2 hybrids yield about 7-10 percent lower than their normal counterparts.

A promising approach to overcoming some of the deficiencies of the homozygous opaque-2 materials involves the visual selection of specific modifiers of opaque-2. It is fairly easy to develop modified, vitreous opaque-2 materials with good ear rot and grain insect resistance. Selections must include endosperm chemical analyses to maintain high levels of protein quality.

Another approach to endosperm textural modification to solve some of the problems associated with opaque-2 corn has been use of the double mutant combination, sugary-2 opaque-2 (*su*2 o2). This modification has improved kernel vitreousness, density and resistance to kernel breakage. The improved vitreous of *su*2 o2 is accompanied by protein quality at least equal to the unmodified opaque-2 materials; however, at this point, yields are 80-85 percent of normal dents.

High-lysine grain can be an important source of high-quality protein in the diets of nonruminants; and nutritional studies have confirmed the potential value of high-lysine corn in helping to meet the world's human and animal nutritional needs. For the present, loss of calories per acre is the trade-off for increased amounts of high-quality protein.

Current U.S. use of high-lysine corn is restricted because of (1) yield differentials compared to normal corn and (2) the corn-to-soybean oil/meal price relationship. Demand for high-quality protein corn in the U.S. is insufficient to command a premium price. However, high-lysine corn is grown to a limited extent as a feed for poultry, swine, dairy cattle and other livestock production needs. In corn-dependent countries where normal corn is a major staple of the human diet, or where high-quality protein supplements for animal feeding are scarce, yield is a secondary consideration. Some high-lysine materials are to the point of development where it may be cost-efficient to grow quality-protein corn as a specialty crop.

Ornamental corn. The so-called ornamental or "Indian" corns commonly show segregation for alleles of several genetic factors that control the production of anthocyanins and related pigments in the aleurone, pericarp and plant tissues of corn. The kernels may be segregating for various color expressions; and varigation of color may even be expressed within a kernel, depending upon the genetic factors involved and their interaction during development of the kernel.

Ornamental corns may be dent, sweet, pop, flint or floury endosperm types. Apart from genetic studies, they are a curiosity and are only grown for ornamental and decorative purposes.

A publication of the National Corn Handbook Project

NEW 6/84 (5M)

Cooperative Extension Work in Agriculture and Home Economics, State of Indiana, Purdue University and U.S. Department of Agriculture cooperating. H. A. Wadsworth, Director, West Lafayette, IN. Issued in furtherance of the Acts of May 8 and June 30, 1914. It is the policy of the Cooperative Extension Service of Purdue University that all persons shall have equal opportunity and access to its programs and facilities without regard to race, color, sex, religion, national origin, age or handicap.

II. TRAITS TO CONSIDER WHEN SELECTING A HYBRID OR VARIETY

Objectives

1. Define plant traits growers consider when selecting a hybrid or variety.
2. Describe corn, soybean, and small grain maturity classification systems.
3. Select proper hybrids and varieties for specific production and management systems.

Improved hybrids and varieties result from incorporating specific desirable traits into seed stocks sold to farmers. The characteristics of the hybrid or variety determine its suitability for specific production conditions, and what is adapted to one set of conditions may not be adapted to another set of conditions. Therefore, plant breeders select plants with specific traits. Some of the most important traits considered by plant breeders are: yield potential, maturity, pest resistance, lodging resistance, drought resistance, ability to withstand population pressures, and crop quality.

A. Yield potential

To be considered desirable, a hybrid or variety must consistently produce high yields of grain, green feed, or fodder. If a hybrid or variety does not have the ability to yield well, a seed company will never release it for sale. High yield is the primary criterion for each hybrid or variety released.

The actual yield of a hybrid or variety may not be the same as its potential. For example, production factors may be such that its full potential under conditions of adequate moisture would not reach its full yield potential during a dry year.

B. Maturity

Adapted, mid- to full-season hybrids or varieties usually yield best. They take full advantage of the entire growing season, available nutrients, and soil moisture.

1. Corn maturity

Given favorable fertility and weather conditions, growers will get the highest corn yields from full-season hybrids. Full-season hybrids usually make optimum use of the length of the growing season, plant nutrients, and available soil moisture. Therefore, it is important to know the maturity ranges of the hybrids you intend to plant.

Corn maturity is classified in the following ways:

a. Days to maturity

Counting days from either planting or emergence to maturity is the oldest and most commonly used method to determine corn maturity. (Some seed companies begin counting from the date of planting, others from the date of emergence). This length of time is frequently expressed as a range. For example, a corn hybrid that matures 98 to 102 days after planting or emergence is called a 98 to 102 hybrid.

The classification of a hybrid also depends on location. A short-season corn hybrid in southern Indiana might be a full-season hybrid in northern Indiana, where the growing season is shorter. Growers often take advantage of the range of maturities available and plant more than one hybrid, as this spreads the risk of stress problems at pollination time.

b. Growing degree days

Counting days to maturity does not take into account differences in growing conditions during the season. Because corn growth directly relates to temperature conditions, a 105-day corn hybrid may require 115 days to mature in a cool summer. On the other hand, it may mature in 100 days or less during a hot summer.

Since corn growth is closely related to temperature, corn maturity can be predicted by counting the accumulated heat units, or growing degree days, from time of planting. Degree days are accumulated at weather stations, and are broadcast over radio stations and printed in newspapers. An example of how growing degree days are used is shown below.

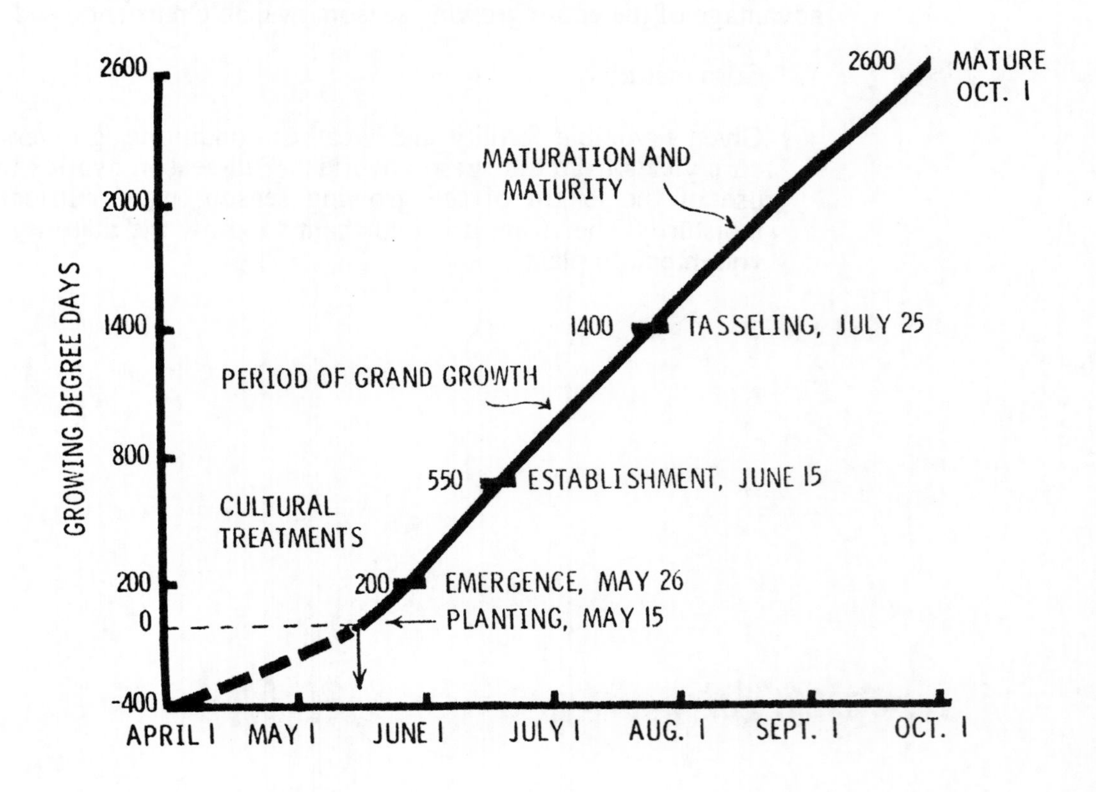

Required Growing Degree Days and Expected Yield Coefficients for Various Corn Hybrid Maturity Groups by Indiana Location.		
Location and maturity group	Growing degree days[1]	Expected Yield coefficient[2]
Northern Indiana		
Full-season	2700	1.00
Medium-season	2550	0.96
Short-season	2400	0.93
Central Indiana		
Full-season	2700-2800	1.00
Medium-season	2550-2650	0.95
Short-season	2400-2550	0.90
Southern Indiana		
Full-season	2800-3000	1.00
Medium-season	2650-2800	0.90
Short-season	2550-2700	0.85

[1] Growing degree days (GDD, base 50) required for maturity. A particular hybrid requiring 2550 GDD would be considered short-season if planted in southern Indiana, and medium-season in central Indiana or northern Indiana. Other than in the north, GDD accumulation is more in western Indiana than in eastern Indiana.

[2] Full season is used as a reference yield (yield = 1.00) for each location. For example, if potential yield for a full-season hybrid in central Indiana is 150 bushels per acre (1.00), then potential yield for a medium-season hybrid would be 142.5 bushels (150 x 0.95) and for a short-season hybrid, 135 bushels (150 x 0.90).

2. Soybean maturity classification

Soybean maturity is dependent primarily upon photoperiod, or hours of daylight per day. Because photoperiod varies as one travels north-south, soybean maturities are classified into narrow geographic zones which stretch east-west. A maturity zone is a region in which soybeans of that maturity group are adapted. Each maturity group has a standard variety with which other varieties in that group are compared. The standard for the Group II maturity classification is the IA2021; for Group III it is Iroquois; and for Group IV it is Stressland. Within each group, varieties are classified as early, mid, or late. Resnik is classified as early Group III because it is better adapted to the northern portion of the Group III maturity area than is Iroquois and it matures earlier than does Iroquois.

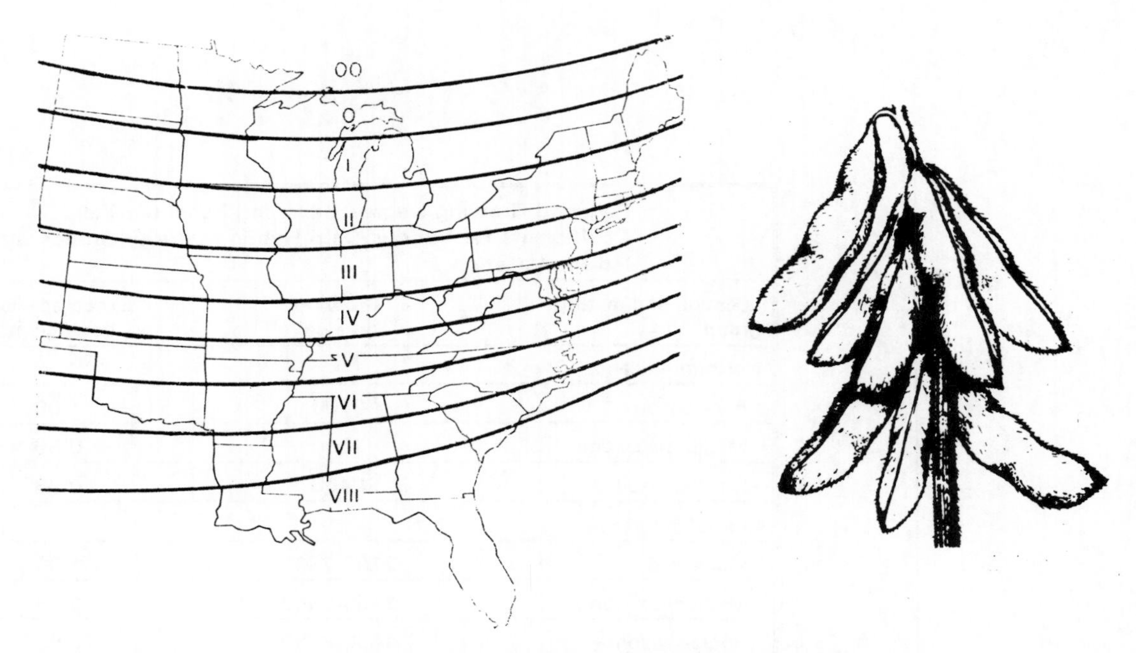

Soybean Maturity Zones for the Ten U.S. Maturity Groups

3. Small grains maturity classification

Small grains fall into two categories, based on when they are seeded. Winter types are seeded and emerge in the fall, and plants lie dormant over winter. Spring types are seeded and emerge in the spring, and plants mature in the summer. Farther north, farmers tend to plant spring types, because winter types are unable to withstand severe winter conditions. Winter types generally out-yield spring types, so farmers grow winter small grains in areas where they are adapted.

C. Pest resistance.

Breeding plants for qualities that contribute to insect and disease resistance is an important way to biologically control these pests. It is especially important to consider racial variations in pests when breeding disease-resistant plants. Certain plant diseases and insects evolve into several races. When plant breeders develop a hybrid that is resistant to the most common races of a disease or insect, it may grow well for several years. However, if a new race of the disease or insect evolves, these plants may have no resistance to it.

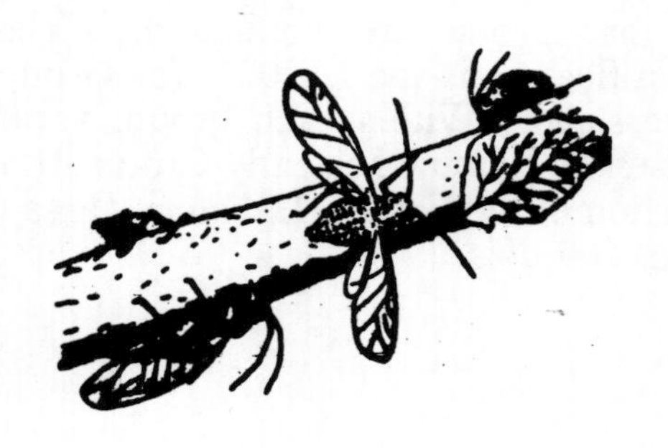

Some plant characteristics which aid in resisting pests:
1. pubescence on stems and leaves
2. production of protective tissue
3. rapid regrowth rates
4. chemical compounds which discourage pest attack

Genetic resistance to Phytophthora root rot in soybeans occurs by inserting alleles for protection against a wide spectrum of prevalent Phytophthora races. The Rps1-k allele is most commonly used, and when the Rps-3 allele is also inserted additional resistance is obtained. The group II variety *Savoy* has both of the alleles.

In corn, resistance to European corn borer has been developed by inserting into hybrids the Bt gene from the bacterium *Bacillus thuringiensis*. The gene causes a natural toxin to develop in the corn plant, which kills the corn borer.

A major concern is that corn borers will one day show resistance to Bt corn. A small percentage of corn borers, perhaps one in a million, may have a natural resistance to the Bt toxin. If Bt corn is planted extensively throughout the corn belt, these resistant individuals will be selected for. Most of the susceptible corn borers would die, while the small number of resistant corn borers would survive. They would then mate with other resistant corn borers, producing resistant offspring. The resistant population would explode, and Bt corn would become useless for controlling corn borer.

A way to prevent resistance from becoming a problem. is to plant at least 20% of corn acres to non-Bt corn, to provide a refuge that would allow a lot of the susceptible corn borers to survive. They would mate with the resistant ones, and the resistant population would never become dominant. Researchers maintain that in order for this to work, a refuge must be maintained on every farm.

D. Herbicide resistance

Corn hybrids and soybean varieties resistant to specific herbicides have recently been developed. Resistance occurs through the plant's ability to either degrade the herbicide metabolically, or by the plant not taking up the herbicide, or sequestering it within the plant. Herbicide degradation within the crop occurs by inserting isolated microbial genes that code for an enzyme that degrades the herbicide. Perhaps the most well known resistance is that of Roundup in corn and soybeans. These are marketed as Roundup-ready corn and soybeans.

E. Lodging resistance

Lodging is a tendency of certain plants to bend at or near soil level and lie nearly flat on the ground. It is a condition found most frequently in cereal grains. When plants lodge, they frequently produce poor grain quality, give low yields, and are difficult to harvest. A lodged plant that partially straightens has <u>elbowed</u>. Elbowing, like lodging, causes reduced yields and makes mechanical harvesting more difficult.

Plant characteristics which contribute to lodging resistance include:
1. Stiff stems or stalks
2. Reduced plant height
3. Disease and insect resistance

F. Adaptation to production conditions

Some production conditions that influence hybrid or variety selection are:
1. Soil fertility
2. Winter conditions
3. Soil conditions
4. Light interception
5. Soil crusting
6. Plant population and spacing

G. Harvestability

Hybrid and variety characteristics which affect harvestability of a grain crop include:
1. Dry-down rate after maturity

   Hybrid characteristics that encourage rapid grain moisture loss include few husk leaves that are thin and that senesce early; loose husks that do not cover the ear tip, and small cob diameter. In early September, grain moisture loss can be as great as one percentage point per day in Indiana.

2. Grain retention
3. Grain separation
4. Grain location
5. Shattering

H. End Use

Plant breeders can develop hybrids and varieties with characteristics suited for a specific end use. For example, a company that extracts corn oil may only be interested in purchasing high oil corn. Read the section entitled Special Purpose Corns in NCH-10-Origin, Adaptation, and Types of Corn.

Maturity Rating Systems for Corn

D. J. Eckert, Ohio State University; and D. R. Hicks, University of Minnesota

Reviewers
G. O. Benson, Iowa State University
T. Daynard, University of Guelph
E. D. Nafziger, University of Illinois
K. R. Polizotto, Potash Corp. of Saskatchewan (IN)

In selecting corn hybrids, the first criterion most farmers are concerned about is maturity—i.e., the approximate length of time for a crop to complete its life cycle in a given environment. Usually, farmers look for hybrids that will take full advantage of the growing season (for maximum yield), will mature before the first killing frost, and will be ready to harvest at a reasonable date and grain moisture content.

Today, most corn hybrids have been maturity-rated by any one of several systems. Proper understanding and use of these systems can greatly improve a producer's chances of choosing the hybrids best suited to his particular farm and production management situation.

This publication discusses the common corn hybrid maturity rating systems, their development, interpretation and various applications. But first, to understand the basis of measurement for all these systems, let's look briefly at what influences maturity in corn and how.

WHAT AFFECTS MATURITY AND HOW

Although maturity is *determined* genetically, it is *influenced* to a great extent by environmental factors, such as soil type, soil moisture, crop nutrition—but especially air temperature. For instance, a given hybrid will usually develop and mature more quickly in warmer years or in warmer locations than in cooler ones. Thus, a hybrid considered short-season in Alabama would likely be an extremely long-season choice in Michigan or New York.

Because maturity is so closely tied to temperature, the following generalizations can be made about desired hybrid maturity relative to geographic location:

- Longer-season hybrids can be planted as one moves further south in the United States, because the growing season becomes longer and warmer.
- The temperature-moderating effect of a large body of water permits use of longer-season hybrids near the shore than could be planted further inland.
- Higher elevations which have cooler prevailing temperatures should be planted to shorter-season hybrids than lower elevations.
- Valleys that warm slowly in the spring and are subject to early frosts in the fall require shorter-season hybrids than the surrounding ridges.

While these generalizations hold true in a majority of cases, be aware that local variations may (and usually do) exist.

GROWING DEGREE DAY (GDD) SYSTEMS

All corn hybrid maturity rating systems are related to temperature effects—some rather loosely, others quite specifically. Presently the one most widely used in the U.S. is based on these two facts: (1) a corn plant must accumulate a certain amount of heat in order to complete its life cycle, and (2) the total amount of heat needed will be relatively constant for a given hybrid. Under the system, the quantity of heat being added is determined from daily temperatures and usually expressed as "growing degree days," "growing degree units" or "heat units."

How GDDs Are Calculated and What They Mean

Growing degree days may be defined as the difference between the mean daily temperature and a chosen base temperature, subject to certain restrictions. The equation and set of restrictions most commonly used in the U.S. to calculate GDDs are as follows:

$$GDD = \frac{H + L}{2} - Base$$

where H is the daily high temperature but no higher than 86°F (30°C), L is the daily low but no lower than 50°F (10°C), and the base is 50°F.

A base is used in order to keep accumulated GDD numbers relatively small; it is set at 50° because corn makes little or no growth below that temperature. The lower cutoff for the daily temperature prevents calculation of negative values. (Corn rarely, if ever, experiences negative growth and development.) The upper cutoff for the daily temperature is based on the fact that at 86°F, corn growth rate begins to decline rapidly due to excessive respiration, moisture stress, etc.

Figure 1 gives examples of how daily growing degree days are determined using the above equation and cutoffs. The accumulated GDDs that identify a corn hybrid's maturity rating is merely the sum of daily GDD values over a given length of time. The time period that seed corn producers normally use to rate their hybrids is from planting to physiological maturity (i.e., the point at which grain filling is essentially completed).

In many states, accumulated growing-degree-day information is being kept for most geographic areas and is available from state Extension Services or the USDA Crop Reporting Service. Figure 2 gives the mean number of growing degree days available in many of the corn growing areas of the U.S. With such information, farmers can make wise hybrid selection decisions, provided they are aware of the following:

$$GDD = \frac{H + L}{2} - Base$$

a. If the high is 80°F and the low is 60°F:

$$\frac{80 + 60}{2} - 50 = \frac{140}{2} - 50 = 70 - 50 = 20 \text{ GDDs}$$

b. If the high is 60°F and the low is 40°F:

$$\frac{60 + 50}{2} - 50 = \frac{110}{2} - 50 = 55 - 50 = 5 \text{ GDDs}$$

c. If the high is 90°F and the low is 70°F:

$$\frac{86 + 70}{2} - 50 = \frac{156}{2} - 50 = 78 - 50 = 28 \text{ GDDs}$$

d. If the high is 45°F and the low is 30°F:

$$\frac{50 + 50}{2} - 50 = \frac{100}{2} - 50 = 50 - 50 = 0 \text{ GDDs}$$

Figure 1. Example of growing-degree-day calculations using 50°F base and 50° and 86°F cutoffs.

- Reliability of a GDD rating system increases with the number of years that local GDD data have been collected.
- Depending on the region, accumulated GDDs can differ significantly within even a rather small area. For example, in hillier parts of Ohio, it is not unusual for GDDs to vary by 200 units over a distance of less than 20 miles. Therefore, GDD values may have to be adjusted, due to the location of one's farm relative to the weather station at which the data were recorded.

How a GDD Rating System Is Used

Because growing degree days characterize several important relationships between corn hybrids and the local growing season, the GDD maturity rating system can be a valuable decision-making aid—and easy to use. For instance, having access to daily GDD information for his area, a farmer can choose hybrids rated to take advantage of the full growing season based on his intended planting date. If planting must be delayed substantially, he can substitute hybrids having a maturity rating within the range of accumulated GDDs left after the new intended planting date. Also, by calculating daily GDD values during the growing season, the farmer who knows his hybrid's maturity rating can often schedule harvesting much more accurately than one who doesn't use such a system.

A word of caution: Although the growing-degree-day maturity rating system is the most common one in the U.S., be aware that not all hybrid corn seed producers use the same set of "givens" to determine GDDs. For instance, some may use temperature cutoffs other than 50° and 86°F in calculating daily GDDs; and some may figure accumulated GDDs over a time period other than from planting to the end of grain filling. Therefore, when using a GDD system to help select hybrids, it's important to find out exactly what scheme was followed. Seed dealers should be able to provide this information.

ONTARIO CORN HEAT UNIT (OCHU) SYSTEM

The OCHU system is quite similar to growing-degree-day systems in that maturity ratings are based on the accumulation of heat units. However, the calculations used to generate daily heat unit values are more complex, and the numerical values are usually 30-40 percent larger.

The OCHU system is used to rate corn hybrid maturity in Canada and is quite accurate under the climatic conditions there.

RELATIVE MATURITY RATING SYSTEMS

Very common in years past and still rather widely used in the U.S. are several relative maturity ranking scales. Using this measure, hybrids are rated according to when they mature in comparison to hybrids of known maturity.

The two most common ones are the Minnesota Relative Maturity Rating System and the Agricultural Experiment Station (AES) Maturity Rating System. Both categorize hybrids into maturity groups—the Minnesota system using a "days" classification (e.g., 95-day hybrids) and the AES system using a numerical series (e.g., 800-series hybrids).

The days or numbers used to identify maturity groups may not always reflect reality (i.e., a 95-day hybrid does not necessarily mature in exactly 95 days); but maturity within and between groups is expected to be constant. For instance, in the Minnesota system, hybrids classed as 100-day should reach physiological maturity approximately 5 days earlier than 105-day hybrids grown under similar climatic conditions and cultural practices. In the AES system, differences between series are generally 4-day intervals—800-series hybrids being 4 days earlier than 900 series, for example.

Rating hybrids relative to those of known maturity permits hybrid maturity recommendations on the basis of "zones of adaptation." This insures that a hybrid adapted to a given zone and planted on a reasonable date will mature before frost. (Climatic zones of adaptation are generally drawn for each state in which a relative rating system is used extensively.)

While it may not provide as much information as a GDD system, hybrids selected by a relative maturity rating system should result in a crop that makes good use of the growing season and is ready for harvest at a reasonable moisture content. However, don't forget that, as with the GDD system, climatic variations in certain localities must be considered and the ratings adjusted if necessary.

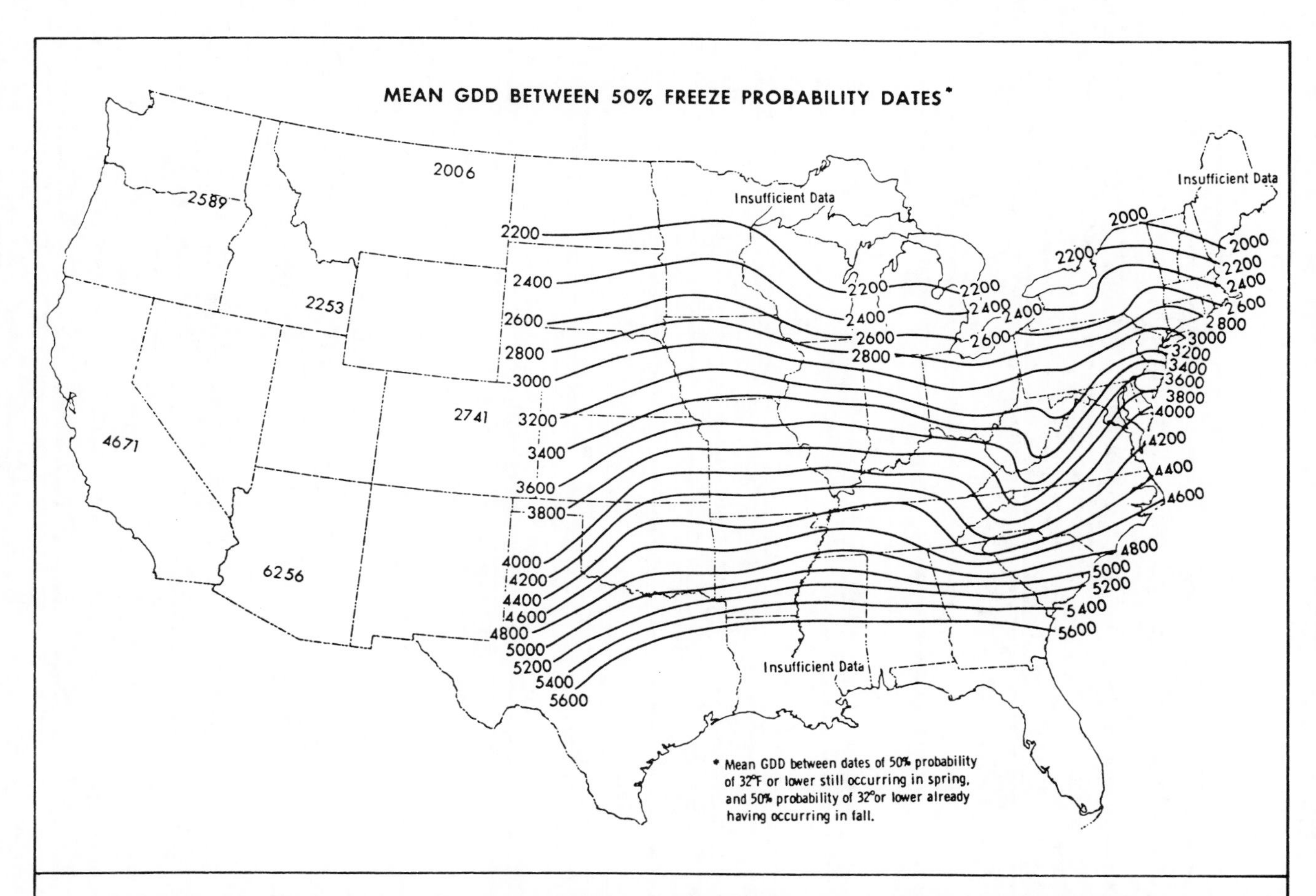

Figure 2. Mean number of growing degree days between April 1 and 50 percent chance of frost in the fall. (Source: USDA Statistical Reporting Service Weekly Weather and Crop Bulletin.)

A publication of the National Corn Handbook Project

NEW 12/85 (5M)

Cooperative Extension Work in Agriculture and Home Economics, State of Indiana, Purdue University and U.S. Department of Agriculture cooperating. H.A. Wadsworth, Director, West Lafayette, IN. Issued in furtherance of the Acts of May 8 and June 30, 1914. The Cooperative Extension Service of Purdue University is an affirmative action/equal opportunity institution.

Corn Hybrid Maturity Management for the Central and Northern Corn Belt

D. R. Hicks, University of Minnesota
G. O. Benson, Iowa State University
D. Bullock, University of Illinois

Reviewers
P. Carter, University of Wisconsin *B. Reiss, Asgrow Seed Co., Inc.*
R. Nielsen, Purdue University *D. Wright, University of Florida*

The objectives of planting corn hybrids of different maturities are to 1) produce maximum profit, 2) minimize risk due to adverse weather conditions, 3) allow a longer harvest period when grain is near optimum moisture content for combining, and 4) allow harvest to begin early.

At one time agronomists recommended planting the short-season hybrids first, followed by mid- and full-season hybrids, successively. However, for corn production in the central and northern Corn Belt states, a better maturity management scheme is to plant full-season hybrids first, follow with mid-season hybrids, and plant early hybrids last. This publication discusses the basis for this maturity scheme.

Data presented are from three years and three locations in southern and central Minnesota. In this article, hybrids differ by 15 relative maturity days (Minnesota relative maturity), which is a large difference in maturity and may be larger than most corn farmers attribute to the terms "short," "mid," and "full." However, while the data presented are from southern and central Minnesota, the concept applies throughout the central and northern Corn Belt. Although hybrids that are termed short-, mid-, and full-season are different hybrids as the location moves from north to south, the effects of planting date on yield, pollination date, and maturity date are similar. Actual calendar dates of optimum planting date and dates of pollination and maturity will, of course, vary with location.

Grain Production and Profit

Production costs and yields are major factors affecting profit from the corn enterprise. An individual's decisions on production inputs will fix production costs. Management to produce the most grain from an individual's corn acres will result in maximum profit from these inputs, at any corn price. Therefore, total grain production from all corn acres should be a producer's objective.

With adequate rainfall or available moisture, highest yields are produced with early planting dates for all corn hybrid maturity groups. However, yield reduction with delayed planting is greatest for full-season hybrids compared with that of earlier hybrids (Figure 1). Therefore, maximum grain production occurs if the full-season hybrids are planted before the earlier maturing hybrids. For maximum production, all groups of hybrids should be planted early, with a major portion of the acreage planted to high-yielding, full-season hybrids.

In Minnesota trials with planting dates from April 24 to June 1, full-season hybrids always yielded higher than mid-season hybrids, which in turn yielded higher than short-season hybrids. For optimum planting dates (late April to early May in Minnesota), the full-season hybrids yielded 18% (20 bu/a) more than the mid-season hybrids, and they, in turn, yielded 16% more than the short-season hybrids. Yield differences due to maturity will be less when a smaller difference in maturity occurs between maturity groups.

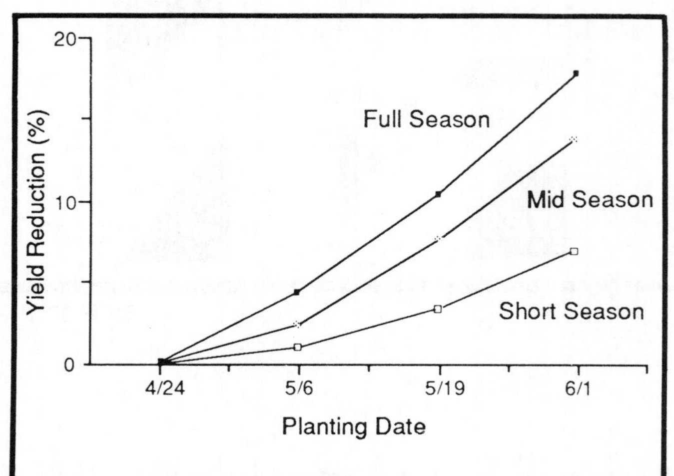

Figure 1. Corn grain yield reduction for three corn hybrid maturity groups planted through the month of May in Minnesota.

Full-season hybrids should not be planted after May 20-25 in most of the Corn Belt because they are unlikely to reach maturity before frost. Grain moisture content of late-planted, full-season hybrids will be higher and drying costs greater compared with that of late-planted, earlier maturing hybrids. Test weight may also be lower on the late planted, full-season hybrids.

The terms "full-," "mid-" and "short-season hybrids" are relative and apply to normal planting dates. For example, 110 RM hybrids are full-season hybrids for southern Minnesota if planted prior to May 20. After May 20, they are not maturity adapted because they are unlikely to reach maturity before the average frost date. Hybrids with lower maturity ratings then become "full season" for the late planting dates. For example, 100 RM hybrids are "full-season" hybrids for southern Minnesota when planted between May 20 and 25. As planting is delayed, they, too, may not be maturity adapted because of the limited remaining growing season.

Many times corn is planted over a period during which soil conditions prevent continuous planting. If the full-season hybrids are planted first, seed exchanges may not be necessary to have hybrids that are maturity adapted for a late planting season. It may be difficult to return seed of full-season hybrids and obtain seed of good-performing, earlier hybrids. Therefore, planting the full-season hybrids first provides a hedge against this problem if planting is partially delayed because of weather conditions in the central and northern Corn Belt.

In the southern and southeastern U.S., grain yields of currently available hybrids are not affected by maturity differences as discussed here. As an average, short- and mid-season hybrids generally yield higher than full-season hybrids when grown under irrigation. However, full-season hybrids often produce higher yields than early and mid-season hybrids when planted late because of summer rains that normally occur during the grain filling period of full-season hybrids. When planted early, short- and mid-season hybrids often go through periods of no rain which may last from 4 to 6 weeks in April, May, and June. Husk coverage and grain quality are often poorer on short- and mid-season hybrids than full-season hybrids used in the southern U.S.

Harvest Maturity Order

Calendar dates of reaching 32% kernel moisture are given in Figure 2 for three hybrid maturity groups planted from late April through May. (32% kernel moisture occurs at or near the same time of physiological maturity.) For all maturity groups, maturity occurs at later dates when planted later. However, late-planted, short-season hybrids will mature before early-planted, full-season hybrids if the early hybrids are planted prior to late May. If mid-season hybrids are planted before mid May, they will mature before early-planted, full-season hybrids. These differences in maturity dates will be reduced when smaller differences exist in short-, mid-, and full-season hybrids. Planting all maturity groups as early as possible maximizes the spread in harvest maturity dates. For example, the greatest spread in harvest maturity dates occurs when all maturity groups are planted in late April.

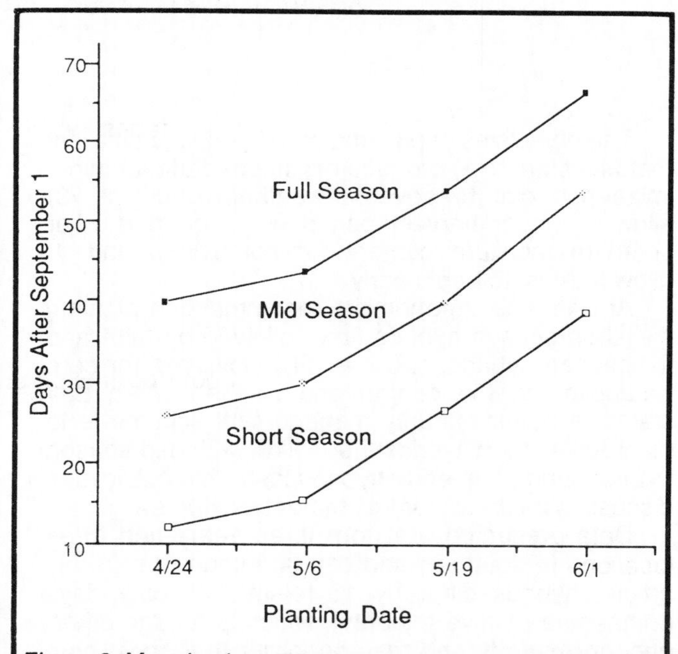

Figure 2. Maturity date (32% kernel moisture) for three corn hybrid maturity groups planted through the month of May in Minnesota.

Scheduled Harvest

Harvesting corn when grain moisture is between 23 and 27% minimizes both harvest loss and kernel damage. The effect of planting full-season hybrids first on pollination period and time when grain is between 23 and 27% moisture is given in Figure 3. The planting scheme in Figure 3 provides for planting the full-season hybrids by May 5, followed by 4 days

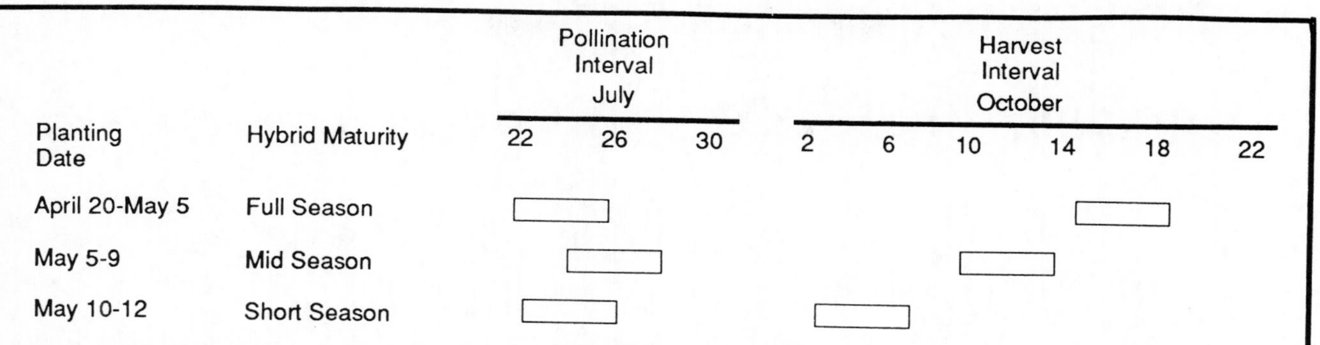

Planting Date	Hybrid Maturity	Pollination Interval July			Harvest Interval October					
		22	26	30	2	6	10	14	18	22
April 20-May 5	Full Season									
May 5-9	Mid Season									
May 10-12	Short Season									

Figure 3. Pollination intervals and fall calendar dates when kernel moisture is in the harvest range of 27 to 23% for three corn hybrid maturity groups when the full-season hybrids are planted first, followed by planting the mid- and short-season hybrids.

of planting mid-season hybrids and 3 days of short-season hybrids. A greater time interval is allocated to planting full-season hybrids because a major portion of the acreage should be planted to the higher yielding, full-season hybrids.

With this example, pollination would occur for all maturity groups during the last part of July. Pollination dates can be spread over a greater number of calendar days by planting the early hybrids first. But pollination for later planted, full-season hybrids then occurs during early August, when there is a greater probability of higher temperatures and soil moisture shortages which adversely affect pollination. And later planting of full-season hybrids results in substantially lower grain yields in the central and northern Corn Belt.

With the scheme in Figure 3, early hybrids are between 23 and 27% kernel moisture content during early October, while mid-season and full-season hybrids are in that moisture range during mid and late October, respectively. Actual calendar dates for both pollination and harvest intervals may vary with years, locations, and hybrids, but the relative order of these events will occur as described.

Summary

Corn maturity management of planting full-season hybrids first, followed by mid- and short-season hybrids, maximizes the grain production and quality of corn for the central and northern Corn Belt. Grain is in the harvestable moisture range for a longer period to facilitate harvest with minimum field loss. Although the data used were from Minnesota and there were large differences (15 RM days) among short-, mid-, and full-season maturity groups, the principles described should apply throughout the central and northern Corn Belt states.

A publication of the National Corn Handbook Project

NEW 9/90 (5M)

Cooperative Extension work in Agriculture and Home Economics, state of Indiana, Purdue University and U.S. Department of Agriculture cooperating. H.A. Wadsworth, Director, West Lafayette, IN. Issued in furtherance of the acts of May 8 and June 30, 1914. The Cooperative Extension Service of Purdue University is an affirmative action/equal opportunity institution.

Agronomy Guide

Purdue University Cooperative Extension Service

(CORN) AY-259

Corn Hybrid Selection for Delayed Planting

Bob Nielsen, Agronomy Department, Purdue University

Corn planting is delayed somewhere in Indiana three out of every five years because of wet spring conditions. If you are facing a later than normal planting date and you haven't planted a large portion of your corn acreage, you need to compare the risks of planting late maturing hybrids to those of planting earlier maturing hybrids. The decision to switch hybrid maturities in a late planting situation is based on: 1) how many days are left from planting until the first killing frost in the fall, and 2) how many growing degree days are expected during that time so that hybrid maturity can be matched to the anticipated accumulation.

The objectives of this publication are: 1) to discuss the impact of delayed planting on the maturation and grain yield of corn, 2) to relate traditional corn maturity rating systems to the newer growing degree day system, and 3) to provide the producer with the necessary hybrid maturity selection information for making decisions when facing delayed planting situations. General guidelines for selecting hybrid maturities at several late planting dates are provided.

Impact of Delayed Planting on Corn Maturation and Grain Yield

Information gathered in Indiana over the last 20 years suggests that delaying corn planting beyond May 10 will cause grain yield losses of about 1 bushel per acre per day. Corn yields decline about 2 bushels per acre per day with delays in planting from May 20 until June 1. These yield losses are due primarily to the shortened part of the growing season from planting to tasselling. This is the time during which the corn plant is building its photosynthetic factory.

Towards the end of May, however, delayed planting can also reduce yield if the grain does not reach **physiological maturity** before the first killing frost in the fall. Physiological maturity is that point in time when the kernels reach maximum dry weight and a black layer of cells forms near the tips of the mature kernels. This **black layer** can be used to identify the occurrence of physiological maturity. Moisture content of the grain is approximately 30-35 percent at this time.

If the corn is frosted before it is physiologically mature, the grain will be lower in test weight and may not dry down properly. Frosted corn usually leads to added harvesting problems and higher drying costs. Proper hybrid maturity selection in a late planting situation will help avoid this potential problem.

Maturity Rating Systems for Corn

Before continuing, working definitions are needed for hybrid maturity and what we mean by early, medium, and full-season hybrids; days to maturity; and growing degree days.

A **full-season** hybrid is defined as one that uses the complete growing season. **Early-** and **medium-maturing** hybrids, therefore, require a shorter growing season than full-season hybrids. Because earlier maturing hybrids do not utilize as much of the growing season to produce grain, they will usually produce lower grain yields than fuller-season corn hybrids. This is why full-season hybrids are normally recommended for planting a major share of your corn acreage. The terms "early," "medium," and "full" are relative to each other and specific to a location and a planting date. For example, a medium-season hybrid in northeast Indiana may be rated as an early hybrid in central Indiana. Similarly, an early-season hybrid planted on May 1 may "become" a full-season hybrid when planted on May 27.

The **days to maturity** rating system is probably best known among corn producers. This type of relative maturity system does not provide you with an absolute number of calendar days from planting to maturity. For example, a 115-day hybrid may require 100 days from planting to physiological maturity during a warm growing season and 130 days during a cool growing season. The days to maturity system instead ranks hybrids in relative order of maturity based on harvest grain moisture percentage compared with check hybrids. For example, a 125-day hybrid would have a greater harvest grain moisture than a 110-day hybrid if harvested on the same day.

We need to be able to accurately measure maturity or growing season requirements to make

West Lafayette, Indiana

more intelligent decisions about hybrid selection in a late planting situation. The terms "early," "medium," and "full" season and "days to maturity" are adequate for general hybrid description, but a fine-tuning adjustment is necessary to more precisely identify the maturity of a hybrid for late-planting decision-making and risk assessment.

This fine-tuning adjustment is the **growing degree day** requirement of a hybrid. It is abbreviated by GDD and is also known by names like "heat unit" and "growing degree unit (GDU)." A GDD is a way of measuring how much heat has accumulated over a certain length of time, usually a 24-hour period. Corn hybrids can differ in their requirements for heat and sunlight for production of maximum grain yield. Hybrid maturities can be associated with definite amounts of GDD's required from planting to physiological maturity. The formula for calculating the number of GDD's accumulated each day is shown below. The 86°F and 50°F limits are used because corn growth is greatly reduced or slowed above 86°F and below 50°F.

$$\text{GDD} = \frac{(\text{Max. Daily Temp.}) + (\text{Min. Daily Temp.})}{2} - 50°\text{F}$$

Note:

1. If the maximum temperature is above 86°F, use 86°F as the maximum temperature in the formula.
2. If the minimum temperature is below 50°F, use 50°F as the minimum temperature in the formula.

Example 1. If the maximum temperature for June 9 was 84°F and the minimum temperature was 60°F, the GDD accumulation for that day would be:

$$\text{GDD} = \frac{(84°\text{F} + 60°\text{F})}{2} - 50°\text{F} = 22$$

Example 2. If the maximum temperature for June 10 was 95°F and the minimum temperature was 45°F, the GDD accumulation for that day would be:

$$\text{GDD} = \frac{(86°\text{F} + 50°\text{F})}{2} - 50°\text{F} = 18$$

Example 3. The total GDD accumulation for June 9 and June 10 would be:

$$\text{GDD} = 22 + 18 = 40$$

The GDD requirements associated with early-, medium-, and full-season maturities are presented in Table 1 for each Crop Reporting District (CRD) in Indiana. **Caution**: Be careful when comparing GDD-based maturity ratings of hybrids from different companies. Some calculate the GDD requirements of a hybrid from *emergence* of the crop to physiological maturity, rather than from *planting*. Such GDD requirements will be about 150 to 200 GDD's less than what is shown in Table 1.

Table 1. GDD requirements of relative maturities for corn hybrids grown in each Crop Reporting District in Indiana.

CRD	Relative	GDD*
NW	early	2400
	medium	2550
	full	2700
NC	early	2400
	medium	2550
	full	2700
NE	early	2400
	medium	2550
	full	2700
WC	early	2550
	medium	2650
	full	2800
C	early	2550
	medium	2650
	full	2800
EC	early	2400
	medium	2550
	full	2700
SW	early	2700
	medium	2800
	full	3000
SC	early	2550
	medium	2650
	full	2800
SE	early	2550
	medium	2650
	full	2800

* GDD accumulations are from planting to physiological maturity.

Calculating when to Switch Maturities as Planting Is Delayed

The Midwest Ag Weather Service at Purdue University (serving Illinois, Indiana, Michigan, Ohio, and Kentucky) monitors GDD accumulations during each growing season and has calculated long-term averages from many years' worth of GDD information. Table 3 is a summary of GDD accumulation normals for each CRD in Indiana. Using this information, we can estimate how many GDD's can be expected from planting to the average first killing frost for any particular planting date. We can then determine whether a particular hybrid maturity (Table 1) will have enough time to mature before frost and, thus, decide whether an earlier maturing hybrid would be safer to plant.

The last safe planting date for any given hybrid maturity allows a hybrid to mature physiologically about two weeks prior to a killing frost. When does the first killing frost occur in your area? Table 2 presents frost dates for each CRD which correspond to probabilities of 75, 50, and 25 percent of a fall frost occurring before these dates. More localized information on frost dates is presented in AY-231, "Determining Spring and Fall Frost-Free Risks in Indiana," and may be more accurate for individual farm locations.

Table 2. Dates of 75, 50, and 25 percent fall freeze risk (chances of occurrence of below 32°F) for each Crop Reporting District in Indiana.

CRD	Chance of freeze occurrence before date 75%	50%	25%
NW	Oct. 14	Oct. 06	Sep. 28
NC	Oct. 14	Oct. 06	Sep. 28
NE	Oct. 14	Oct. 06	Sep. 28
WC	Oct. 21	Oct. 13	Oct. 05
C	Oct. 21	Oct. 13	Oct. 05
EC	Oct. 14	Oct. 06	Sep. 28
SW	Oct. 28	Oct. 20	Oct. 12
SC	Oct. 21	Oct. 13	Oct. 05
SE	Oct. 21	Oct. 13	Oct. 05

Source: Midwest Ag Weather Service, Purdue University

The following example will illustrate how the information in the three tables can be used to decide when to switch hybrid maturities if planting is delayed:

Example 4. Let's assume that spring rainfall has prevented you from planting a large portion of your corn acreage in central Indiana. The short-term weather forecast suggests that you may be able to get back in the field during the week of May 27. Do you switch from full to medium- season hybrids, or perhaps from medium to early-season?

1. From Table 1 you know that a full-season hybrid for central Indiana requires 2800 GDD's from planting to physiological maturity. You also know that medium- and early-

Table 3. Growing degree day accumulation normals from April 1 for each Crop Reporting District in Indiana.

Week dates	NW	NC	NE	WC	C	EC	SW	SC	SE
Apr 01 - Apr 07	31	31	30	38	38	35	50	47	47
Apr 08 - Apr 14	69	69	67	85	84	78	110	104	104
Apr 15 - Apr 21	115	115	112	141	139	129	180	170	170
Apr 22 - Apr 28	170	170	167	207	204	189	260	246	246
Apr 29 - May 05	234	234	231	282	278	258	350	331	331
May 06 - May 12	308	308	305	367	362	336	449	426	426
May 13 - May 19	391	391	388	461	455	423	558	530	530
May 20 - May 26	487	486	483	568	560	522	679	645	645
May 27 - Jun 02	596	594	590	687	677	633	812	772	772
Jun 03 - Jun 09	718	714	709	819	806	756	957	910	910
Jun 10 - Jun 16	852	846	841	963	948	891	1115	1060	1060
Jun 17 - Jun 23	993	985	979	1114	1096	1032	1279	1216	1216
Jun 24 - Jun 30	1139	1129	1122	1269	1248	1177	1447	1376	1376
Jul 01 - Jul 07	1290	1278	1270	1429	1405	1327	1619	1540	1541
Jul 08 - Jul 14	1446	1432	1423	1593	1566	1481	1795	1708	1710
Jul 15 - Jul 21	1604	1588	1578	1759	1729	1637	1973	1879	1881
Jul 22 - Jul 28	1760	1742	1731	1923	1890	1791	2149	2048	2051
Jul 29 - Aug 04	1914	1893	1882	2085	2048	1943	2323	2215	2219
Aug 05 - Aug 11	2066	2042	2031	2245	2204	2093	2495	2381	2385
Aug 12 - Aug 18	2216	2188	2178	2402	2357	2240	2665	2545	2549
Aug 19 - Aug 25	2358	2326	2317	2551	2502	2379	2827	2701	2705
Aug 26 - Sep 01	2490	2455	2446	2691	2638	2509	2980	2848	2853
Sep 02 - Sep 08	2613	2574	2566	2821	2765	2629	3123	2985	2992
Sep 09 - Sep 15	2726	2684	2676	2942	2883	2740	3257	3113	3122
Sep 16 - Sep 22	2829	2784	2776	3052	2990	2841	3380	3230	3241
Sep 23 - Sep 29	2921	2873	2865	3151	3086	2931	3490	3335	3348
Sep 30 - Oct 06	3002	2951	2943	3238	3170	3010	3588	3428	3443
Oct 07 - Oct 13	3072	3018	3010	3314	3243	3078	3674	3509	3525
Oct 14 - Oct 20	3132	3075	3076	3378	3305	3136	3747	3579	3595
Oct 21 - Oct 27	3182	3122	3114	3432	3357	3184	3809	3638	3655
Oct 28 - Nov 03	3222	3160	3152	3475	3399	3223	3860	3687	3704
Nov 04 - Nov 10	3251	3189	3181	3507	3430	3253	3900	3725	3742
Nov 11 - Nov 17	3270	3208	3200	3529	3451	3273	3929	3753	3770
Nov 18 - Nov 24	3285	3223	3215	3546	3468	3289	3952	3776	3793
Nov 25 - Dec 01	3296	3234	3226	3559	3481	3301	3971	3794	3811

Source: Midwest Ag Weather Service, Purdue University

season hybrids require 2650 and 2550 GDD's, respectively.

2. From Table 2 you know that the average date (50% risk) for the first killing frost in central Indiana is October 13.

3. From Table 3 you observe that, from April 1 until October 13 in central Indiana, you can expect to receive, on average, 3243 GDD's. But, you're not planting on April 1. How many GDD's can you expect to receive from May 27 until October 13? From Table 3, again, you see that by May 27 in central Indiana, 560 GDD's have generally accumulated since April 1. By subtracting those 560 GDD's from the 3243 GDD's, you are left with 2683 GDD's to be accumulated from May 27 until October 13.

4. Now that you know that only 2683 GDD's are expected for the rest of the growing season, which hybrid maturity should you plant the week of May 27? A full-season (2800 GDD) hybrid probably will not mature before a fall frost. A medium-season (2650 GDD) would just barely mature. You must decide whether the risk (50% chance) of an early fall frost is low enough to warrant growing the medium-season hybrid. An early-season hybrid (2550 GDD) will probably mature before a fall frost.

Suggestion: If you want to be more conservative (less risk) about your hybrid maturity selection, use the date for which there exists a 25 percent (1 year out of 4) chance of a first killing frost. If you can assume more risk, use the date for which there exists a 75 percent (3 in 4) chance of a first killing frost.

General Guidelines for Switching Hybrid Maturities

The following suggestions for switching hybrid maturities with delayed planting were developed using the method outlined above and assuming 50 percent frost risk dates. Keep in mind that these were developed using long-term averages and should be used only as guidelines. If temperatures were below normal throughout the season, the GDD accumulations would also be below normal for the season. Maturity would be delayed and yield would be reduced.

For Southern Indiana

1. After June 3, switch from full- to mid-season hybrids
2. After June 10, switch from mid- to early-season hybrids
3. After June 17, switch to soybeans

For Central Indiana

1. After May 20, switch from full- to mid-season
2. After June 3, switch from mid- to early-season
3. After June 7, switch to soybeans

For North Central and Northeastern Indiana

1. After May 13, switch from full- to mid-season
2. After May 20, switch from mid- to early-season
3. After June 3, switch to soybeans

For Northwestern Indiana

1. After May 13, switch from full- to mid-season
2. After May 27, switch from mid- to early-season
3. After June 3, switch to soybeans

Summary

Corn planting is delayed somewhere in Indiana three out of five years because of wet springs. When planting is delayed until late May and early June, the potential for yield reduction from frost damage on immature corn increases dramatically for late maturity hybrids. The possibility of late planting dictates that the corn grower know how and when to switch hybrid maturities to avoid this potential yield loss.

Determining a safe hybrid maturity to use in late planting situations is based upon the average date for the first fall frost in your area and how many Growing Degree Days (or heat units) you can expect from planting until that average frost date. The expected GDD accumulation can then be matched with the GDD requirements of different hybrid maturities.

The information and guidelines in this publication will help the corn grower make accurate and timely decisions about hybrid selection and delayed planting. Planning ahead for late planting situations is one part of being a top corn production manager.

For additional information and publications on other aspects of crop production systems, contact your local Cooperative Extension Service agent.

(Crop-Corn) **RR 7/87 (2M)**

Cooperative Extension work in Agriculture and Home Economics, state of Indiana, Purdue University, and U.S. Department of Agriculture cooperating; H. A. Wadsworth, Director, West Lafayette, IN. Issued in furtherance of the acts of May 8 and June 30, 1914. The Cooperative Extension Service of Purdue University is an affirmative action/equal opportunity institution.

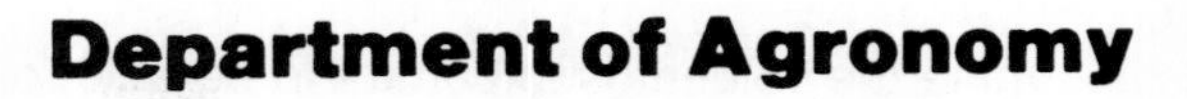

Department of Agronomy

Turfgrass Varieties, Cultivars, and Seed Labels

Jeff Lefton, Extension Turfgrass Specialist

This bulletin covers three areas:

- Selection of turfgrass varieties and cultivars for various sites.
- Turfgrass varieties and cultivars for Indiana.
- Turfgrass seed label interpretation.

Specific Site Situations and Seed Recommendations

- Sunny area (irrigated; mowed properly; fertilized two or more times per year).

1. 100% Kentucky bluegrass blend with at least 3 varieties.
   Use 1.5 lb. of seed per 1,000 sq.ft.
2. 80% or more Kentucky bluegrass plus 20% or less turf-type perennial ryegrass.
   Use 3 to 4 lb. per 1,000 sq.ft.
3. 100% Turf-type tall fescue.
   Use 6 to 8 lb. of seed per 1,000 sq. ft.

- Sunny area (limited irrigation; mowed properly; fertilized at least once, preferably 2 or more times).

1. 100% Turf-type tall fescue.
   Use 6 to 8 lb. of seed per 1,000 sq.ft.
2. Kentucky bluegrass plus turf-type perennial ryegrass (various percentages).
   Use 3 to 4 lb. of seed per 1,000 sq.ft.

- Shade (dry; lower maintenance)

1. 30 to 50% Kentucky bluegrass (blend of 2 or 3 shade-tolerant varieties) plus 30 to 60% fine fescue and/or 0 to 20% turf-type perennial ryegrass.
   Use 4 to 5 lb. of seed per 1,000 sq.ft.
2. 100% turf-type tall fescue.
   Use 6 to 8 lb. of seed per 1,000 sq.ft.

- Shade (wet; lower maintenance)

1. 70% or more Sabre rough bluegrass (*Poa trivialis*) plus a blend of shade-tolerant Kentucky bluegrasses.
   Use 2 to 4 lb. of seed per 1,000 sq.ft.

- Athletic Fields

1. 100% turf type tall fescue.
   Use 6 to 8 lb. of seed per 1,000 sq.ft.
2. 80% or more Kentucky bluegrass blend at least two varieties plus 20% or less turf type perennial ryegrass.
   Use 3 to 4 lb. of seed per 1,000 sq.ft.
   (NOTE: Overseeding with a turf-type tall fescue is not recommended in a bluegrass or bluegrass-ryegrass turf situation.)

- Alkaline and saline (salt) soil conditions.

1. Use a mixture containing 'Fults' alkaligrass. Two examples of mixtures containing alkaligrass appear in Table 1.

Table 1. Two Alkiligrass Seed Mixtures.

	Percent
Tall Fescue	30
Perennial Ryegrass	10
'Dawson' Red Fescue	10
'Scaldis' Hard Fescue	10
'Fults' Alkaligrass	40
Kentucky Bluegrass	30
Perennial Ryegrass	10
'Dawson' Red Fescue	10
'Scaldis' Hard Fescue	10
'Fults' Alkaligrass	40

Recommended Varieties and Cultivars

Table 2 lists various varieties and cultivars recommended by Purdue turfgrass researchers. They are representative of products available on the market.

Interpreting Turfgrass Seed Labels

A good turf area starts with the selection of the correct turfgrasses for a given set of conditions. Quality seed is usually more expensive. To determine the value it is important to be able to correctly interpret a turfgrass seed label.

In Indiana the seed law is a "truth in lending" law. If the percentage is correctly stated on the label, it meets the intent of the present law. However, one needs to look at the label much more closely to properly evaluate quality to establish value. An example of a typical seed label follows.

18.8% Adelphi Kentucky bluegrass	85% germination
32.1% Merit Kentucky bluegrass	83% germination
31.0% American Kentucky bluegrass	85% germination
14.9% Fiesta perennial ryegrass	90% germination
0.0% Crop	
0.2% Weed	
3.0% Inert	

Purity

The purity figure indicates the percent, by weight, of pure seed of each component in the mixture. Not all the pure seed is live seed. The purity for Adelphi Kentucky bluegrass in the sample seed label is 18.8%.

Table 2. Recommended Varieties and Cultivars.

Kentucky Bluegrass	
America	Touchdown*
Mystic	Majestic
Eclipse*	Bristol*
Ram I	Nassau
Plush	Merit
Sydsport	Birka*
Adelphi	Victa*
Glade*	Enmundi

Fine Leafed Fescue	
Reliant*	Banner*
Biljart*	Koket*
Scaldis*	Jamestown*
Waldina*	

Perennial Ryegrass	
Palmer*	Citation II*
Prelude	Birdie II*
Yorktown	Premier
Diplomat	Regal*
Pennfine	Manhattan II*
Fiesta II*	Repell
All Star	Blazer
Derby	

Turf-Type Tall Fescue	
Rebel*	Jaguar*
Falcon*	Arid
Houndog*	Apache
Olympic*	Finelawn*
Rebel II	Pennant
Adventure	Galway
Mustang	

* Reported to be shade tolerant. (Refer to AY-14, "Improving Home Lawns in Shade.")

Germination

The germination figure indicates the percent of pure seed that will grow. Table 3 contains guidelines for establishing reasonable germinations for various cool season grasses. The germination percents for Kentucky bluegrass in the sample seed label all fall within the 80-85% range.

The other factors that must be considered in evaluating a seed mixture or blend are the % crop, % weed, and % inert.

Table 3. Germination Ranges for Various Cool Season Grasses.

Turf-type	Minimum%	High Quality%
Kentucky bluegrass	80	85
Fine fescue	85	90
Perennial ryegrass	85	90
Tall fescue	85	90

Crop

The crop figure indicates the percent, by weight, of seeds in a package that are grown as a cash crop. This includes many of the coarse hay or pasture grasses. Examples of crop may include orchardgrass, timothy, redtop, clover, and bentgrass. The crop percentage for high quality bluegrass should be as close as possible to 0.00%, as it is in the sample seed label. As a general guideline, anything over 0.5% should be avoided.

A label indicating 0.1% crop is approximately 1-1/2 oz. in 100 lb.. A seeding of 50 lb. of Kentucky bluegrass per acre with 0.1% crop on the label is like spreading 60,000 timothy seeds or 235,000 bentgrass seeds. The law states that if the percent crop is greater than 5% by weight, the crop seed must be listed by its specific name.

Weed

The weed figure indicates the percent, by weight, of weed seeds in the package. A weed is any seed that has not been included in pure seed or crop. As a general guideline, anything over 0.3% should be avoided.

A seeding of 50 lb. of Kentucky bluegrass per acre with 0.05% weeds on the label is like spreading 18,000 knotweed seeds, 140,000 chickweed seeds, or 37,000 *Poa annua* seeds. The maximum by law in Indiana is 2.5%. Unfortunately, this does not protect the purchaser of high quality turfgrass seed blends or mixtures.

Noxious Weeds

This defines the number per pound or per ounce of weed seeds considered legally undesirable. Most of those listed for Indiana are not problems in turf areas. The list is designed for farm fields.

Inert

Inert indicates the percent, by weight, of material in the container that will not grow. This includes corn cobs or chaff for bulk to make the package larger. Sand may be added so that the package can meet weight requirements.

Seed with more than 8% inert should be avoided. As a general guideline, higher quality bluegrass seed should be less than 4% inert. The percent inert by weight on the package x 2 equals the approximate percent inert by volume in the package. For example, the inert figure in the sample seed label is 3%. This would be approximately 6% volume. For a seed label reading 12%, this would approximate 24% by volume.

Date Tested

Check the date on the seed label. This date should be within the last nine months.

The best bet on getting a good mixture or blend of seed is to buy name brands from a reputable seed house or your local garden shop.

REV 7/88 (5M)

Cooperative Extension work in Agriculture and Home Economics, state of Indiana, Purdue University, and U.S. Department of Agriculture cooperating; H.A. Wadsworth, Director, West Lafayette, IN. Issued in furtherance of the acts of May 8 and June 30, 1914. The Cooperative Extension Service of Purdue University is an affirmative action/equal opportunity institution.

CROP GROWTH AND DEVELOPMENT

I. THE PLANT CELL

Objectives

1. Identify the key components found in living and nonliving portions of a plant cell.
2. Describe the functions of cell parts identified.

The cell is the basic structural unit of plants. Cells reproduce, assimilate, respire, respond to changes in the environment, absorb water and minerals from their surrounding environment and release by-products into their external environment. Activities of the plant result from activities of the cell.

A cell is comprised of both living and nonliving matter. The living substance of a cell is called **protoplasm** where all metabolic and physiological processes occur. The protoplasm is differentiated into **cytoplasm** and **organelles**. Some nonliving structure is the cell wall, a rigid envelope surrounding the cytoplasm (Fig. 1). You should become familiar with the functions of the following structures:

A. Middle lamella

B. Cell wall

C. Nucleus

D. Cytoplasmic membrane

E. Vacuole

F. Chloroplast

G. Mitochondria

H. Cytoplasm

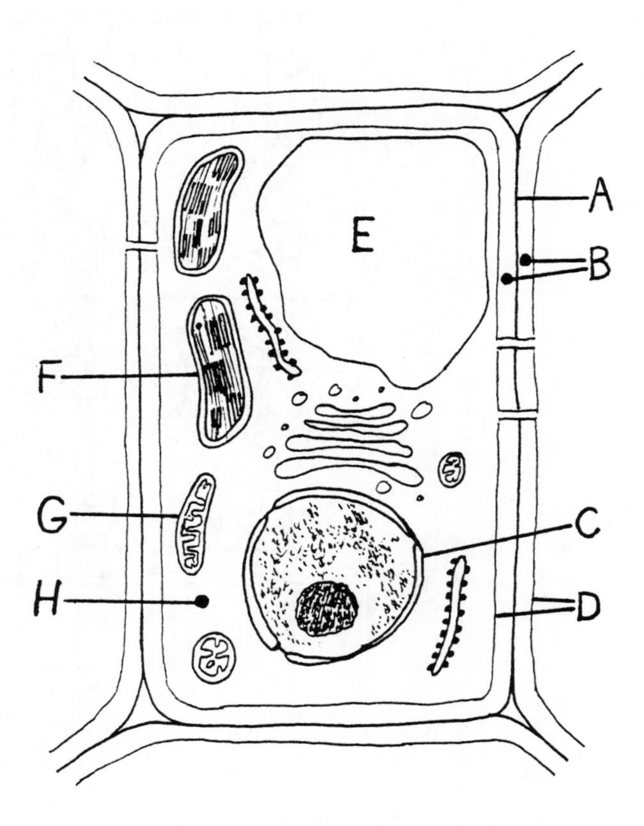

Fig. 1. The plant cell

II. THE LEAF

<u>Objectives</u>

1. List 8 parts of a typical leaf and describe their functions.

Typically, the blade (lamina) of a leaf is a flat, thin, much expanded structure that exposes a large surface to light and air. For example, a single stem of mature alfalfa plant may bear at most 90 leaves having a combined surface area of 16,000 cm^2. The primary function of leaves is photosynthesis, for which they are well adapted. You need to become familiar with the location and function of the internal leaf parts. (Fig. 2).

A. Cuticle
B. Epidermis (upper)
C. Palisade mesophyll
D. Spongy mesophyll
E. Intercellular space
F. Epidermis (lower)
G. Guard cell
H. Stomate
I. Xylem
J. Phloem

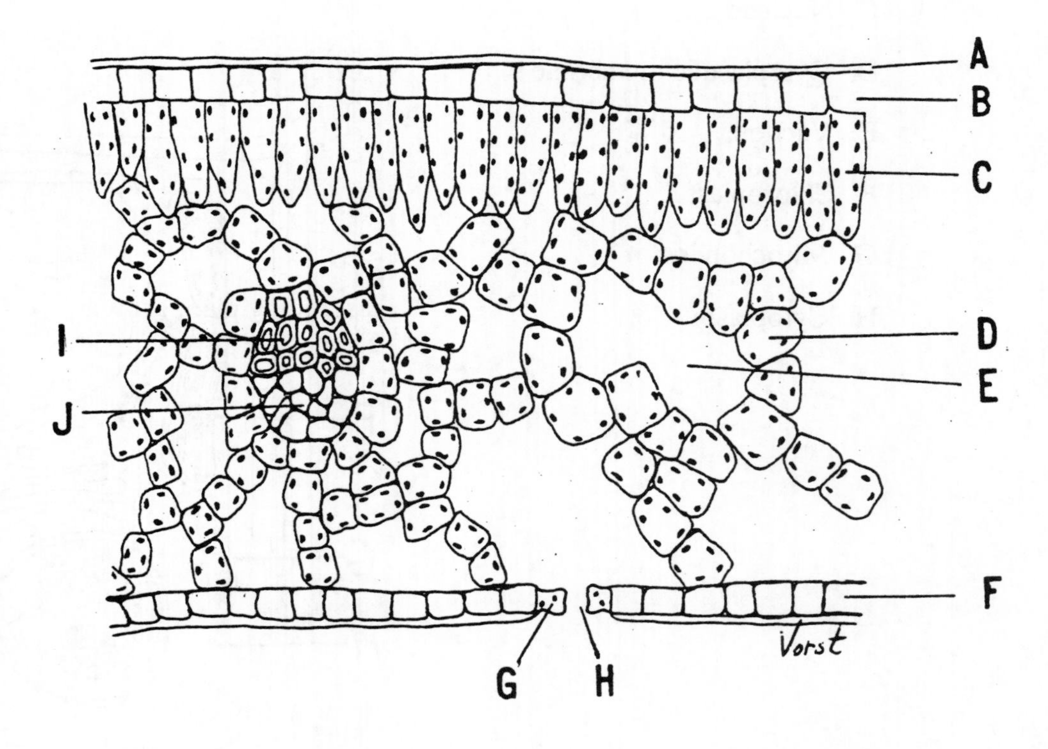

Fig. 2. A leaf cross-section

III. CROP GROWTH

Objectives

1. Define the pattern of crop growth during the growing season.
2. Describe von Leibig's "Law of the Minimum".

Growth involves an irreversible increase in size which is usually accompanied by an increase in dry weight, length, area or volume. This process involves the formation of protoplasm by **assimilation**. Growth rate is slow when a plant is young, then accelerates until a maximum is reached, followed by a decline (Fig. 3).

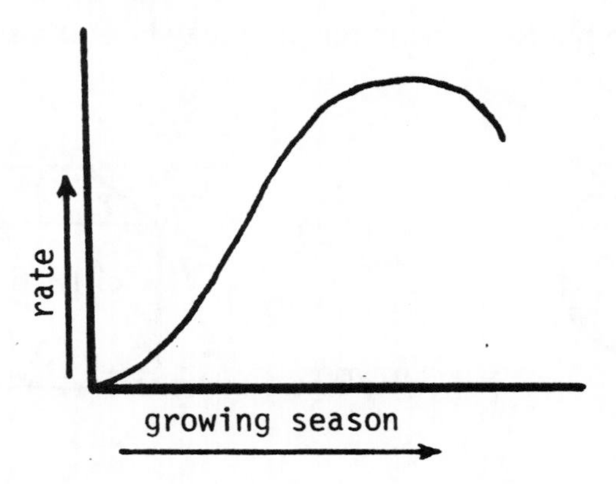

Fig. 3. Pattern of plant growth during the growing season.

Growth is more than just an increasing amount of plant matter. Differential growth of plant organs results in a characteristic plant shape. This shape is genetically controlled, yet the environment can alter the rate and pattern of growth. One theory, the Law of the Minimum, proposed by Justus von Liebig in 1840 states that "A deficiency of one necessary constituent, all others being present, renders the soil barren for crops for which that nutrient is needed". This has been restated to include all production factors and is sometimes referred to as the barrel stave law (Fig. 4). This theory, while not absolutely correct, is still useful in understanding plant growth.

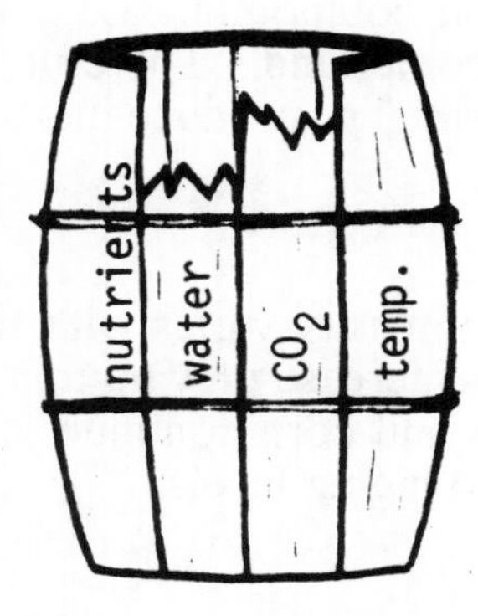

Fig. 4. Each stave represents a production factor. The lowest factor limits production.

IV. PHOTOSYNTHESIS

Objectives

1. State the general chemical equation for photosynthesis.
2. Identify 8 factors affecting the process of photosynthesis.

Approximately 375 billion tons of sugar are synthesized annually by plants. It is estimated that 90% of this is produced by aquatic plants, while the remaining 10% is produced by terrestrial plants. About 2% of the food made by plants is used by animals of the earth. The human race consumes directly or indirectly, 0.2% of the total.

The photosynthetic reaction may be summarized as follows:

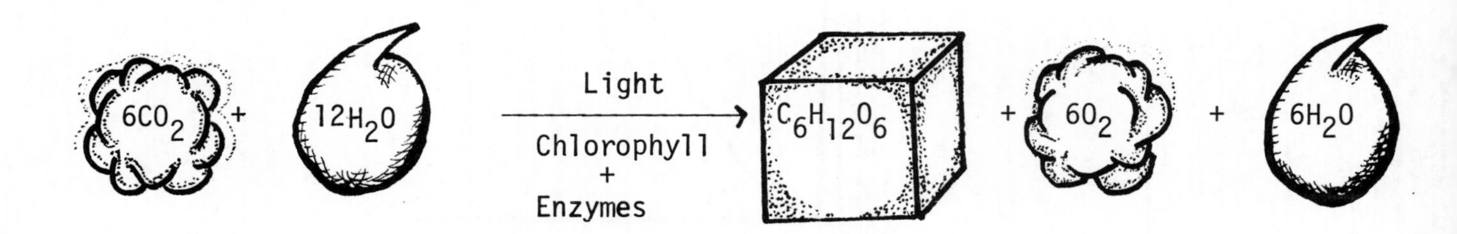

The sugar, glucose, can be changed into starch, cellulose, fats, or amino acids. These organic compounds can be stored as carbohydrates, (dry matter), or oxidized through respiration which releases energy. The fate of photosynthates is a function of plant growth and development, and environmental factors. To obtain optimum crop yields a farmer must maintain conditions favorable for a high rate of photosynthesis. Listed below are eight factors which influence rate of photosynthesis, and dry matter production by agronomic crops.

A. Type of plant

There are two different biochemical pathways by which plants carry on photosynthesis and "fix" carbon into sugar. C4 plants (corn and sorghum) will "fix" an efficient 4 carbon compound. C3 plants (soybeans and alfalfa) will "fix" a less efficient 3 carbon compound. Therefore, in terms of photosynthetic efficiency, C4 plants have the potential to produce more dry matter per acre than do C3 plants.

B. Light intensity

The optimum light intensity varies with the species. For shade plants and C3 plants photosynthetic rates increase only up to a certain intensity and then level off. Full sun plants like sorghum and corn continue to increase their photosynthetic rates as light intensity increases to higher levels.

C. Temperature

Dry matter production of most crops increases from the minimum value of about 10°C to 30-33°C. At high temperatures dry matter production decreases because of increased respiration rates.

D. Carbon dioxide concentration

a. In the atmosphere

Atmospheric CO_2 levels have been rising steadily since the mid 1800s, primarily due to burning of fossil fuels. This has resulted in higher photosynthetic rates than in the past.

b. In the canopy

Concentration of CO_2 in the atmosphere above the crop canopy is usually 300-350 ppm (Fig. 5). Wind movement brings CO_2 down into the crop canopy, and it then moves by diffusion through the stomates into the leaf.

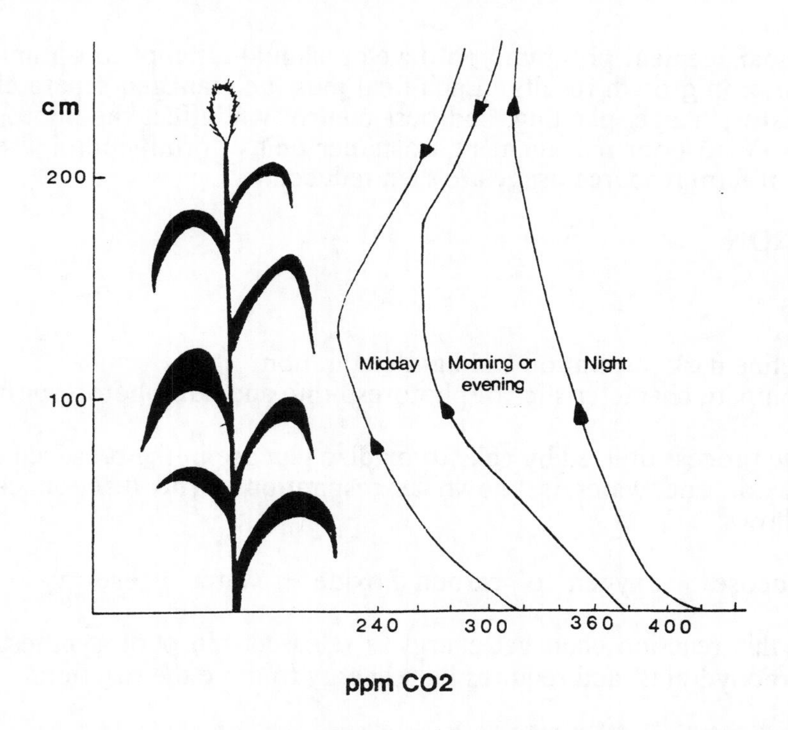

Fig. 5. CO_2 concentration changes with time, day and distance above the soil surface.

E. Leaf water potential

Water is necessary not only as a reactant in the photosynthetic reaction, but also to keep the plant hydrated. The water status of plants can be expressed as leaf water potential. Leaf water potential is the strength the plant exerts to hold water in the leaf. When adequate water is available leaf water potential is zero. As water availability decreases, leaf water potential drops, and is expressed as a negative value. As leaf water potential drops, dry matter production is more seriously limited.

F. Leaf carbohydrates

The carbohydrate level of the leaf affects rate of additional carbohydrate formation. If the leaf contains a high level of carbohydrates the rate of photosynthesis is suppressed until the carbohydrates are either used or translocated out of the leaf. This phenomena is called feed-back inhibition.

G. Leaf area (LAI)

LAI is the ratio of leaf surface to land surface, or the acres of leaves per acre of land. Dry matter production increases as LAI increases to between 3 and 4 for most agronomic crops. At higher LAI values, the production level decreases because the lower leaves are overly shaded resulting in mutual shading.

H. Plant maturity

Seedlings produce less dry matter than actively growing plants at or near vegetative maturity. As plants approach senescence, rate of dry matter production decreases. A typical "S" curve characterizes the factor.

The crop management program you develop should attempt to optimize production factors so efficient crop growth results. Each field must be managed separately. How you manage nutrient status, tillage, planting, and pest control will affect the photosynthetic efficiency of your crop. With poor management, costs per unit of production increase. Both profit and efficiency of farm resource usage are then reduced.

V. RESPIRATION

<u>Objectives</u>

1. Define dark respiration and light respiration.
2. Compare characteristics of photorespiring and non-photorespiring plants.

The process utilized by cells to oxidize glucose in the presence of oxygen into carbon dioxide and water is known as respiration. The reaction can be summarized as follows:

Glucose + oxygen ó carbon dioxide + water + energy

In this reaction chemical energy is released. In photosynthesis, energy is stored in carbohydrates, and requires light energy to drive the reaction.

There are two types of respiration that may occur in plants:

1. Normal (dark) respiration, which results in the release of energy used in plant processes.

2. Photorespiration, which is a wasteful release of energy occurring only during the day, and which serves no useful purpose for the plant. Photorespiration can occur in some plants at rates that reduce yield potential as much as 50%.

Characteristics of non-photorespiring and photorespiring plants.

	Non-photorespiring plants	Photorespiring plants
1.		
2.		
3.		

Net Assimilation Rate

In plants, photosynthesis stores chemical energy in the form of carbohydrates, and respiration releases this energy by converting the energy stored in the carbohydrates to carbon dioxide and water. The **net assimilation rate** (NAR) is used to measure the rate of photosynthesis minus respiration losses. NAR is also referred to as the **apparent photosynthesis rate**.

Increasing NAR provides more carbohydrate that can be used for storage and yield. NAR could be increased by increasing photosynthetic rates, or decreasing the rates of respiration. An obvious way to decrease respiration losses in C_3 plants would be to reduce or eliminate photorespiration. In C_4 plants, such as corn, some plants have lower rates of normal respiration than others. Environmental factors such as temperature, light, water, carbon dioxide, and mineral nutrients all affect NAR.

VI. TRANSPIRATION

Objectives

1. Describe the process of transpiration.
2. List 5 factors that influence the rate of transpiration.

Agronomic crops usually contain at least 85-90% water by weight. The internal relative humidity of a plant organ is 100%. Yet the wet walls of mesophyll cells in the leaf are constantly exposed to air. Water evaporates from the walls and accumulates as vapor in the intercellular spaces. Slowly this vapor diffuses out of the leaf. This water vapor loss is called **transpiration**. A single corn plant may transpire 400 pounds of water during the growing season. An acre of corn plants at a population density of 20,000 transpires in excess of 2,000 tons of water during a season. Approximately 99% of the water taken up by a crop plant is lost through transpiration.

Evaporation of water may lower the temperature of leaves 2°- 6°C. This is of some value on very hot days. However, the leaf loses heat by conduction, convection and radiation. For example a leaf receiving 2.1 calories of radiant energy/cm^2/min. removes about 35% of the heat by transpiration, 60% by radiation and 5% by convection. Therefore, understanding the factors affecting transpiration is important to maintain a high rate of photosynthesis and a moist environment for cells.

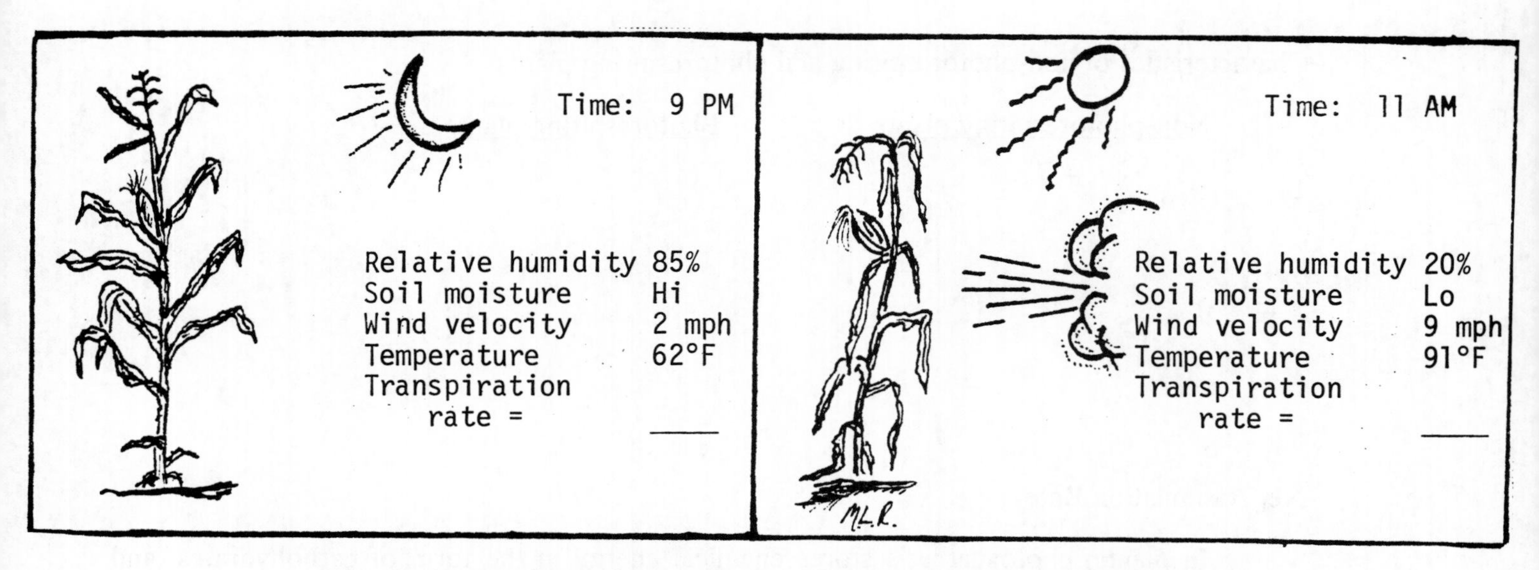

Fig. 6. Comparison of some factors affecting the rate of transpiration.

Using the information in Fig. 6, summarize the effects of the following factors on crop transpiration rates. Fill in the blanks with no change, increase, or decrease. For example, if transpiration rates increase as the factor increases write "increase".

Increasing level of factor	Effect on transpiration rate
Light	____________
Temperature	____________
Relative humidity	____________
Wind velocity	____________
Soil moisture	____________

VII. EVAPOTRANSPIRATION

Objective

1. Define evapotranspiration.

   Evapotranspiration is an indication of total amount of water lost from a field by both evaporation from the soil and plant transpiration. Early in the growing season more water is lost from a field by evaporation, whereas transpiration becomes more important as the crop becomes larger later in the season.

TILLAGE AND PLANTING MANAGEMENT

I. OPTIMIZING YIELDS WITH CORRECT TILLAGE MANAGEMENT

Objectives

1. Describe the essential characteristics of a desirable seedbed.
2. Differentiate between primary and secondary tillage operations.
3. Compare moldboard plowing and chisel plowing.
4. Outline the advantages and limitations of conservation tillage systems.
5. Illustrate how the choice of a tillage system effects the management of pesticides and fertilizers.

The primary purpose of tillage is to create a favorable environment for crop germination, establishment, and growth. For row crops, the field is a seedbed for only a short period, and is a rootbed for most of the growing season. Therefore the most desirable tillage system will provide for a good seedbed for the crop at planting, as well as prepare a rootbed that enhances crop growth during the rest of the growing season. It is important to note that conditions for an ideal seedbed are different than those for an ideal rootbed. Rootbeds need to be fertile, loose and friable, and have a rough surface to aid in water infiltration and weed control.

Good seedbed preparation will provide the following:

- Proper moisture and temperature for rapid crop seed germination.
- Favorable environment for seedling root growth.
- Small soil aggregates for good soil-seed contact
- Good surface tilth that prevents crusting and aids water infiltration and aeration.

A. Tillage operations

Tillage operations can be categorized as primary or secondary.

1. Primary

Primary tillage refers to the first operation which is done to the soil. Primary tillage can be done in either the spring or the fall, depending on soil and climatic conditions.

Primary tillage is usually done either with a moldboard plow or a chisel plow. Each of these has certain advantages and disadvantages, and the farmer should carefully consider the following characteristics of each when deciding which to use.

a. Moldboard plow

1. greatest incorporation of residue which may aid in control of certain insects, weeds, and diseases. However, complete residue incorporation also increases erosion and runoff potential.

2. time consuming operation.

3. not adapted to erosive soils, especially if all of the residue is incorporated.

4. on most soils, better soil structure results after fall plowing than after spring plowing, due to the freezing-thawing and wetting-drying that occurs over the winter in fall plowed fields.

b. Chisel plow

1. leaves more residue on soil surface. If a twisted point is used approximately 50% of the residue is incorporated, whereas approximately 25% residue incorporation occurs when straight points are used.

2. faster than moldboard plowing. Depending on soil type, a 100 hp tractor can cover approximately twice as much land with a chisel plow than with a moldboard plow.

3. can be done on more sloping soils.

4. may aid in breaking up subsurface compaction.

2. <u>Secondary</u>

Secondary tillage aids in seedbed preparation, after primary tillage is completed. Frequently secondary tillage is done to aid in efficient operation of planting equipment, or to aid in pest control.

Five common secondary tillage methods are:

a. Disc - creates a fine seedbed to depth of disc operation, but also may cause subsoil compaction if the soil is wet.

b. Field cultivator - because of the stirring action of the tires, compaction is less likely to occur. The field cultivator leaves a rougher soil surface than does the disc.

c. Harrow - used to level the soil surface and create a fine, smooth seedbed.

d. Soil conditioners (do-all) - usually a combination of rolling coulters or discs, field cultivator tines, and a soil leveling device, consisting of rolling baskets or a harrow of some type.

e. Pulvi-mulcher - consists of two sets of fluted rollers, with a series of tines inbetween. The tines may be raised or lowered. This creates a fine, level seedbed particularly well suited to forage seeding.

Fig. 1 Nitrogen applicator and nurse tank

Fig. 2 Tandem disc used for secondary tillage

Fig. 3 A Conventional planting system

Fig. 4 Planting in a No-till system

B. Tillage systems

Tillage systems can be grouped according to the number of operations and type of equipment used. Historically, the corn belt farmer first plowed, and then made 2-5 secondary tillage trips over the field before planting. Recently, however, many farmers have reduced the number of field operations in an effort to save both time and money. Refer to NCH-22, **Definitions of tillage systems for corn** for a more detailed discussion of tillage systems used for row crop production.

Types of tillage systems:

1. Conventional tillage

   Conventional tillage systems involve several operations to prepare the seedbed for planting. A typical conventional tillage system for corn production might include plowing, one or more diskings, or field cultivating and harrowing before planting. Reduced tillage systems combine operations or tillage tools, thus requiring less energy per unit area are used.

2. Conservation tillage

   Conservation tillage systems are based on tillage operations that incorporate little or no crop residue, thus minimizing wind and water erosion. Many systems exist, and many combinations of equipment are used. While soil conservation is emphasized, additional benefits include savings in moisture, labor or equipment. Read CT-1, "What is Conservation Tillage?" to obtain more detailed information on conservation tillage.

3. No-till

   The no-till system involves using a no-till planter to plant directly into untilled soil. Tillage is accomplished with a fluted coulter that rolls directly in front of the openers on the planter which place the seed in the soil. CT-1 provides additional information.

4. Ridge-tillage

   Ridge-tillage is a conservation tillage system in which the soil is left undisturbed prior to planting. The crop is planted on 10-15 cm high ridges that were created at time of lost cultivation the previous year, and maintained each year thereafter.

Major advantages of conservation tillage or no-till systems are reduced soil erosion and reduced cost of production because fewer trips over the field are required. (Table 1)

Table 1. Effect of Surface-Applied Crop Residue on Runoff and Soil Loss.

Residue in tons/acre	Runoff as pct. of rainfall	Soil loss in tons/acre
0	45	12
0.25	40	3
0.50	25	1
1	0.5	0.3
4	0	0

C. Pesticide and fertilizer management under different tillage systems.

1. Pesticide management

   Conventional tillage which incorporates residues aids in control of many soil borne pests. This leads to reduced needs for chemical pest control.

   Under reduced and no-till systems, rates of chemical pesticide application may be higher, because of a better environment for pest buildup, and the additional organic material on the soil surface "ties up" more of the pesticide.

2. Fertilizer management

   Time of application may be spring or fall for any of the systems, however, incorporation is difficult or impossible with some reduced or no-till systems. Because of residue on soil surface, sidedressing corn after emergence may be more difficult with no-till systems.

Definitions of Tillage Systems for Corn

J. C. Siemens, University of Illinois; E. C. Dickey, University of Nebraska; and E. D. Threadgill, University of Georgia

Reviewers
J. Ackley, Deere & Company, IL
M. R. Gebhardt, University of Missouri
D. R. Griffith, Purdue University

If "tillage" is defined as the mechanical manipulation of soil, it follows, then, that a "tillage system" would be the sequence of soil-manipulation operations performed in producing a crop. Today, however, such a definition is recognized as inadequate. We know, for instance, that the management of nonharvested plant tissue (i.e., residue) affects both crop production and soil erosion, and that field operations in which the soil is not tilled have a marked influence on soil condition.

Therefore, in this publication, a tillage system is the sequence of *all* operations involved in producing the crop, including soil manipulation, harvesting, chopping or shredding of residue, application of pesticides and fertilizers, etc. But before describing and comparing the various tillage systems for corn, some terminologies and possible points of confusion need to be addressed. These have to do with primary vs. secondary tillage and the different ways in which similar tillage systems could be defined.

PRIMARY AND SECONDARY TILLAGE

For many tillage systems, the specific operations can be separated into "primary" and "secondary." Primary tillage loosens and fractures the soil to reduce soil strength and to bring or mix residues and fertilizers into the tilled layer. The implements ("tools") used for primary tillage include moldboard, chisel and disk plows; heavy tandem, offset and one-way disks; subsoilers; and heavy-duty, powered rotary tillers. These tools usually operate deeper and produce a rougher soil surface than do secondary tillage tools; however, they differ from each other as to amount of soil manipulation and amount of residue left on or near the surface.

Secondary tillage is used to kill weeds, cut and cover crop residue, incorporate herbicides and prepare a seedbed. The tools include light- and medium-weight disks, field cultivators, rotary hoes, drags, powered and unpowered harrows and rotary tillers, rollers, ridge- or bed-forming implements, and numerous variations or combinations of these. They operate at a shallower depth than primary tillage tools and provide additional soil pulverization.

Equipment that permits primary and/or secondary tillage *plus* planting in a single operation is also available.

TILLAGE SYSTEMS FOR CORN

Throughout the United States, many different tillage systems are used in producing corn. Most provide for planting the crop into a flat seedbed, but some prepare a raised seedbed or ridges. Frequently, the tillage system dictates the design of the corn planter and its attachments. These attachments can range from a complete secondary tillage implement operating ahead of the entire planter to coulters or flexible harrow teeth that just till in front of each row of a planter.

Because of the number and diversity of corn tillage systems, it is difficult to give each one a meaningful name or even a precise definition. For instance, a system may be identified according to its ultimate objective (e.g., conventional, clean, reduced, minimum or conservation tillage); *or* according to the primary tillage tool used (e.g., moldboard plow, chisel plow, disk or rotary-till system); *or* according to its basic function (e.g., subsoil-plant, slot-plant, no-till, zero-till, strip-till or ridge-plant system).

COOPERATIVE EXTENSION SERVICE • PURDUE UNIVERSITY • WEST LAFAYETTE, INDIANA

The name problem is further compounded by the fact that definitions differ between regions of the country and even within regions. Different names may be used to identify the same tillage system, or the same name may refer to different tillage systems, all depending on where you live.

Thus, to accurately define or describe a given tillage system, all the operations that make up the system must be listed, including tillage, chopping or shredding residue, and application of pesticides and fertilizers. The American Society of Agricultural Engineers has compiled definitions and illustrations for most agricultural implements in *ASAE Standard S414*. This standard, revised periodically to include major new tillage implements, is available from ASAE, 2950 Niles Road, St. Joseph, MI 49085.

Recognizing that a confusion in terminology does exist, here are general definitions for the major tillage systems grouped according to ultimate objective, primary tillage implement and basic function. (Some of these systems are discussed in detail in other NCH fact sheets.)

Tillage Systems as Defined by Objective

Conventional tillage. This refers to that set of tillage operations most commonly used in a given geographic area to produce a given crop. Since operations vary considerably under different climatic, agronomic and other conditions, the definition of "conventional tillage" likewise varies from one region to another—and even within a region.

For instance, on the prairie soils of central Illinois, conventional tillage for corn means: (1) in the fall, phosphorus, potassium and lime are applied and the ground moldboard plowed; (2) in the spring, the ground is disked, nitrogen and herbicides applied before field cultivating twice and then the crop planted; and (3) the crop is rotary hoed if necessary and cultivated once. On the sandy Coastal Plain soils of south Georgia, on the other hand, conventional tillage usually involves the following sequence of operations, all done in the spring: disking, moldboard plowing, application of fertilizer and herbicides, and field cultivating or disking before planting, with the crop subsequently cultivated two or more times.

Conventional tillage is often used as the "standard" or "check" in experiments to assess the potential of other tillage systems in a given area. Factors usually compared in such studies include: soil conditions, plant emergence, growth and yield costs, erosion control, energy use and labor requirements.

Clean tillage. Another defined-by-objective system is clean (or residue-free) tillage. It involves a set of operations that prepares a seedbed having essentially no plant residue on the surface. Many conventional tillage systems are also clean-till, particularly those based on use of the moldboard plow. However, a clean soil surface can be achieved with other tools, depending on the previous crop, amount of surface residue, and number and timing of tillage operations.

Reduced tillage. This refers to any system that is less intensive than conventional tillage. Either the number of operations is decreased or a tillage tool that requires less energy per unit area is used in place of a tool used in the conventional tillage system.

Minimum tillage. The definition of minimum tillage has varied over the years and from region to region; thus, the term is today little used and not very meaningful. Perhaps the best definition has been: "The minimum soil manipulation necessary for crop production or for meeting tillage requirements under existing conditions." The way most people use the term minimum tillage, they mean reduced tillage as defined above.

Conservation tillage. Although specific operations or sequence of operations might differ, most would agree that the objective of conservation tillage is to provide a means of profitable crop production while minimizing soil erosion due to wind and water. The emphasis is on soil conservation; but moisture, energy, labor and even equipment conservation are sometimes additional benefits.

To be considered conservation tillage, a system must produce, on or in the soil, conditions that resist the erosive effects of wind, rain and flowing water. Such resistance is achieved either by protecting the soil surface with crop residue or growing plants, or by increasing the surface roughness or soil permeability.

How effectively these conditions control erosion depends on certain related factors, including soil type, degree and length of slope, rainfall pattern and intensity, and cropping sequence. The Universal Soil Loss Equation (USLE), which takes all these factors into consideration, may be used to estimate the extent to which a particular conservation tillage system will control erosion. Local Soil Conservation Service personnel can provide assistance in using the USLE.

Tillage Systems as Defined by Major Implement

Moldboard plow system. The moldboard plow is used in the fall or spring for primary tillage, with various secondary tillage operations performed ahead of planting. Often before plowing, the previous crop residue is chopped or disked, and fertilizers and lime (if necessary) are applied.

Secondary tillage varies widely in type and number. *If herbicides are incorporated,* the plowed soil is often leveled, then herbicides applied and incorporated with one or two passes of a tillage tool that also prepares the seedbed. Fertilizers may be applied after the leveling operation or after planting (especially nitrogen in the form of anhydrous ammonia for corn). A portion of the fertilizers may be applied during the planting operation. A rotary hoe or row-crop cultivator is also frequently used for additional weed control and soil loosening after crop emergence.

If herbicides are not incorporated, one or more secondary tillage operations are used to level the soil and prepare a seedbed, with the herbicides usually applied after planting. Fertilizing may be done before, during or after planting.

Chisel plow system. Except, of course, for the primary tillage tool, the chisel plow system is essentially the same as the moldboard plow system. The chisel plow generally leaves a rougher soil surface and considerably higher percentage of residue on or near the surface than does the moldboard plow.

Chisel plows come with a variety of chisel point shapes and sizes—from straight to twisted and from 2.5-inch to 4-inch widths. Chisel plows are also available with coulters or a disk gang mounted in front to cut through residue, thereby eliminating the need to disk or chop the residue beforehand.

Disk system. Tillage is usually not as deep with the disk system as with the moldboard plow or the chisel plow system. The number and type of operations in the disk system may vary considerably. An offset disk or tandem disk is usually used for primary tillage in the fall or spring, whereas secondary tillage may include tandem disking, field cultivating, harvesting and/or other operations. Fertilizers and herbicides are applied the same as with the moldboard plow system.

Rotary-till system. The rotary tiller may be used for primary tillage, secondary tillage or both. Sometimes the tool is mounted ahead of the planter to prepare the seedbed and incorporate herbicides in a one-pass tillage-planting operation. There are also rotary tillers available that only till narrow strips into which the corn is planted.

Tillage Systems as Defined by Function

Subsoil-plant system. This is a type of reduced tillage system that has become popular in areas having severe compaction problems. The system uses a combination tillage tool. In a single operation, the area beneath each row is subsoiled and tilled by coulters or other devices, and the crop then planted in the tilled strip.

Till-plant (or ridge) system. With this system, the corn is grown on pre-formed ridges or beds. The ridged seedbeds are made initially either (1) when cultivating the previous crop, using a cultivator equipped with sweeps and/or disk hillers, or (2) after harvesting the previous crop, using a moldboard lister bottom or short disk gangs.

In the spring, the residue is chopped, if necessary. Till or ridge planters usually have a sweep or double disk mounted in front of each planting unit. These sweeps or disks, set to operate 1-2 inches deep, transfer the disturbed soil and residue to the area between the rows. This leaves a clean strip into which the crop is planted.

With the till-plant system, preemergence herbicides are often banded or broadcast after planting. Several alternatives are available for applying fertilizers.

No-till (zero-till or slot-plant) system. This is a system whereby seed is planted in previously undisturbed soil. A no-till planter is designed and equipped to plant through residue and into firm soil. The only "tillage" done is the making of a narrow strip or slot into which the seed is dropped.

Generally, fertilizer has been broadcast on the soil surface before planting; however, equipment is available to apply fertilizer below the surface in otherwise no-tillage situations. Preemergence herbicides are usually applied after planting, with contact herbicides applied, if necessary, before the crop emerges to kill growing weeds.

A publication of the National Corn Handbook Project

NEW 1/85 (5M)

Cooperative Extension Work in Agriculture and Home Economics, State of Indiana, Purdue University and U.S. Department of Agriculture cooperating. H. A. Wadsworth, Director, West Lafayette, IN. Issued in furtherance of the Acts of May 8 and June 30, 1914. It is the policy of the Cooperative Extension Service of Purdue University that all persons shall have equal opportunity and access to its programs and facilities without regard to race, color, sex, religion, national origin, age, or handicap.

Purdue University Cooperative Extension Service

What Is Conservation Tillage?

C. Janssen and P. Hill
Department of Agronomy, Purdue University

As its name implies, conservation tillage conserves soil by reducing erosion. In the Midwest, erosion by water is the primary concern, whereas western regions of the country are more susceptible to wind erosion.

Soil erosion removes the productive layer of topsoil, reducing crop yields and land value. Soil removed from fields eventually ends up as sediment in streams, rivers, or lakes. Sediment collects in surface waters, reducing their water-holding capacity. Some crop nutrients and pesticides attach to soil particles and are carried and deposited in waterways along with the soil.

Factors affecting soil loss by water erosion include:

- rainfall patterns
- erodibility of soil
- slope length and height
- soil cover
- cropping patterns and management

Conventional tillage, such as moldboard plowing, leaves the soil surface bare and loosens soil particles, making them susceptible to the erosive forces of wind and water. Conservation tillage practices reduce erosion by protecting the soil surface and allowing water to infiltrate instead of running off. Table 1 shows the relationship between residue cover and soil loss. The Conservation Technology Information Center (CTIC) defines conservation tillage as any tillage and planting system that leaves at least 30 percent of the soil surface covered by residue after planting.

Table 1. Effects of surface residue cover on runoff and soil loss.

Residue Cover	Runoff	Soil Loss
%	% of rain	tons/acre
0	45	12.4
41	40	3.2
71	26	1.4
93	0.5	0.3

Conservation tillage practices are grouped into three types: no-till, ridge-till, and mulch-till. Keep in mind that no one conservation tillage method is best for all fields. Decisions should be based on the severity of the erosion problem, soil type, crop rotation, latitude, available equipment, and management skills. Before adopting a conservation tillage system, first seek advice from:

- Nearby farmers who are successfully practicing conservation tillage.
- Seed, chemical, and other agri-business dealers with experience in serving the needs of conservation tillage.
- Representatives from your local Soil and Water Conservation District, Natural Resources Conservation Service (formerly the Soil Conservation Service), or Cooperative Extension Service.

No-till

No-till leaves the soil undisturbed from harvest to planting. Planting is done in a narrow (usually 6 inches or less) seedbed or slot created by coulters, row cleaners, disk openers, in-row chisels, or roto-tillers. A press-wheel follows to provide firm soil-seed contact.

No-till planting can be done successfully in chemically-killed sod, in crop residues from the previous year, or when double-cropping after a small grain. Herbicides are the primary method of weed control, although cultivation may be used for emergency weed control.

Soil conservation results from the high percentage of surface covered by crop residues.

Ridge-till

Ridge-till involves planting into a seedbed prepared on ridges with sweeps, disk openers, coulters, or row cleaners. The ridges are rebuilt during cultivation. Except for nutrient injection, the soil is left undisturbed from harvest to planting.

Ridge-till works best on nearly level, poorly drained soils. The ridges speed up drainage and soil warm-up. Cultivation controls weeds along with some herbicides.

Ridge-till systems leave residues on the surface between ridges. Soil conservation depends on the amount of residue and the row direction. Planting on the contour and increased surface coverage greatly reduce soil loss.

Mulch-till

Mulch-till uses chisel plows, field cultivators, disks, sweeps, or blades to till the soil before planting. The tillage does not invert the soil but leaves it rough and cloddy. Various chisel points or sweeps attached to the shanks affect the amount of residue cover left on the soil surface.

Fall chiseling should be done to a depth of 8-10 inches, and spring chiseling should be no deeper than 6 inches. Disking or other shallow tillage operation can be used in seed bed preparation. A standard, or tandem, disk does not till as deep and leaves more residue on the surface compared to heavy (offset) disks. Herbicides and/or cultivation control weeds in a mulch- till system.

The effectiveness of mulch-till systems in reducing erosion depends on surface roughness, amount of residue, and tillage direction.

References:

1994 National Crop Residue Management Survey,
Conservation Technology Information Center.
West Lafayette, IN.

11/94 (2M)

Cooperative Extension work in Agriculture and Home Economics, state of Indiana, Purdue University, and U.S. Department of Agriculture cooperating; H. A. Wadsworth, Director, West Lafayette, IN. Issued in furtherance of the acts of May 8 and June 30, 1914. The Purdue University Cooperative Extension Service is an equal opportunity/equal access institution.

II. THE PLANTING OPERATION

Objectives

1. List the proper depth of planting for corn and soybeans.
2. Explain how corn and soybean yields are affected by planting pattern and plant populations.
3. List the general guideline for dates of planting corn and soybeans in Indiana.
4. Describe management factors to consider when double cropping in Indiana.

Crop management of planting time can make the difference between profitable and nonprofitable grain farming.

A. Planting depth

1. Corn

Getting kernels as little as 1/4 to 1/2 inch too deep can make a big difference in emergence in cool, early season soils. Planting depth should be between 1.5 and 2 inches under most soil conditions. With very early planting, 1.25" depth may be justified. With shallow planting, root development can be seriously affected and result in "floppy" or rootless corn syndrome. These conditions can be particularly damaging if sidewall compaction or insect feeding also restricts root development. Shallow planting can also increase risks associated with certain pre-emergence-applied herbicides.

2. Soybeans

For normal conditions, 1.5 - 2.0 inches is recommended for conventional tilled soybeans, and 1.0 inch is recommended for no-till soybeans. When no-tilling soybeans, planting too shallow may result in seed lying on the residue and not germinating. Under dry conditions planting deeper is recommended, but planting deeper than 2 inches frequently delays emergence and reduces final population.

B. Plant population and row spacing.

1. Plant populations tend to be too low for optimum corn yields, and too high for optimum soybean yields. Consider increasing plant population for early plantings. Normal stand loss is approximately 10% but may be 15% under cool, wet planting conditions.

2. As plant population is increased more fertilizer is needed. (Fig. 1). Higher populations are recommended where soil fertility and water holding capacity are high.

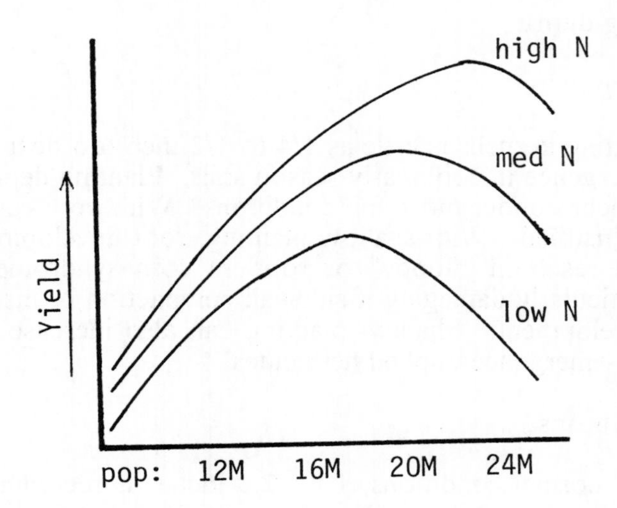

Fig. 1. As fertility level increases, so does optimum plant population. (1M = 1,000 plants)

3. Variable rate planting

Varying plant populations within fields is again receiving serious attention because of the availability of GPS technology. It seems logical to reduce plant populations on soils with lower yield potential. However, economic analysis of variable rate planting by Purdue University indicate it is not always profitable.

Variable rate planting has less profit potential for soybeans than for corn, because of the greater ability of soybeans to adapt to high or low populations. For corn, research has shown that variable seeding rate tends to be most profitable for land with less than 400 bu/A yield potential. On land with 120-140 bu/A expected yields the profit potential is reduced, and on land with over 180 bu/A expected yields, farmers are probably better off with uniform rate seeding. While seeding lower than optimum can reduce yields higher than optimum seeding rates carry little yield penalty.

Variable rate seeding may be most valuable in areas of the Cornbelt where some of the low yield potential soils are used to produce corn.

C. Date of planting

1. Determined mostly by soil temperature and calendar date. When soil temperature is at 50°F at a 5 cm depth at 8:00 AM, it is time to plant. However, if the season is late because of unfavorable spring weather, corn should be planted as soon as possible.

2. Generally, earlier planting date results in greater yield potential (Fig. 3).

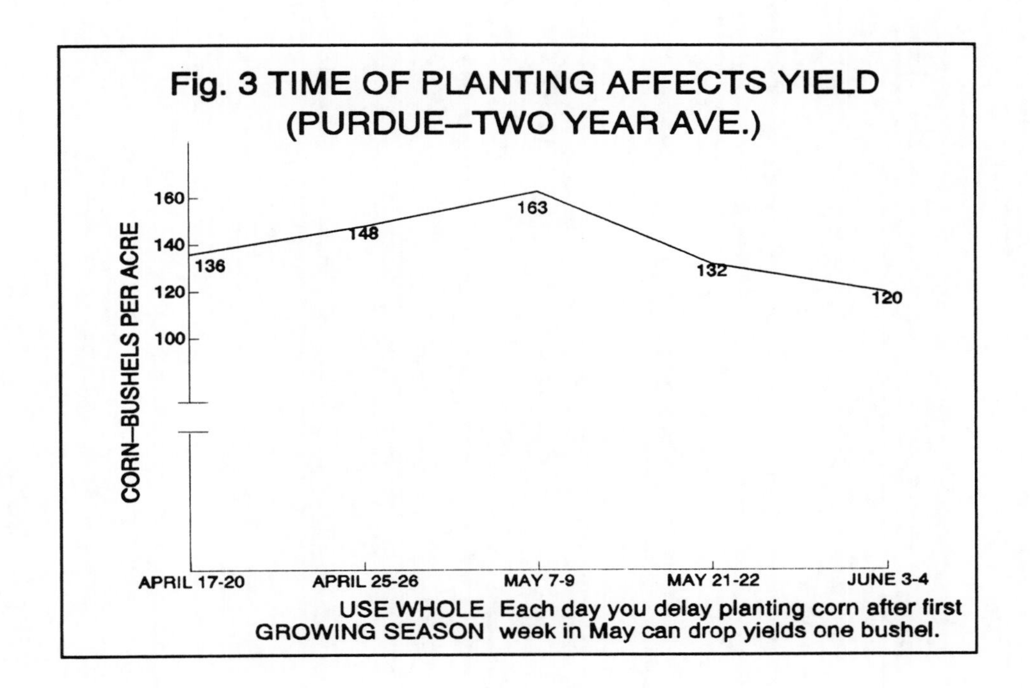

Agronomy Guide

Purdue University Cooperative Extension Service

CROPS (SOYBEANS) AY-217

Plant Populations and Seeding Rates for Soybeans

E. P. Christmas
Agronomy Extension Specialists, Purdue University

As a result of widespread adoption of highly productive management practices such as solid seeding or narrow rows, soybean growers have become more aware of the importance of optimum plant populations and seeding rates in soybean production systems. Besides following new management practices, growers also have been able to control plant populations with considerable precision because of the availability of high quality seed and improved planting equipment. New management practices and seed quality improvement indicate growers should reevaluate their current seeding practices. This publication is a guide designed to help growers determine optimum soybean population levels and seeding rates needed to reach those levels.

IMPORTANCE OF PROPER PLANT POPULATIONS

High soybean yields are possible with a wide range of plant populations because single plants of most varieties will utilize a 7- to 9-inch area in all directions around the main stem. Plants adjust to low populations by producing more branches per plant and by increasing the number of pods on both the main stem and branches. There is, however, little change in seed size and in seed number per pod. While the production of more branches and pods per plant maintains the yield potential for soybeans, harvest losses may be greater in thin stands since the pods on the lateral branches will be close to the soil surface and branch lodging is apt to occur. Leaves on plants in a thin stand also take longer to produce a ground-covering canopy. This allows more weed competition and soil moisture evaporation. In contrast, a stand that is too thick may result in excessive early lodging which means reduced yields as well as increased harvest loss.

When grown under high populations, individual plants produce fewer pods, fewer branches, grow taller, and pod higher off the soil surface than when grown at low populations. Yield potential is maintained with high populations since there are more plants per acre. Soybean populations that are too high also undergo a natural thinning process due to the intense competition between plants, which reduces the stand to a more acceptable level. In other words, plants are eliminated after emergence. In summer, soybean populations can vary perhaps as much as 50 percent from recommended levels without affecting yields, as long as missing plant spaces are not too large and weeds are controlled.

There are varietal differences in soybean response to over- or under-population. Taller varieties that are lodging prone are likely to have reduced yields if populations are too high. Shorter varieties are more likely to have reduced yields if populations are too low. In general, fewer problems occur when stands are established at or near recommended levels.

SELECTING OPTIMUM PLANT POPULATIONS

Plant population refers to the number of soybean plants emerged and established in the field which can contribute to overall crop performance (yield, competition with weeds, moisture use, etc.). It is usually expressed in terms of plants per acre or plants per linear foot of row.

Soybean plant population recommendations for Indiana are shown in Table 1 (columns 3 and 4) for various row widths. At the wider row spacings (20- to 36-inch), stands that vary ± one plant per foot will differ little in yield. However, as row width is narrowed (20-inch and below), establishing the stand to within ± 1/2 plant per foot becomes more important.

The population guidelines in Table 1 are based on responses of different public soybean varieties. Most varieties grown in Indiana will produce maximum yields at these populations but there are exceptions.

For determinate varieties, you may need to increase plant populations to realize full yield potential. An extreme example is the determinate variety, Hobbit 87, which is recommended for solid seeding (6-, 7-, or 8-inch rows) at populations of 200,000 to

West Lafayette, Indiana

Table 1. Suggested Per-acre Plant Populations and Seeding Rates for Soybeans Planted at Various Row Widths.

Col. 1 Row width	Col.2 Feet of row/a.	Col. 3 Suggested number plants /ft. of row	Col.4 Recommended total plant population (Col.2 x Col.3)	Col.5 Required seeding rate at 90% germination and 90% emergence	Col.6 Lb. seed needed if 2500 seed/lb. (81% final emergence*)
in.	ft.	plants/ft.	plants/a.	seeds/ft.	lb./a.
36	14,520	7.0	101,640	8.6	50
30	17,424	6.0	104,540	7.4	52
20	26,136	5.0	130,680	6.2	65
18	29,040	4.5	130,680	5.6	65
16	32,670	4.0	130,680	5.0	65
15	34,848	3.7	130,680	4.6	65
14	37,336	3.5	130,680	4.3	65
12	43,560	3.0	130,680	3.7	65
10	52,272	2.8	143,750	3.4	71
8	65,340	2.5	163,350	3.1	81
7	74,674	2.2	168,020	2.8	83
6	87,120	2.0	174,240	2.5	86

* Final emergence = percent warm germination x percent expected emergence.

250,000 plants per acre, or about three plants per foot of row. You can see from the table that this is 50 percent greater than the standard recommendations. When planting semi-dwarf types, follow your seedsman's population recommendations. These suggested populations have a large built-in safety factor for unusual conditions. There is no need to adjust the seeding rate for soil type or planting date. Recent studies show that soybeans can produce a normal seed yield with populations down to about 60,000 plants per acre when planted by June 7 (80,000 required for later planting) so long as the plant stands are reasonably uniform and weeds are not a serious problem.

DETERMINING PROPER SEEDING RATES

"Seeding rate" refers to the number of seeds planted and is expressed in terms of "seeds per foot of row" or "pounds of seed per acre" (Table 1, columns 5 and 6). The relationship between the number of seeds planted and the eventual plant population depends primarily on seed quality and expected emergence of live seed.

The basic formula for calculating the seeding rate necessary to establish a particular desired population is as follows:

$$\text{Seeding rate} = \frac{\text{Suggested plants/ft. of row (Col. 3 from Table 1)}}{\text{Pct. germination} \times \text{Pct. expected emergence}}$$

Example: A farmer is planning to plant Resnik soybeans in 30-inch rows. He has purchased seed labeled "90 percent germination." Previous experience indicates that, with his soil type and equipment, he can expect to lose about 10 percent of the seedlings from crusting problems, leaving a 90 percent live seed emergence. What should be his seeding rate to establish the desired stand?

$$\frac{\text{6 plants/ft. for 30-in. rows}}{\text{.9 (germ.)} \times \text{.9 (emerg.)}} = \frac{6}{.81} = \text{7.4 seeds/ft.}$$

Thus, our farmer would set his planter to drop about 74 seeds per 10 feet of row.

Information about seed drop, soybean plates, and sprocket settings is given in the owner's manual for planting equipment. Some manuals use pounds per acre when referring to seeding rates. Therefore, to select desired settings corresponding to seeds per foot, it is necessary to convert to an acre basis. This is done using Table 1 (column 2) and Table 2 in conjunction with the following formula:

$$\text{Lb. seed per acre} = \frac{\text{Ft. of row/acre (from Table 1)} \times \text{Seeding rate (from above)}}{\text{Seeds/lb. (from Table 2)}}$$

Referring to our previous example, the farmer's seeding rate on a pounds per acre basis using Resnik soybeans would be:

$$\frac{\text{17,424 ft. of row/acre} \times \text{7.4 seeds/ft. of row}}{\text{2,830 seeds/lb. for Resnik}} = \frac{128,938}{2,830} = \text{45.6 lb. seed}$$

Regardless of the planter used, seed drop should be field-verified regularly. The number of seeds per pound also should be calculated for each lot of every variety planted because the number of seeds per pound will vary from lot-to-lot depending upon growing conditions.

The most difficult figure to accurately determine

when using the above seeding rate formula is percent expected emergence. Percent warm germination is the standard measure for seed quality; every bag of commercially-processed seed in Indiana must be labeled with this information. The warm germination test is run under ideal conditions (i.e., 7 days at 70°F and high humidity) and thus is essentially a test for live seed.

Table 2. Average Number of Seeds per Pound for Various Public Soybean Varieties.

Maturity group	Variety	Ave. seeds per lb.*
Group II	Archer	2600
	Burlison	2330
	Century 84	2520
	Chapman	2170
Group III	Bass	2800
	Edison	3070
	Harper 87	2280
	Hobbit 87	2650
	Linford	2380
	Pella 86	2320
	Resnik	2830
	Williams 82	2570
	Winchester	2330
Group IV	Flyer	3170
	Corsica	2530
	Spencer	2600
	Delsoy 4210	2320

*Number of seeds per pound will vary from year-to-year, depending on growing conditions.

FACTORS AFFECTING EMERGENCE

Emergence in the field is dependent upon a number of factors. The main ones are soil conditions, weather, date of planting, cultural practices including planting depth, and seed treatment. Emergence response to these factors often is referred to as "seedling vigor"; and all the factors must be considered in predicting percent emergence.

Soil Conditions

Soil conditions are determined by soil type, weather, and tillage practices. A cloddy, compacted, or crusted soil will generally reduce soybean emergence. Personal knowledge of and experience on individual fields is the best guide for estimating the effect of soil conditions on percent emergence.

Weather

In addition to influencing soil conditions, rainfall and temperature play a critical role in the germination and growth of soybeans. Soil and air temperatures of 55-60°F are necessary for seed germination and seedling growth; and as temperatures increase (up to about 90°F), rate of germination and growth likewise increase. Adequate soil moisture is needed to initiate seedling growth, but too much or too little can adversely affect soybean emergence.

Date of Planting

Planting date effect on emergence is related to weather and soil conditions. As soil and air temperatures increase during the planting season, the percent of emergence also should increase.

Purdue studies show that there will likely be less difference in emergence between early-and late-planted soybeans with high quality seed than with low quality seed. For instance, at 90 percent germination, we might expect a 3-8 percent emergence difference between a May 10 and a June 10 planting date; but for seed testing 80-85 percent, this difference increases to 8-13 percent.

Cultural Practices

Soybeans should be planted 1-1½ inches deep. Deeper planting will reduce emergence, because the distance a soybean can grow up through the soil is limited.

Secondary tillage affects the condition of the seedbed and influences the accuracy of planting depth, crusting tendency of the soil, and moisture retention. Therefore, review your specific cultural practices, and consider if and how they might influence—positively or negatively—seedling emergence when calculating your seeding rate.

Seed Treatment

The effect of a fungicide seed treatment on soybean emergence depends on two things—seed quality and weather conditions. Treating high quality seed (90 percent or greater germination) has little effect on laboratory germination or field emergence. Treatment of lower quality seed, however, may increase field emergence 5-10 percent.

The weather during germination also can affect the seed treatment-emergence relationship—i.e., as planting is delayed and conditions for germination improve, there is less emergence response to seed treatment. Remember, too, that fungicide treatment rarely increases yields, so long as adequate minimum populations are established.

POPULATIONS AND SEEDING RATES FOR DOUBLE CROP SOYBEANS

The recommended plant populations shown in Table 1 are valid for double crop soybeans. However, seeding rates for soybeans following wheat may be greater, because of added difficulty in establishing desirable stands. For seeding rate and population information regarding double crop soybeans, see Purdue Extension publication ID-96, "Double

Cropping Wheat and Soybeans in Indiana," available from your county Extension office.

SOYBEANS PLANTED WITHOUT TILLAGE

It is not necessary to increase seeding rates for no-till soybeans if a properly weighted and adjusted drill or planter is used and soil conditions are acceptable. Special care should be taken to assure correct planting depth, good slot closure, and soil firming around the seed. Most problems occur when trying to plant when the soil is either too wet or too dry. If too wet, it is difficult to obtain good slot closure and soil firming because of "slabbing out" of soil by the no-till coulters. If too dry, it is difficult to obtain adequate coulter penetration and seed placement at the correct depth. Uniform stands and solid seeding (rows 10" or less) are important, valuable aids in providing good, early canopy closure and enhanced weed control for no-till soybean production.

SUMMARY

Many factors affect the relationship between seeding rate and the established stand for soybeans. Fortunately, the growth characteristics of most varieties are such that establishing precise populations is not critical. Only when stands are 50 percent thinner or 50 percent thicker than the optimum populations suggested in Table 1 would serious problems be expected.

Calculating desired plant populations can be approached several ways. Table 1 goes through the exercise of determining seeding rates, expressing them on the basis of seeds per foot of row and pounds of seed per acre.

The accuracy of your seeding rate decision depends on your ability to predict percent emergence. Initially, your prediction as to emergence probably will be based on personal observation. But for future decisions, do a few field checks to confirm or improve your accuracy.

REV 2/93 2.5M

Cooperative Extension work in Agriculture and Home Economics, State of Indiana, Purdue University, and U.S. Department of Agriculture cooperating; H. A. Wadsworth, Director, West Lafayette, IN. Issued in furtherance of the acts of May 8 and June 30, 1914. The Cooperative Extension Service of Purdue University is an affirmative action/equal opportunity institution.

D. Double cropping management

Double cropping consists of producing two successive crops on the same land during one growing season. There are two important advantages of double cropping to the crop producer. First, it allows him to make intensive use of his land, and secondly, he can spread production and price risks over two crops in the same year.

While several crops are suitable for double cropping, the most popular combination in Indiana is soybeans following winter wheat. To be successful, a double cropping system must be properly managed. Read the article, "Managing Double Cropped Soybeans". Then list the points to consider in each of the following areas.

1. Management factors to consider

    a. Variety selection

    b. Row width

    c. Seeding rate and plant population

    d. seeding date

    e. seeding method

MANAGING DOUBLE CROPPED SOYBEANS

Choosing a Variety

Because of the strong photoperiodism inherent in the soybean (i.e., response to shortening days), later-maturing varieties can be grown in a double cropping program than one might otherwise think possible.

A good rule-of-thumb is to plant the latest variety that can be safely matured (leaves begin to yellow) before a killing freeze. This generally is a medium-season variety for the normal planting time in an area (see Figure 3). Such varieties will give the most vegetative growth, height, canopy closure, weed suppression and yield, and still mature before frost.

Varieties too early for a given area tend to be short, difficult to harvest and lower yielding. Conversely, varieties too late may have good height and canopy development but will still be green at the first killing freeze. The result is lower bean quality and yield than for a variety that would mature safely.

Remember in considering the 'right' maturity that, as one moves east in Indiana, average date of the first frost comes earlier because of higher elevation. For instance, the first-freeze date is the same for Lafayette and Richmond, even though Richmond lies on a line about 60 miles further south.

Row Width

Because of the delayed planting date, double crop soybeans have only about half their normal vegetative growth period remaining. They don't have as much time for lateral branching and canopy development as May-planted beans.

For this reason, rows should be 20 inches or less. Rows 30 inches or wider will yield about 25 percent less and rarely, if ever, develop a closed canopy, which is important for good space utilization and shading out of late weeds and volunteer small grain.

If your planter is 30 inches or wider, plan to double back and split the middles. Missing 'splitting the middles' by a few inches won't matter, since you don't cultivate no-till double cropped soybeans.

Seeding Rate and Plant Population

A seeding rate of 8-9 seeds per foot is suggested for 20-inch rows. This means 80-100 pounds of seed per acre for most varieties (see Table 5). The table also indicates a population of 4-5 plants per foot is adequate for optimum yields in 20-inch rows, while 6-7 plants is about right if you must use 30-inch rows.

Table 5. Suggested Plant Populations and Seeding Rates for No-Till Double Crop Soybeans Planted at Various Row Widths.

Row width	Final stand desired	Seeding rate required*
in.	plants/ft.	lbs./acre
30**	7	81
20	5	87
15	4	93
7-10	2½	109

*Assumes 60% germination and emergence rate, and average seed size (2500 seeds/lb.).

**Avoid if possible.

Within-row population is not as important as row width in influencing yield. Percent emergence will tend to be lower when planting in small grain stubble than in a well-prepared seedbed, thus the higher-than-normal seeding rates for double crop soybeans.

PLANTING EQUIPMENT FOR NO-TILL DOUBLE CROPPING

A no-till planter must do six things: (1) cut through small grain straw and stubble, (2) open a furrow or planting slit, (3) place the seed in the soil, (4) regulate depth of the seed, (5) cover the seed, and (6) firm the covering soil. Any planter that does all this will plant 'no-till.'

Coulters, Disc Openers and Press Wheels

The double disc openers on a conventional planter usually do not have sufficient downward force to cut through straw and stubble, open a furrow and provide enough loose soil for covering. Therefore, you will need some type of fluted or rippled-edged coulter ahead of the planter to accomplish these things. The tilled strip need not be very wide, especially if the planter units are close-coupled.

The original 2-inch wide fluted coulter works well in many situations. However, in heavy soils that are a bit wet, it tends to throw out to the side slabs of soil which are then not available for covering. A 1-inch wide fluted coulter avoids this problem and works well over a much wider range of conditions. The coulter need only to loosen enough soil to insure good soil-seed contact in the planting slit.

During planting, the straw must be dry and the soil reasonably firm to enable the coulter to cut cleanly through the straw rather than weaving it into the slit (Figure 4). Disc openers behind the coulter are better for planting through straw than straight runners (Figure 5). More important, however, is the excellent depth control provided by the 1¾-inch exposure of disc-opener depth bands.

The object is to place the soybean seed directly into the slit, then cover it with 1 inch of firm, moist soil (Figure 6). A ribbed press wheel helps close the slit, firm the soil and minimize moisture loss around the seed (Figure 7).

Figure 4. A fluted coulter, properly weighted, cuts through small grain stubble and tills a narrow band of soil to provide an excellent seedbed.

Modifying Pull-Type and Mounted Toolbar Planters

Pull-type toolbar planters (with three toolbars) have been used most for no-till double cropping in Indiana. The coulters are mounted on the front toolbar. To provide extra weight for good coulter penetration in firm, fine-textured soils, farmers often add iron rails, tractor wheel weights or tractor front-end weights. These are not as bulky as water-filled barrels or other means.

To use a mounted toolbar planter, you'll need to add a second toolbar up front for mounting the coulters. Toolbar spacers are available from most toolbar equipment manufacturers. However, this modification 'stretches out' the planter and may soon tax the lifting capability of your tractor's three-point hitch hydraulic system, especially if the planter is very wide and if much extra weight is added for coulter penetration.

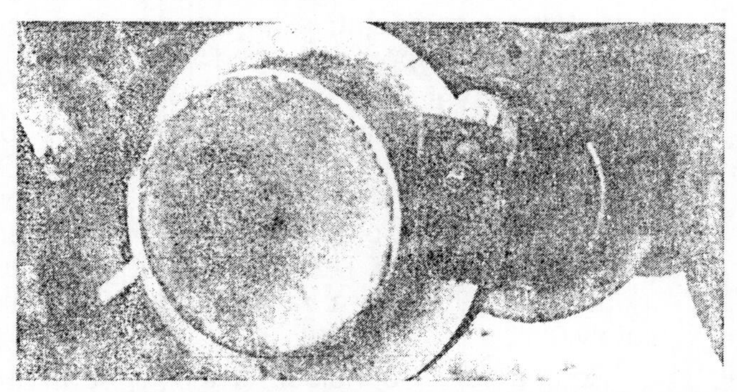

Figure 5. Double disc openers cut through and shred straw better than runner shoes. The depth bands shown expose 1¾ inches of the disc opener and place the seed about 1 inch deep after firming.

Modifying Conventional Pull-Type Planters

To use a conventional pull-type planter, you may be able to mount the coulters on the bar used to mount fertilizer openers; or it may be necessary to add an extra toolbar in front of the fertilizer bar. This will increase the permissible turning radius of the unit to avoid tractor tire injury.

With either coulter mounting, you might be exceeding the structural strength limits at the planter hitch and other frame members. Therefore, be prepared to assume the risks of a major breakdown during that critical double crop planting period.

Figure 6. The goal of no-till double cropping is to place the soybean seed about 1 inch deep in firm, moist soil in the small grain stubble.

Modifying Grain Drills

Some farmers have modified grain drills for no-till planting of double crop soybeans. Purdue's preliminary results with grain drills suggest an additional 4 bushel-per-acre yield advantage for 7- to 10-inch wide rows vs. 20-inch rows.

If you go this route, you'll need a coulter of some kind mounted in front of each opener on the

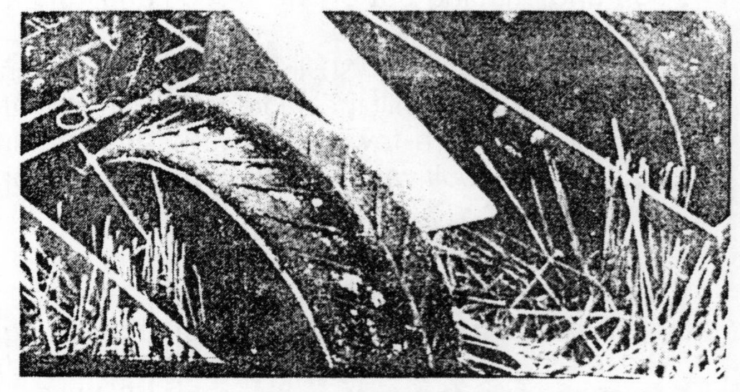

Figure 7. A ribbed press wheel is an effective aid in firming the soil around the soybean seed in the narrow tilled strip.

drill. Consider using depth bands on the drill openers. And for conventional end-wheel drills, use individually-mounted seed firming wheels or 'gang' press wheels.

The same precautions hold for modifying a grain drill as for modifying conventional pull-type planters. That is, you might exceed the structural limits of the drill hitch and other frame members; and you may not be able to turn as sharp if a coulter toolbar must be added across the front of the drill.

Do-it-yourself plans are available from a Canadian university for adding no-till coulters to a grain drill, and one U.S. manufacturer markets a press wheel drill. Currently, there are only two grain drills available designed for no-till planting.

OTHER METHODS OF DOUBLE CROPPING

Conventional Method

'Conventional' double cropping means planting soybeans in a fitted seedbed following the plowing, discing or chiseling of the small grain stubble. This has always been considered a rather high-risk practice because of soil moisture loss.

Double cropping soybeans in a plowed and fitted seedbed is feasible, however, when there is ample soil moisture and steps are taken to conserve this moisture. A number of farmers have obtained consistent stand establishment by plowing and working the ground in the evening immediately after wheat harvest, then planting that night with a cultipacker behind the planter or drill. This greatly reduces the moisture loss from conventional tillage by reducing the time the soil is 'open' for drying under the hot summer sun.

Conventional double cropping should probably be limited to those years when soil moisture is plentiful at wheat harvest time, and when planting can be done the same day as harvest. If planting is delayed after plowing until the top 2 inches of soil are dry, risk of poor stand establishment is high.

Planting in a conventional seedbed, of course, has certain advantages over the no-till method. No special equipment or modifying of existing equipment is necessary. Also, conventional weed control techniques, including rotary hoeing and row cultivating, can be done if beans are planted in rows. And there is less expense per acre for herbicides.

With a conventional seedbed, use normal soybean seeding rates (see Extension publication AY-217, "Plant Populations and Seeding Rates for Soybeans"). If using a grain drill, the rate should be about 2-2½ seeds per foot in 7-inch rows.

Intermediate Methods

There are many 'intermediate' forms of double crop tillage between conventional and no-till, including chiseling the stubble, discing once or twice, etc. All of these open up the soil somewhat to drying. They are essentially a compromise—i.e., enough tillage to prepare a seedbed for conventional equipment, but hopefully, not enough to permit excessive drying on top.

Intermediate systems have not been researched to any extent in Indiana, but most will work *if* there is plenty of soil moisture. A rolling cultivator may be helpful in allowing row cultivation where considerable wheat stubble remains on top.

SUMMARY OF DOUBLE CROPPING MANAGEMENT GUIDELINES

Dependable second crop stand establishment and weed control are the items critical to consistent success. And adequate soil moisture is the key to stand establishment, regardless of planting method.

With existing technology, double cropping is best adapted to the southern 2/3 of Indiana. It is difficult, but sometimes possible, in the northern third of the state, with special early wheat harvest techniques. The western half of the state has more favorable climatic conditions for double cropping than the eastern half because of lower elevation and a later killing frost date.

No-till double cropping is an intensive crop management system that offers high-management-ability farmers an opportunity to reduce erosion, conserve moisture at a critical time, cut labor and production costs, and lessen the risks of growing two crops in one year.

Plowing or tillage in any form is a drying operation. Thus, elimination of tillage except in the planting strip conserves soil moisture, while the straw mulch on top retards loss of any rainfall received. These features aid in the more consistent establishment of double crop soybeans.

Once established, the tap-rooted soybean is quite drought-tolerant; and if weeds are controlled, yields will likely average about 60 percent of normal bean yields in a given environment (or rarely be below 15-20 bushels per acre). The secret to effective weed control with herbicides in soybeans is good weed suppression in the small grain.

All management systems have limitations, and no-till double cropping is no exception. Research and farmer experience both suggest that, when the soil is dry in the top 2 inches or weeds are not suppressed, no-till double cropping should not be attempted.

HARVEST AND STORAGE OF GRAIN AND FORAGE CROPS

I. HARVESTING AGRONOMIC CROPS

Objectives

1. List 3 major considerations for the harvest and storage of a crop.
2. Define physiological maturity.
3. Define harvest maturity.
4. List the factors used to determine harvest maturity for a forage crop.

Growers often do not obtain the maximum yield from a crop due to losses incurred during harvest and storage. Proper management minimizes losses which may occur during crop harvesting and storage.

There are 3 key factors affecting crop harvest: (1) time of harvest, (2) method of harvesting, and (3) how the crop is to be utilized. In the case of grain crops to be used for livestock feed, a farmer wants a high grain yield which contains high levels of carbohydrate and protein. When harvesting forage crops to be used for livestock feed, a farmer will try to maximize forage production which is high in protein and digestibility. Farmers involved in seed production for commercial companies harvest when optimum yields are comprised for seed with high vigor and germination ability. Other crops, like barley, that are to be used for malting, must be low in protein and high in starch at harvest, so high quality malt is obtained.

A. Maturity

Physiological maturity (P.M.) occurs when a plant or an organ of the plant reaches maximum dry weight. In grain crops this type of maturity is associated with:

1. Formation of abscission layer (black layer in corn)

2. Moisture content of the grain

3. Accumulated growing degree days (heat units)

Harvest maturity (H.M.) is reached when the crop has dried sufficiently so that it can be harvested with minimal grain damage. At this point the grower will be able to harvest the greatest amount of dry matter with the highest market value. Harvest maturity will vary, depending on harvesting method and equipment available. For example a corn grower who does not have adequate drier facilities may wish to delay harvest so the grain will contain less moisture when it is harvested.

The general relationship between dry matter accumulation and loss of kernel moisture for corn is illustrated below. This illustration shows the importance of understanding the difference between P.M. and H. M.

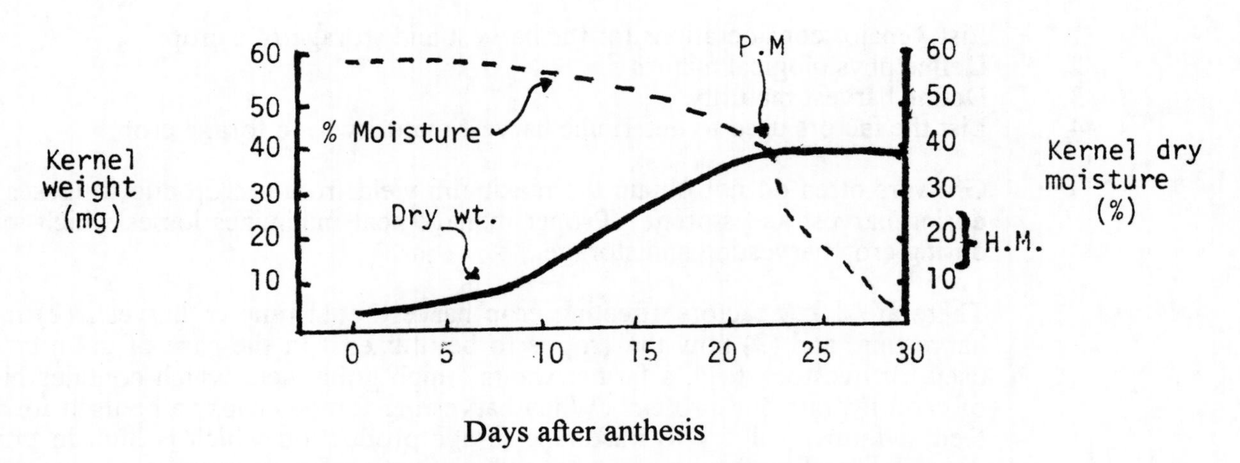

Fig. 1. General relationship between kernel development and water loss in the kernel during development.

Harvesting grain before **physiological maturity** may have adverse effect upon:

1. yield -

2. quality -

3. test weight -

Harvesting grain before or after **harvest maturity** may have adverse effect upon:

1. yield -

2. quality -

3. test weight -

Harvest maturity for many forage crops depends upon:

1. Stage of maturity -

2. Leafiness -

3. Digestibility -

II. STORAGE

Objectives

1. State the proper moisture content for storing grains, silage, haylage and hay.
2. Describe 3 methods of storing hay.
3. Decribe how quality silage is maintained.

Once a crop has been harvested the means of storage is the next consideration. A farmer will try to reduce or minimize respiration, heating, and storage pest infestation in an effort to maintain a high quality product. Quality will be a function of moisture, temperature, and method of storage.

A. Moisture and temperature conditions

In order to store a crop safely while maintaining quality, it must be stored at the proper moisture content.

1. corn

2. soybean

3. silage

4. haylage

5. hay

Maintaining the proper temperature range for both the crop and storage facility are important in reducing fungal growth and insect damage. High storage temperatures increase both the respiration rate of the crop and the amount of storage pests, both of which reduce quality.

An excessive moisture content will encourage fungal activity, which increases the storage temperature. However, when high moisture conditions exist a farmer can lower the storage temperature to control storage pest and reduce spoilage. Storage fungi grow slowly at 40-50°F and rapidly at 80-90°F.

B. Forage storage methods

Once a forage crop has reached harvest maturity the method of harvest and storage often determines its eventual use. For example, a forage crop harvested as hay could be stored as:

1. Baled hay

   a. square

   b. round

2. Long loose hay

3. Field chopped hay

   a. cubes

   b. pellets

Forages such as corn, alfalfa, or sorghum can also be harvested for silage and stored in silos. Silage or ensilage is a method of storage which preserves the crop at a high moisture content under anaerobic conditions. This allows the forage to ferment. To obtain high quality silage, the silos must be air-tight.

Hybrid Maturity-Energy Relationships in Corn Drying

D. J. Eckert, Ohio State University; R. B. Hunter, University of Guelph; and H. M. Keener, Ohio Agricultural Research and Development Center

Reviewers

P. H. Grabouski, University of Nebraska *R. Krech, Houston Co. Extension Office, MN*
J. H. Herbek, University of Kentucky *S. J. Murdock, Creston, IA*
G. J. Bossaer, White Co. Extension Office, IN

Because energy prices will continue to rise in the future and corn is an energy-intensive crop, efficient energy management is a major factor in maintaining a profitable production program. Proper hybrid selection is an important part of such management.

High-yielding hybrids whose maturities take fullest advantage of the available growing season are generally the most energy-efficient choices. A hybrid which matures far in advance of anticipated harvest may not make full use of available solar radiation, and therefore not realize the full yield potential of the growing season and the energy-related inputs provided by the farmer. Conversely, a hybrid which is not mature at the time of frost can impose an increased artificial drying cost on the farmer, in addition to not achieving full yield potential because it was killed before grain filling was complete.

MAJOR ENERGY INPUTS OF CORN PRODUCTION

Nitrogen fertilizer is normally the major energy input of a corn production program. Maximizing the efficiency of applied nitrogen involves optimizing all production factors, including date of planting, tillage, soil fertility, hybrid selection, and pest control, in addition to proper management of the nitrogen itself.

Artificial drying is the other major component of energy consumption in corn production. Proper hybrid selection is extremely important in managing the drying load. Farmers should strive to harvest corn at as low a grain moisture content as possible (without incurring yield losses due to lodging, ear drop, or weather delays) to minimize their drying loads. Planting hybrids of several maturities can help to minimize both drying load and harvest losses by extending the length of the optimum harvest period.

FACTORS AFFECTING KERNEL MOISTURE

At pollination and during the period of kernel establishment, kernel tissue is over 80% water. During the periods of grain filling and post-physiological maturity, the kernel loses moisture (Figure 1). The rate of moisture loss declines with time, and the factors affecting drydown vary depending on the stage of kernel development.

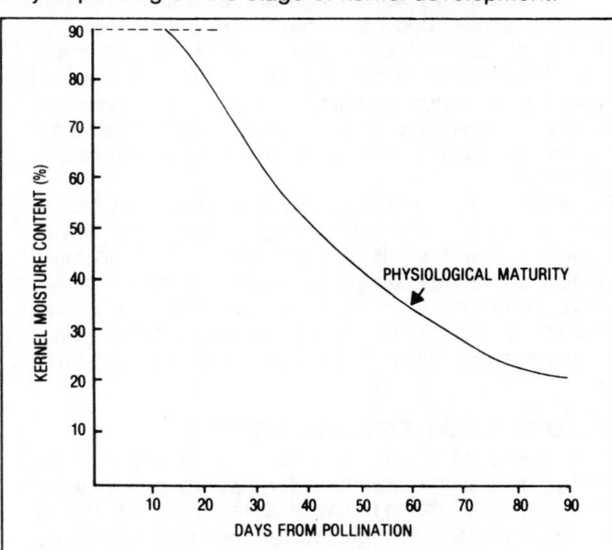

Figure 1. Relationship between kernel moisture and stage of growth.

During Grain Filling

In the grain filling period, rate of drydown is related primarily to the rate of kernel development and the specific hybrid being grown. Since air temperature is the primary external factor affecting development, there is a relationship between kernel moisture and air temperature. A corn hybrid grown in a cool environment takes longer to develop than when grown at higher temperatures. Since kernel moisture loss during grain filling is related to development, kernels filling under cool growing conditions will dry at a slower rate than kernels filling under warmer conditions.

At Physiological Maturity

Hybrids can differ significantly in kernel moisture content at physiological maturity. Kernel moisture content may vary from less than 30% to more than 40% moisture. Adverse weather late in the grain filling period may increase kernel moisture at maturity. A hybrid which normally matures at a kernel moisture content of 30% may be so affected by adverse conditions that it ceases filling at kernel moisture levels of 35% or higher.

After Physiological Maturity

Following physiological maturity, the reduction in kernel moisture content is closely related to the physical factors which affect moisture evaporation. Climatic factors such as temperature, humidity, wind speed, and rainfall all affect drydown.

Hybrid characteristics also play an important role in drydown after maturity. Characteristics reported to affect drydown include husk coverage, husk size and number, kernel shell permeability, and ear inclination. In general, kernels of hybrids with a small number of short, narrow husks which open to expose the ear as the kernels mature will tend to dry more rapidly than those on hybrids whose ears are covered by thick, tight husks. Kernels with thin shells dry more quickly than those with thick shells, and those on drooping ears tend to dry more quickly than those on erect ears.

EFFECTS OF FROST ON FIELD DRYING

When a killing frost occurs prior to normal physiological maturity, the dry-matter accumulation pattern in the kernels is altered; however, the rate of change in kernel moisture content may not change appreciably.

There is some suggestion that the rate of kernel drydown may decrease following such a frost. The decrease may be due to husks remaining tight around the ear and not opening in a normal manner. Other data, however, have shown increasing rates of drydown after frost, indicating that drydown response to frost may be related to hybrid type.

Regardless of its effect on drydown rate, frost before maturity normally halts grain filling, leaving the farmer with partially filled kernels and excess water which must be removed prior to storage, an expensive and energy-intensive proposition. In addition, the crop then may experience cooler, more humid conditions which slow field drying, particularly if the frost was not unusually early.

THE COST OF ARTIFICIAL DRYING

That drying corn after harvest is expensive is illustrated by the fuel requirements for operating high temperature dryers (Table 1). These data show that drying from 25% down to 15% moisture requires only two-thirds as much fuel as drying from 30% down to 15%. At fuel prices of $0.70 per gallon of LP gas and $0.05 per kilowatt-hour of electricity, the drying cost savings accrued by harvesting at the lower moisture content would be nearly $0.083 per bushel in this particular high-temperature dryer. Other drying systems could be expected to show similar cost relationships.

Table 1. Fuel Consumed and Costs of Drying Corn from Several Harvest Moistures Down to 15% Moisture.

Harvest	Fuel consumed		
moisture	LP gas	Electricity	Cost*
%	gal./bu.	kwh./bu.	$/bu.
35	0.472	0.066	0.334
30	0.337	0.049	0.238
25	0.219	0.033	0.155
20	0.109	0.017	0.077

* Based on fuel costs of $0.70 per gallon of LP gas and $0.05 per kilowatt hour of electricity.

Savings of this magnitude can affect overall profitability. In the example above, farmers harvesting 120 bushels of corn per acre would incur drying charges $10.00 per acre less by harvesting at the lower rather than higher grain moisture content. At corn prices of $2.50 per bushel, this provides the same return as increasing yields by four bushels per acre, at no additional cost.

SELECTING HYBRIDS TO MINIMIZE ARTIFICIAL DRYING

Whether one is concerned with energy conservation or simply profit, hybrid maturity makes a difference. To reduce drying loads, farmers should choose high-yielding hybrids which are as dry as practical at harvest.

Harvesting at 20 to 25% grain moisture is often cited as a reasonable compromise between drying load and harvest loss, but many farmers find themselves forced to harvest at higher moistures to beat the onset of winter weather. A different hybrid selection often will solve this problem.

The following guidelines will aid in making proper hybrid choices:

1. Know the growing season, including when planting can be accomplished normally, and when a killing frost normally occurs.
2. Plant the longest-season hybrid first, and choose its maturity to ensure that it will mature before frost in an average year.
3. Plant shorter-season hybrids next. Try to stagger maturities so that the hybrids mature in series in the fall.
4. Harvest each hybrid when its moisture content falls to 25% or below.
5. If planting is delayed, switch to shorter-season hybrids.
6. When using performance trial data to select hybrids, consider both yield and harvest moisture rather than yield alone. Use anticipated grain prices and drying charges to figure the overall profit potential for different hybrids.

A publication of the National Corn Handbook Project

NEW 1/87 (5M)

Cooperative Extension work in Agriculture and Home Economics, state of Indiana, Purdue University and U.S. Department of Agriculture cooperating. H.A. Wadsworth, Director, West Lafayette, IN. Issued in furtherance of the acts of May 8 and June 30, 1914. The Cooperative Extension Service of Purdue University is an affirmative action/equal opportunity institution.

SUSTAINING WORLD AND U.S. AGRICULTURE

I. THE WORLD FOOD SITUATION

Objectives

1. Describe characteristics of developing and developed countries.
2. Describe food production and consumption patterns in developed and developing countries.
3. Compare energy use and environmental conditions between developing and developed countries.

A. Developing and Developed Countries

The distinction among groups of countries in the world is arbitrary. A traditional method of classifying countries as developed (sometimes called industrialized), and developing (sometimes called less developed) has been made strictly on economic terms. Development meant an increase in per capita national income, or an increase in its total amount of goods and services produced (GNP). However, defining development in strictly economic terms grossly simplifies the true nature of the status of the countries of the world. In some cases, social changes associated with economic growth may not be desirable. Health care, crime rates, housing, and other components of quality of life may not be associated with increased economic growth. Development therefore encompasses both economic growth and the social changes that economic growth brings about. A developing country is one in which agriculture and mineral resources dominate the economy. About 80 percent of the world's population live in such nations. World population is growing, to a projected 8 billion people in 1998. The largest growth rates will occur in the developing countries (fig. 1).

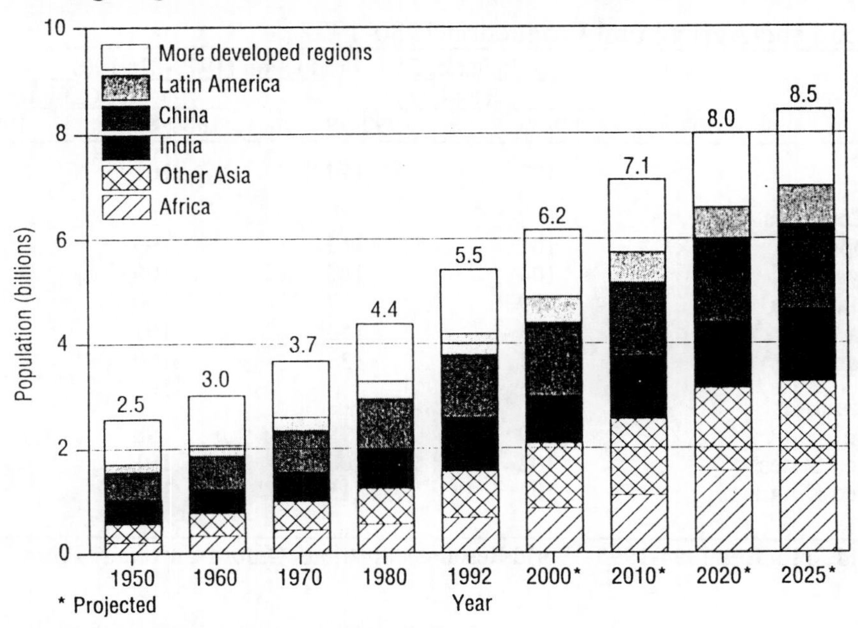

Fig. 1. Distribution of world population 1950-2025.
Source: Current Demographic Situation and Future Change: Selected Pages from World Population Prospects, 1992.

The gap between the rich and the poor is increasing. United Nations estimates that the gap between the richest 20 percent and the poorest 20 percent doubled between 1965 and 1995. The 1990's saw the richest 20 percent receiving 150 times the income of the poorest 20 percent of the world's population.

Table 1. Population size, growth rate, and nutrition characteristics of selected countries.

	Population (millions) 1990	1995	Average annual % population increase 1990-1995	% of required calories	% income spent on food 1980-85
World	5295	5759	1.68		
Developing Countries					
Ethopia	50	58	3.05	73	50
Kenya	24	28	3.35	89	39
Afghanistan	16.5	23	6.74	72	Unknown
India	931	1394	1.91	101	52
Honduras	5	6	3.00	98	39
Developed Countries					
Japan	123.5	126	0.38	125	16
United States	250	263	1.03	138	13
United Kingdom	57.5	58	0.24	130	12

Source: World Resources 1994-1995, World Resources Institute, Oxford University Press, 1994.

B. Food Production and Supplies

Lack of adequate food is a worldwide problem. An estimated 0.5 to 1 billion people in the world are hungry today. About 80 percent are women and children living in small villages in developing countries.[1] Food production has increased worldwide, but has not kept up with demand in many places (Table 2). Many factors affect food production and consumption worldwide (Table 3).

Table 2. Food and Agricultural Production 1980-1992

	Index of Food Production (1979-81=100) Total		Per Capita	
	1980-82	1990-99	1980-82	1990-92
World	103	127	101	105
Developing Countries				
Ethopia	101	113	99	85
Kenya	104	145	100	99
Afghanistan	99	77	100	69
India	103	153	101	122
Honduras	104	133	100	92
Developed Countries				
Japan	99	99	98	93
United States	102	108	101	97
United Kingdom	102	115	102	112

Source: World Resources 1994-1995, World Resources Institute, Oxford University Press, 1994.

[1] Seitz, John, Global Issues, Blackwell Publ., 1995.

Table 3. Developed and Developing Countries

	U.S.	Japan	Afghanistan	Ecuador	Ethiopia
Index of Agricultural Production per capita (1979-81=100)	97	90	69	109	85
Life Expectancy	76	79	43	67	47
Food Expend. %	13	16	Unknown	30	50
Purch. Power (~income)	$22,000	19,000	Unknown	4,100	370 (est)
% of labor force in agriculture	2	6	55	30	75
Cropland per capita ha	.74	.04	.46	.25	.27
% Grain Fed to Livestock	69	48	0	31	0

Source: World Resources 1994-1995, World Resources Institute, Oxford University Press, 1994.

C. Energy and the Environment

The ability of the earth to support its population is determined by more than the amount of food that can be produced. Consumption of natural resources, waste generation, and environmental management also play a role, and vary widely (Table 4).

Table 4. Causes of land degradation in selected countries, 1945-1991.

Region	Over-grazing	Defores-tation	Agricul-ural Misman-agement	Degraded Area as Share of Total Vegetated Land
		(million hectares)		(percent)
World	679	579	552	17
Asia	197	298	204	20
Africa	243	67	121	22
South America	68	100	64	14
Europe	50	84	64	23
North & Cent. Amer.	38	18	91	8

Source: World Resources 1994-1995, World Resources Institute, Oxford University Press, 1994.

Worldwide energy use has increased dramatically over the past 50 years, with most of the increase occurring in the developing countries (Table 5). In 1990, the United States, with about 5% of the world's population, consumed about 25% of the world's energy. (seitz). The world is also entering another period of energy transition, away from oil to other sources. Renewable, non-polluting sources of energy are needed worldwide. Potential sources include: wood, water power, wind, wastes, and solar energy. Not all of these are feasible sources; the US Government's *Global 2000 Report* projects demand for wood will exceed supplies by 25% by the year 2000.

Table 5. Per capita commercial energy consumption, selected countries, 1991.

	Gigajoules consumed	% change since 1971
World	60	+2
Developing Countries		
Ethopia	1	13
Kenya	3	-10
Afghanistan	5	179
India	9	36
Honduras	5	-20
Developed Countries		
Japan	140	52
United States	320	-4
United Kingdom	157	17

Source: World Resources 1994-1995, World Resources Institute, Oxford University Press, 1994.

II. SUSTAINABLE AGRICULTURE

Maintaining a health food fiber production system for future generations should be a goal of every agriculturalist. This means our agricultural production systems mustmaintain or improve the natural resource base, preserve environmental quality, ensure profitability, a maintain a viable social infrastructure for rural America. The American Society of Agronomy has defined Sustainable Agriculture as "one that, over the long term, enhances environmental quality and the resource base on which agriculture depends, provides for basic human food and fiber needs, is economically viable, and enhances the quality of life for farmers and society as a whole." Sustainable agriculture can best be viewed as a philosophy, rather than a specific set of practices to follow.

In order for agriculture to be truly sustainable it must be sustainable in three areas:

1. Economic

   The farmer needs to make a profit, and agriculture will be economically sustainable as long as supply matches demand at reasonable food prices, and with normal profits.

2. Environmental

   Natural resources are used in food production, but our resource base can be protected through its wise use. Undue contamination or destruction of soil, air, and water resources for short term profits will not maintain sustainable agricultural systems.

3. Sociological

   Agriculture also has a social context, which has largely been ignored in our efforts to maximize production. Vital, healthy rural communities which recognize the importance of quality of life are needed to maintain a sustainable agriculture.

U.S. agriculture is faced with many economic, environmental, and food safety issues. Increased production costs and low farm prices have forced many small to medium sized operators to quit farming. This has created both an economic and social problem. Peaceful rural communities are becoming "urbanized" as uncontrolled development places housing subdivisions on prime farmland. Along with this development comes social problems that many small communities aren't equipped to handle.

Agriculture is the largest single nonpoint source of water pollutants, resulting from sediments, salts, fertilizers, pesticides and manures. In some areas, irrigation is depleting natural aquifers at rates exceeding recharge. Improper cultivation causes excessive wind and water erosion, and improper use of pesticides has contaminated natural resources in some areas. The American consumer, while having the greatest diversity and safest food supply in the world, is becoming increasingly concerned about the continued availability of a safe food supply.

Future agricultural production systems will need to address these and other issues as they arise in the future. The need for environmentally sound production systems that are profitable and maintain an acceptable quality of life will demand a constant reassessment of agriculture, both nationally and internationally.

LABORATORY

CONTENTS

EXAMINING SOILS: IN THE FIELD

OBJECTIVES

1. Use a soil survey report to locate a farm, determine soil type, and explain the soil's characteristics.
2. Define soil horizons and list characteristics of horizons you will see.
3. Use soils information to decide suitability of land for various uses.
4. Explain how a soil cover affects soil temperature.
5. List factors affecting soil erosion.

Characteristics of the soil are important factors to consider when deciding how land can best be used. Much can be learned about suitability for grain production, pasture, home or industrial construction, roadways, and other uses by carefully examining soil reports and critically evaluating the soil's characteristics.

To be able to describe a soil, one must be familiar with terms such as **soil profile**, **color**, **texture**, **structure**, and **horizons**. With practice one can look at a soil profile and relate its characteristics to yield potential, drainage problems, water holding capacity for crop growth, erodibility and many other factors. Management decisions concerning what type of crop will grow best on a particular soil, the environmental consequences of tillage methods, when one should plant, or suitability of soil for home sites or industrial uses can be aided by a basic knowledge of soil.

Examining soils in the field provides a perspective of how the soil fits into, and interrelates with, the total environment. Use the following study guide to help you understand the major objectives at each of the stops.

FIELD TRIP STUDY GUIDE

1. Using the soil survey report, locate the site where you are standing, and answer the following questions about the soil.

   a. Soil symbol: ________________ Soil name ________________

   b. Characteristics:

      - Topography

      - Parent material

      - Natural drainage

      - Susceptibility to erosion

      - General productivity

2. Examine the soil profile and answer the following questions:

   a. How many horizons can you see?

      What are they?

      At what depth are they?

   b. What is the color of the A horizon?

      What is the predominant color of the B horizon?

      Is there any mottling?

      What type of drainage does mottling indicate?

   c. How does soil texture differ in the A and B horizons?

      Which horizon has the greatest clay content?

d. How does structure differ between horizons?

3. After examining the soil at this location, what are desirable and undesirable factors for each of the following land uses:

   a. row crop production

   b. home sites

   c. road construction

   d. permanent pasture

4. How does the surface temperature of the soil at the location with soil cover differ from the location without soil cover?

   a. Factors which affect soil temperature are:

   b. Would the present soil temperature be desirable for:

      - Corn germination
      - Weed seed germination
      - Organic matter decomposition
      - Root growth

   c. What could the farmer to do encourage desirable soil temperature for crops being grown?

5. The Universal Soil Loss Equation

How do each of the factors in the Universal Soil Loss Equation contribute to the amount of soil loss?

R The rainfall factor represents potential erosion based upon an area's seasonal distribution, intensity and total rainfall.

K The soil erodibility - factor deals with the actual soil type. Infiltration capacity and structural stability are considered in determining this factor. These properties are affected by the soil texture and organic matter content, as well as several other characteristics.

LS The length and steepness of the slope is important in determining the soil loss because a greater degree incline will increase the velocity of the water, giving it more force to detach soil particles. An increase in the length of slope will provide a greater volume of flooding water, which also affects soil detachment.

C When crop managment factor is not considered, one is determining annual soil loss on bare soil. The crop management factor is based upon the type of crop planted in an area (some crops are better than others at holding soil in place) and the cropping system used.

P The erosion control factor shows the effect of using different tillage systems as well as other conservation practices, such as terracing or contour cropping.

EXAMINING SOILS: IN THE LABORATORY

OBJECTIVES

Part A

1. Define soil texture and soil structure.
2. Recognize how soil texture affects crop production.
3. Mechanically analyze soil texture using the Bouyoucos Method.
4. Given the percent of sand, silt and clay in a soil, use a textural triangle to place the soil in the correct textural class.

Part B

1. Define macropore, micropore and capillary action.
2. Describe the relationship between soil texture and plant available water.
3. Describe how soil macropores are created and destroyed, and recognize their impact on drainage.
4. Recognize the principles of tile drainage.

Part A

1. Soil Texture and Soil Structure

Soil texture refers to the relative proportion of sand, silt and clay in the soil. As seen in Figure 1, sand particles are the largest in diameter (0.05 - 2.00 mm), clay particles are the smallest in diameter (< 0.002 mm), and silt is between sand and clay (0.002 - 0.05 mm).

Soil structure refers to the way the soil particles are arranged to form structural units called **peds**.

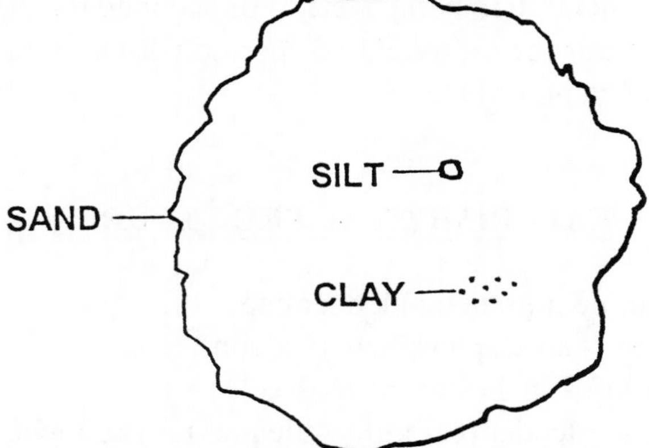

Fig. 1. The relative sizes of sand, silt, and clay.

2. Influence of Soil Texture on Crop Production

Soil texture affects the amount of particle surface area contained in a soil. Particle surface area is closely linked with soil fertility and water-supplying ability. Soils with clayey textures have the greatest particle surface area, while sandy-textured soils have the least. A gram of silt loam soil contains 75 to 185 square meters of surface area on its sand, silt, and clay particles.

Soil texture also affects the size and amount of pore space. Pore space determines the rate of air and water movement through the soil.

By affecting pore space and particle surface area, soil texture greatly influences soil internal drainage, water-supplying ability, fertility, aeration, and erosion.

Soil scientists need to identify and classify soils just as crop scientists must do with agronomically important plant species. For soils this includes determining the texture. Texture determination can be made by the ribbon or feel method, and by laboratory analysis using a hydrometer: the Bouyoucos Method.

3. Mechanical Analysis for Soil Texture Determination

The Bouyoucos hydrometer method of determining soil texture measures the concentration of soil particles in suspension at a given time, and depends on the rate of settling of sand, silt, and clay. This method is based on the fact that large soil particles will settle out of suspension faster than smaller particles. The reason for this is that large particles have less surface area in proportion to their weight than do small particles. This underlies a principle known as Stokes Law. Sand will settle first, followed by silt, and finally, under certain conditions, the clay.

The hydrometer gives an indication of the density of the suspension. As more material settles out of suspension, the density of the suspension becomes less. The hydrometer is calibrated to give a reading directly in grams of material in suspension per liter of suspension.

Soil particles tend to cling together due to the effect of organic and inorganic colloidal material, inorganic cementing materials, etc. For accurate results of any mechanical analysis, the particles must be thoroughly dispersed with a chemical dispersing agent such as sodium hexametaphosphate.

EXPERIMENTAL PROCEDURE

1. Pour 50 g soil sample into metal blender cup.
2. Add 1/4 teaspoon hexametaphosphate (Calgon).
3. Add water (to within 2 inches of the top).
4. Attach metal cup to blender (carefully--blender turns on when cup is attached).
5. Blend for approximately 10 minutes.

6. Pour soil suspension from metal cup into 1000 ml cylinder.
7. Rinse remaining soil from metal cup into cylinder with squirt bottle.
8. Put hydrometer into cylinder (carefully).
9. Add water to cylinder to 1000 ml mark.
10. Remove hydrometer.
11. Stretch parafilm across the end of cylinder.

	LAB READING	**ADJUSTED**	**FINAL RESULT**
12. Shake cylinder end to end (to bring soil into suspension)			
13. Set down cylinder. NOTE TIME.	___min___sec		
14. Put hydrometer back in cylinder.			
15. Read hydrometer AT 40 SECONDS PAST STEP 13.	15. _____g (silt and clay)		
16. Measure temperature of water.	16. _____°C	17._____g (silt and clay)	
17. Adjust silt and clay reading. (For each degree above 20°C, add 0.4 g. For each degree below, subtract 0.4 g)			
18. Calculate weight of sand fraction (50g-#17).		18._____g (sand)	**19.____% sand**
19. Calculate % sand (based on 50g total).			
20. Read hydrometer AT 2 HOURS PAST STEP 13.	20._____g (clay)	21._____g (clay)	**22.____% clay**
21. Adjust clay reading for water temperature.			
22. Calculate % clay.			
23. Calculate weight of silt fraction (#17-#21).		23._____g (silt)	**24.____% silt**
24. Calculate % silt.			

25. When completed, rinse out cylinder and blender cup in lab sink.

26. **From percentages of sand, silt, and clay, determine the soil texture classification using the textural triangle:**

Soil Textural Classification: ______________________________

4. <u>The Textural Triangle</u>

The texture of a soil is expressed by the use of class names which are found within the textural triangle. The class name is a function of the percentage of sand, silt and/or clay found within the soil. Once these percentages are correctly determined by mechanical analysis the textural triangle is used to describe the soil's class. (Fig. 2).

Determine the class of soil containing 30% sand, 60% silt and 10% clay.

a. Using percent sand, draw a straight line parallel to the silt axis, from the percent sand axis across to the percent clay axis.

b. Using the percent silt, draw a straight line parallel to the clay axis, from the percent silt axis to the percent sand axis.

c. Using the percent clay, draw a straight line parallel to the sand axis, from the percent clay to the percent silt axis. All three lines should intersect at one point. The name of the textural class is as stated within the area where the lines intersect.

d. Therefore the textural class of the above soil: silt loam.

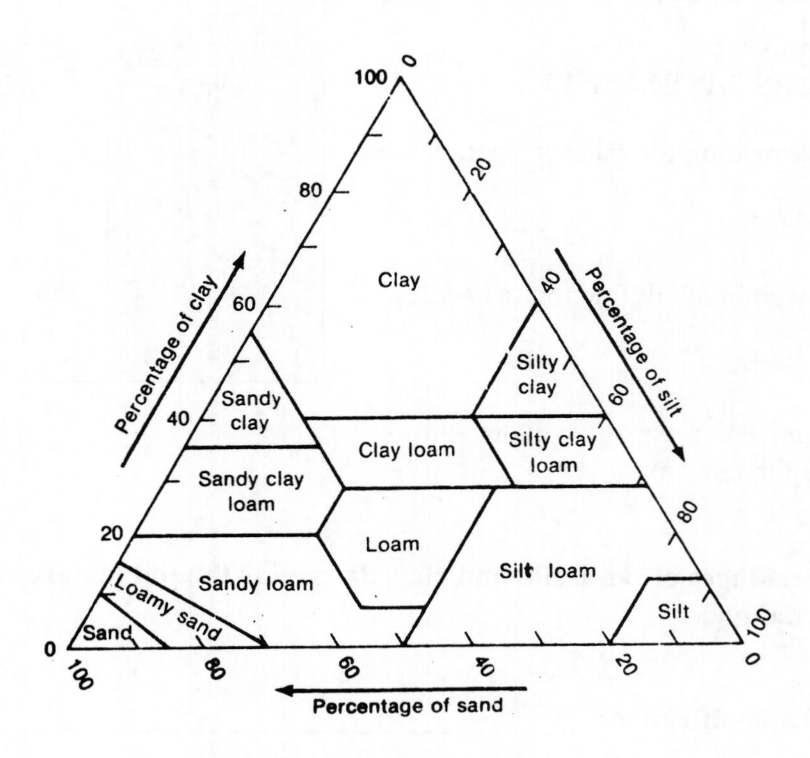

Fig. 2. Soil Textural Triangle

Part B

1. Soil Pores and Water Movement

All soils contain pore spaces of various sizes. Micropores are so small that they restrict air movement. Water moves by capillary action in micropores, which is the natural attraction for water into openings the approximate size of a hair.

Macropores are pores large enough to allow rapid movement of water. Large volumes of water can travel through macropores very quickly, resulting in rapid infiltration rates.

2. Soil Texture and Plant Available Water

Commonly, water held by the soil remains there under negative pressure known as **tension**. This tension is referred to as **Soil Moisture Tension (SMT)**. Water becomes less available for crop use as the water layer on a soil particle decreases because the tension exerted by the soil on the water increases (Fig. 3).

Not all water in the soil is available for plant use. After sufficient rain, a soil becomes saturated. Water will quickly drain out of the macropores, leaving the soil holding as much water as it can in the micropores. At this point, the soil is said to be at **field capacity**. Plants will remove water from the micropores until the **wilting point** is reached. At the wilting point, the water remaining in the soil is held so tightly by tension that plant roots cannot extract it and the plants will die unless soil water is replenished.

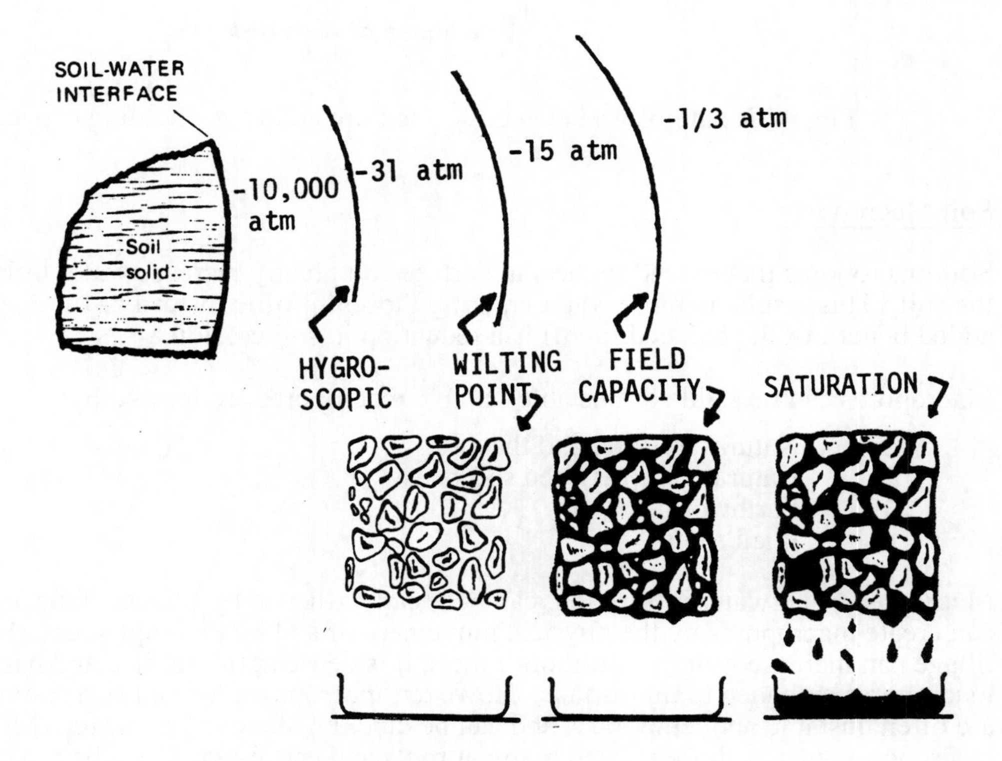

Fig. 3. Water tension and field conditions.

Soil structure, organic matter content, and texture all influence a soil's water-holding capacity, but texture has the greatest influence on the amount of water a soil can hold. Plant-available water is that water held at a tension between wilting point and field capacity. Soils high in clays hold the most water overall, but silt loam soils hold the greatest amount of plant-available water (Fig. 4).

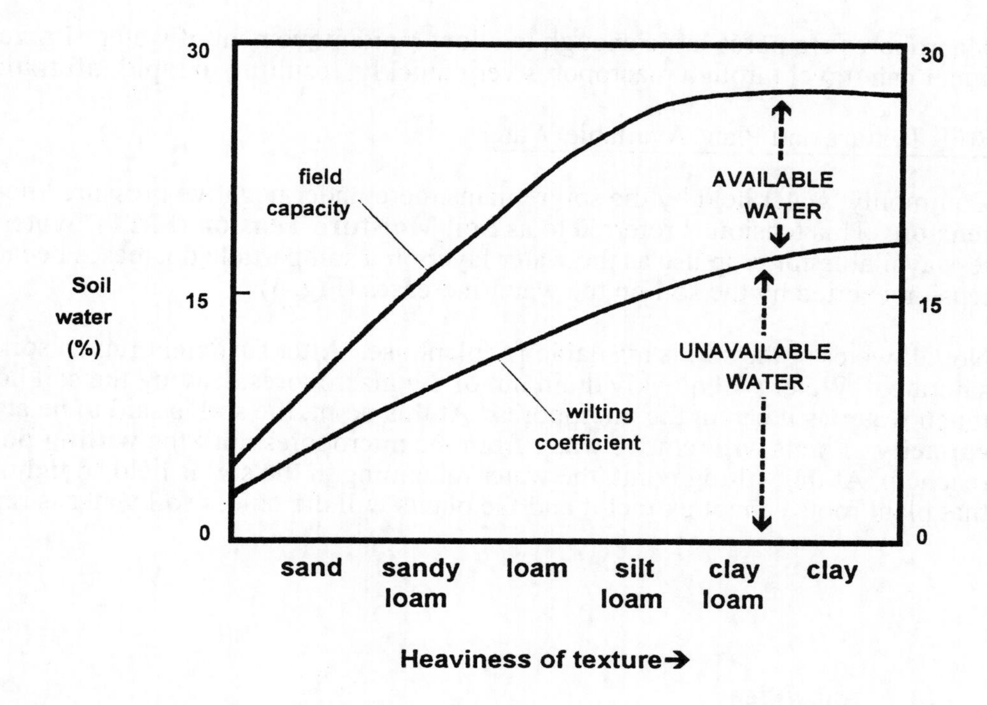

Fig. 4. Relationship between soil texture and water-holding capacity

3. Soil Macropores

Soil macropores that extend to the soil surface can greatly increase water infiltration into the soil. This results in more water entering the soil profile instead of running off. An added benefit of the reduced runoff is a reduction in soil erosion.

Macropores can be stable or unstable. Stable macropores are formed by:

1. Natural freezing and thawing
2. Natural gaps between soil peds
3. Earthworms
4. Plant roots

Macropores, and water infiltration, can be greatly affected by tillage. Tillage operations can create macropores by the physical movement of soil by the implement. Sometimes, tillage can increase water infiltration if the soil is dense at the surface and has few macropores extended to the surface. However, macropores formed as a result of tillage are often unstable and temporary, and can be quickly destroyed by water. More stable macropores such as those created by plant roots and earthworms, can be sealed off by tillage, thus reducing water infiltration. If tillage is repeated and intense, soil peds are broken down, resulting in a further loss of macropores. Thus, a field that is intensely tilled often has reduced infiltration rates.

4. Tile drainage

Systems of underground tile are designed to remove excess water from crop fields, and usually provide an effective means of improving soil drainage. Clay, cement, and perforated plastic tubing are the most common materials from which modern tile are made. Regardless of what material the tile is made from, they all function by the same principles. The tile is usually placed 1 meter below the soil surface. The tile line spacing varies, depending on soil type (Table 1). Water moves from the surrounding soil to the tile, enters through perforations in the tubing or through open joints between the tiles, and then moves through the tile to an outlet ditch.

Table 1. Suggested Spacing Between Tile Laterals for Different Soil and Permeability Conditions.

Soil	Permeability	Spacing (feet)
Clay & clay loam	Very slow	30-60
Silt & silty clay loam	Slow to moderately slow	60-100
Sandy loam	Moderately slow to rapid	100-300
Muck & peat	Slow to rapid	50-200

To facilitate movement of the water, tile systems are commonly designed with 3-6 inches of fall per 1000 feet of tile.

Advantages of improved soil drainage include warmer soil temperatures in the spring, better soil aeration, more days to perform field operations with heavy machinery, and reduced risk of soil heaving.

DEMONSTRATIONS

1. The container filled with layers of silt and sand simulates a stratified soil profile and has been filled according to the diagram below.

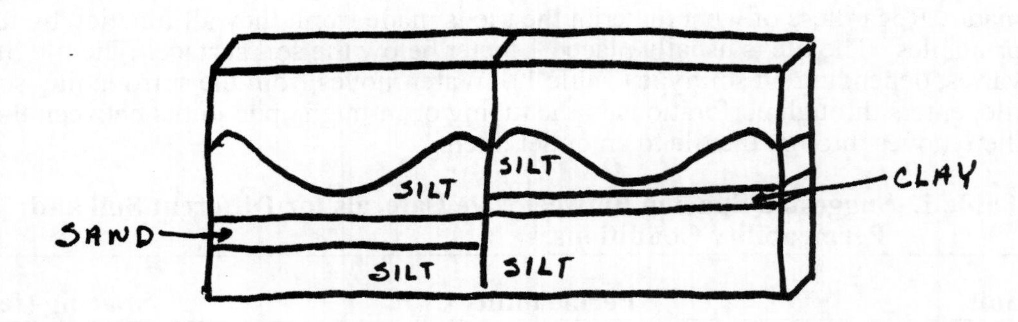

As the water moves through the soil observe the wetting front movement. Notice any change that occurs when the wetting front reaches the stratified layers.

2. a. The rates of water movement through cultivated and pasture soils are demonstrated with the large glass tubes. The soil aggregates used passed through 4-mesh screen but were held on a 20-mesh screen. List any differences in water movement between the two soils.

   b. Select soil peds of similar size from the cultivated and pasture soil. Place each in a wire basket and briefly swirl in a beaker of water. Explain what happened and why. What is the effect of tillage on aggregate stability?

DEFINITIONS

Adhesion: Attraction of molecules to unlike surfaces.

Aeration: The movement of air into and out of pore space in a soil.

Available water: Water which plant roots are able to take up and use.

Capillary rise: The attraction of water into openings of the approximate diameter of a hair, wick action.

Cohesion: Attraction of like molecules to one another.

Clay: Soil particles less than .002 mm in diameter.

Field capacity: Condition when water is thought to be most available to plants, held with 1/3 atmosphere of tension.

Gravitational water: Held with less than 1/3 atmosphere of tension; drains from soil profile by gravity.

Horizon: Layer of soil with somewhat uniform color, texture, structure.

Hygroscopic water: Water held so tightly to soil colloids that it is unavailable to plants.

Infiltration: Movement of water into a soil.

Inorganic soil component: Composed of weathered mineral matter.

Macropores: Pores large enough to allow rapid movment of water.

Micropores: Pores small enough to restrict air movement: water moves by capillary action in micropores.

Organic soil component: Composed of organic residues, plants and animals.

Percolation: Movement of water through the soil.

Pore space: Space between soil particles which can be occupied by air and water.

Profile: Vertical section of soil extending down from the surface and exposing the underlying layers of development.

Sand: Soil particles larger than 0.05mm, which provides coarse characteristics to the soil.

Silt: Soil particles between 0.05 - 0.002mm, which feels smooth and floury when wet.

Soil aggregate: A group of soil particles bound together by organic matter.

Structure: Arrangement of soil particles.

Unavailable water: Water held by the soil into aggregates with so much tension that plants cannot use it.

EARLY GROWTH AND DEVELOPMENT OF SEEDLINGS

OBJECTIVES

1. Identify the legume seed and grass caryopsis structures, explain their functions, and describe characteristics of good quality seed.
2. List the environmental factors that influence germination and explain how good management can influence these factors.
3. Explain dicot and monocot germination and seedling development.
4. Explain how method of emergence and environmental conditions influence planting depth of corn, small grains, and soybeans.

To be able to judge good quality seed one must know the various parts of the seed or kernel and their function in germination. The harvest yield you get in September can be limited by poor quality seed before planting even begins that Spring. A farmer wouldn't buy a tractor without judging its quality first. Buying seed should require the same careful judgment. In both cases looking at the individual parts is important in determining the overall performance.

A. Legume Seed and Grass Caryopsis Identification

1. Dicotyledons

Legumes are **dicotyledonous** plants since they have two "seed leaves" called cotyledons in their embryo. Some common legume crops are soybeans, alfalfa, red clover, and peas.

Carefully remove and examine the outer covering of a soaked soybean seed. This thin covering is called the **testa** (seed coat) and is actually the mature ovule wall. The seed coat is a valuable protective covering against soil borne diseases and insects. On the seed coat there is an oblong mark or scar called the **hilum**. This is the point where the seed was attached to the pod. Hilum color differs considerably and is useful in identifying different varieties.

Separate the two remaining halves. These are the **cotyledons** which serve as the food source for the germinating legume seedling. Therefore, a characteristic of high seed quality is disease-free cotyledons. Notice that these two cotylendons are attached at one end of the cotyledonary node. Above the cotyledonary node is the **epicotyl** (plumule) consisting of the growing point and embryonic leaves. Below the cotyledonary node is the **hypocotyl** which is important in emergence, and the **radicle** which develops into the root system. This distinction between the hypocotyl and radicle is difficult to see without the use of a microscope. The epicotyl, hypocotyl, and radicle must all be functioning properly for successful germination and establishment.

Label the following diagram of a legume seed.

Seed coat
Hilum
Epicotyl
Cotyledonary node
Cotyledon
Hypocotyl
Radicle

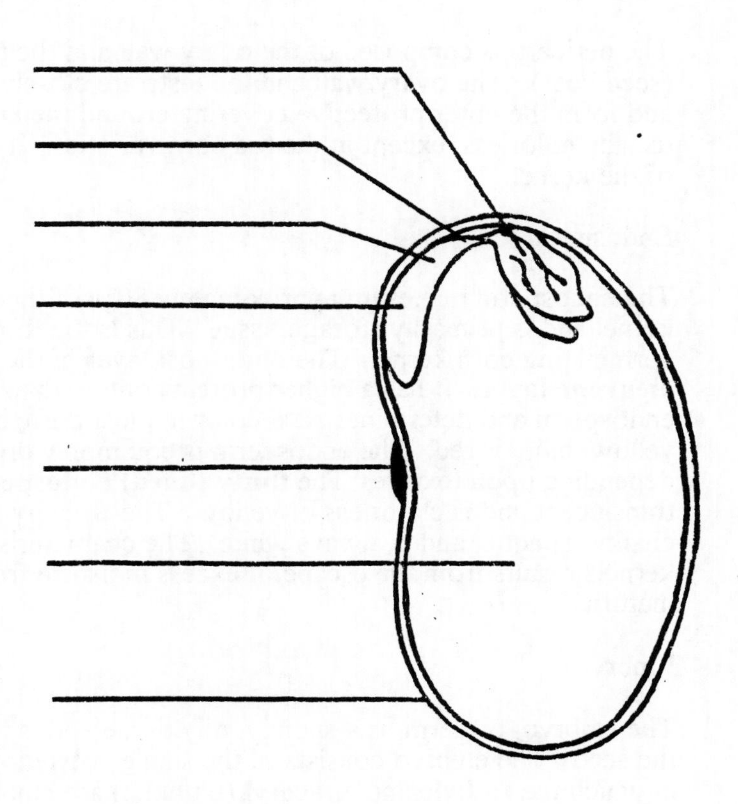

Fig. 1. Legume Seed

2. Monocotyledons

Members of the grass family are classified as **monocotyledons** since they have only one cotyledon or "seed leaf" in the embryo. All of the cereals and many of the forage crops are classified as monocotyledons. The "seed" of grasses are not true seeds, but actually one-seeded fruits known as a **grain kernel** or caryopsis. The term "seed" is commonly used for these one-seeded fruits as a matter of convenience.

The corn kernel, because of the size of its parts, will be used as an example of the structure of the grass caryopsis. Their three major components of the grass caryopsis are the **pericarp**, **endosperm**, and **embryo**.

a. Pericarp

The pericarp is composed of the ovary wall and the remnants of the testa (seed coat). The ovary wall and the testa are closely adherent to all points and form the outer protective covering around the kernel. The pericarp is usually colorless, except in the red corn varieties. It is the protective covering of the kernel.

b. Endosperm

The endosperm makes up approximately 80% of the dry weight of the mature kernel and is primarily storage tissue. This is the food source for the germinating corn kernel. The outermost layer of the endosperm is called the **aleurone layer**. It has a higher protein content than the rest of the endosperm and determines seed color in most corn, being either colorless, yellow, blue or red. The endosperm is commonly divided into two parts, depending upon texture. The **flinty (hard) endosperm** is hard and translucent and is clolorless or yellow. The **starchy (soft) endosperm** is soft, chalky, opaque, and is always white. The characteristic depression in all dent kernels results from the escape of excess moisture from the soft endosperm at maturity.

c. Embryo

The embryo, or germ, is a spongy, oily tissue and is the only living portion of the seed. The embryo consists of the single cotyledon or first leaf (scutellum) in which the **radicle** and **epocotyl** (plumule) are embedded. The **coleoptile** (second leaf) is a modified protective covering for the epicotyl. The epicotyl is a dome-shaped structure pointing toward the dented end of the kernel and develops into the stem and leaves. The radicle points toward the base of the kernel and is enclosed in the **coleorhiza** or root sheath.

Dissect two moistened corn kernels so you obtain cross sections as shown in Fig. 2. Carefully examine the cross sections and identify the structures listed above the diagrams. After you have seen these structures in your corn kernels, label them in the diagrams. If you have any questions consult your laboratory instructor.

Silk scar
Pericarp
Radicle
Cotyledonary node (scutellar node)

Hard endosperm
Soft endosperm
Aleurone layer

Cotyledon (scutellum)
Coleoptile
Epicotyl
Coleorhiza

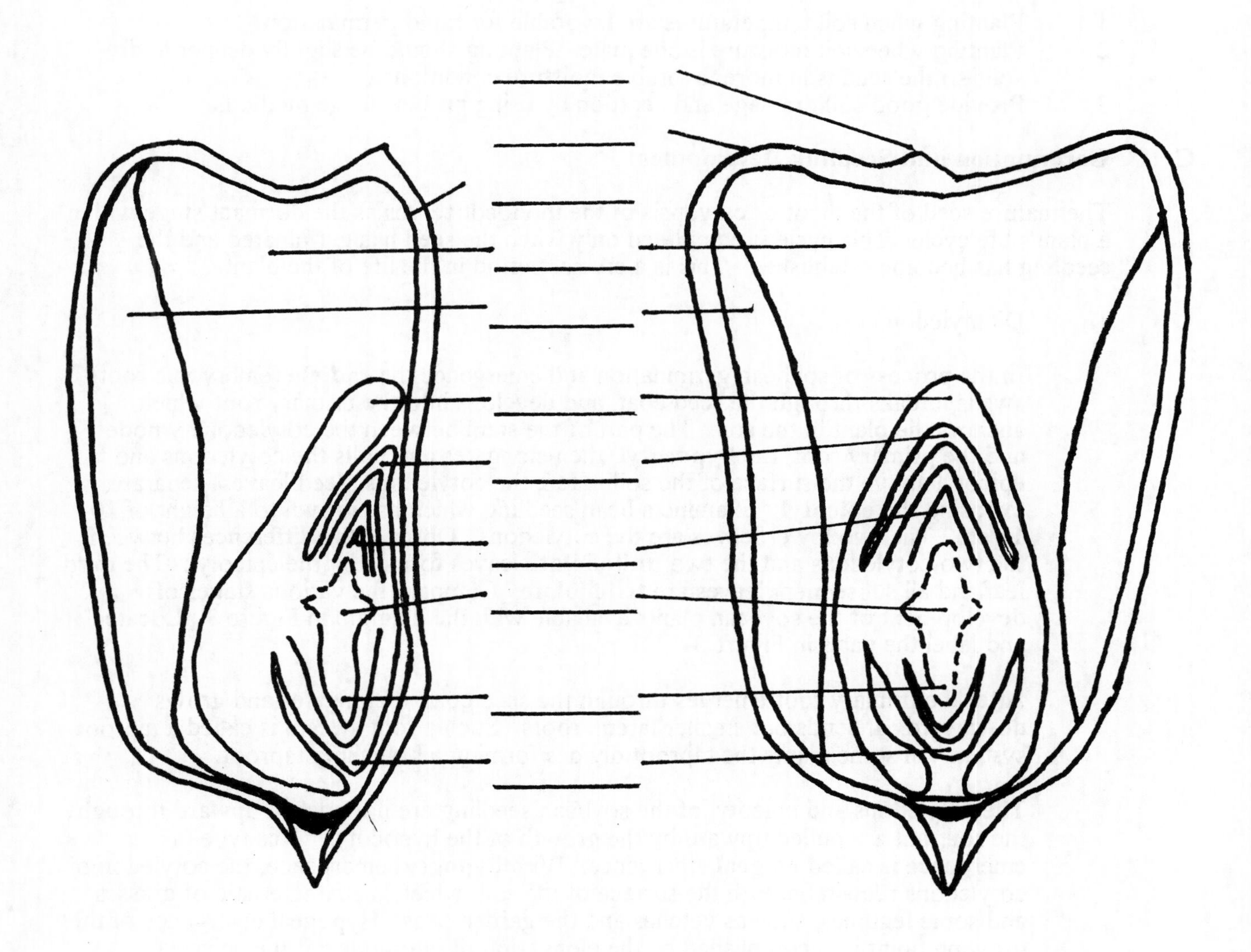

Fig. 2. Parts of a corn kernel

B. Environmental Factors and Managment Practices Influencing Germination

Growth activity is resumed by the embryo when a viable seed is planted under favorable conditions of temperature, oxygen, and moisture. In the germination process, water is imbided by the seed, which activates the enzymes in the cotyledon, converting starch, fats, and proteins into forms usable by the germinating seedling.

The following management practices will help alleviate poor germination:

1. Planting when soil temperatures are favorable for rapid germination.
2. Planting when soil moisture is adequate. Planting should be slightly deeper in dry soils so the seed is in more favorable moisture conditions.
3. Provide good soil drainage and aeration by using proper tillage methods.

C. Germination and Seedling Development

The mature seed of the dicot or caryopsis of the monocot serves as the dormant stage within a plant's life cycle. This cycle is completed only when the seed has germinated and the seedling has become established. This is a crucial period in the life of the plant.

1. Dicotyledon

In the process of soybean germination and emergence the **radicle** (embryonic root) swells, breaks through the seed coat, and develops into the primary root which anchors the plant in the soil. The part of the stem between the cotyledonary node and the primary root, the **hypocotyl**, then elongates and pulls the cotyledons and epicotyl above the surface of the soil. Here the cotyledons (seed leaves) separate, exposing the **epicotyl**. Examine a bean seedling which has grown to a height of 10-15 cm. The two lower leaves are the cotyledons. Observe the differences between the two cotyledons and the two **unifoliolate** leaves formed by the epicotyl. The third leaf and all subsequent leaves are **trifoliolate**. Compare the various stages of development of the soybean plants available with the diagram in Figure 3. Locate and label the parts in Figure 3.

After the primary root emerges through the seed coat, it elongates and grows downward. It sends out smaller lateral roots. Such a root system is called a taproot system. In some plants the taproot divides forming a branching taproot.

The cotyledons and epicotyl of the soybean seedling are not pushed upward through the soil, but are pulled upward by the growth of the hypocotyl. This type of emergence is called **epigeal** emergence. With **hypogeal** emergence, the cotyledon or cotyledons remain beneath the surface of the soil, which is characteristic of grasses and some legumes, such as vetches and the garden peas. Hypogeal emergence of the growing point is accomplished by the elongation of internodes of the epicotyl.

Label the following parts in Fig. 3.

hypocotyl	cotyledons
cotyledonary node	primary root
epicotyl	secondary roots
unifoliolate leaf	nodules

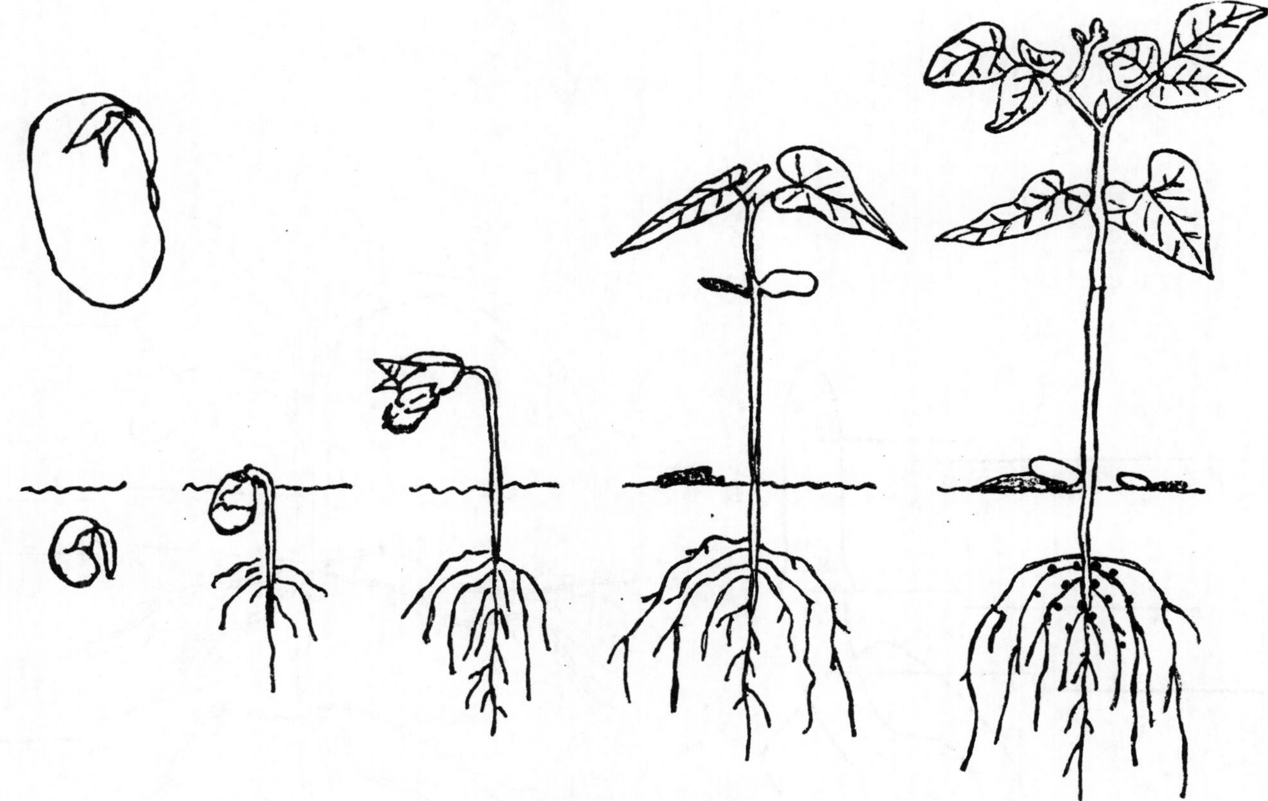

Fig. 3. Epigeal emergence of bean

2. Monocotyledon

In the process of germination and emergence of a monocot, the **radicle** and **coleorhiza** elongate and break through the pericarp first. The radicle then breaks through the coleorhiza and grows downward, anchoring the seedling in the soil. Shortly after emergence of the radicle, the **coleoptile** emerges from the pericarp and grows upward. Both the **coleoptile** and the **first internode**, located above the **cotyledonary node**, elongate until the coleoptile is above the soil surface. After the coleoptile emerges, light causes the coleoptile and first internode to stop elongating. The **epicotyl** continues to elongate, breaking through the tip of the coleoptile and exposing new leaves as it develops.

Compare the various stages of development of the corn plants available with the diagrams in Fig. 4 and then label the drawings with the seedling parts given.

coleoptile	coleoptilar node
crown roots	first internode of stem
cotyledonary node	radicle or primary root
	seminal roots

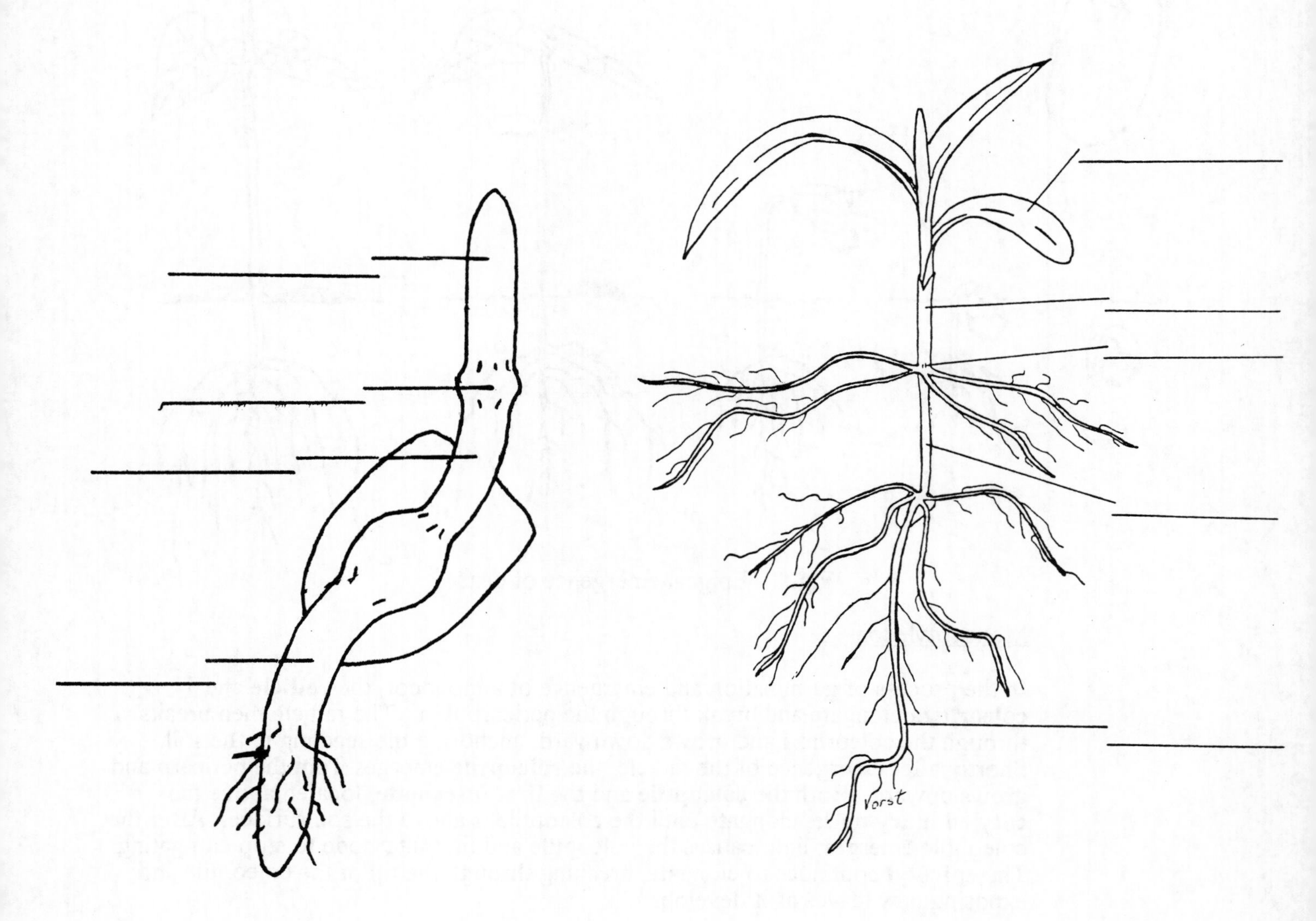

Fig. 4. Corn seedlings

NATIONAL CORN HANDBOOK

GROWTH & DEVELOPMENT NCH-3

Corn Plant Growth—from Seed to Seedling

G. O. Benson, Iowa State University and H. F. Reetz, Jr., Potash & Phosphate Institute, IL

Reviewers

D. R. Hicks, University of Minnesota
K. R. Polizotto, Lilly Research Labs, IN
E. D. Nafziger, University of Illinois
S. Ritchie, University of Wisconsin

This publication discusses growth of the corn plant from seed through germination and emergence to the established seedling stage. Emphasis is given to describing the major structures present at each stage, but with mention also made of how environmental factors influence the plant's physiology and, in turn, structural development at that stage.

Understanding how the corn seed develops into a seedling has important implications both for making wise production management decisions and for early detection of field problems. Unless you know what *should* be happening in the early life of the corn plant, you are apt to overlook what *shouldn't* be happening—until too late!

The Kernel

The main parts of the corn kernel are the pericarp (seed coat), endosperm and embryo (germ). These structures are identified in Figure 1.

The *pericarp* seems to protect the enclosed endosperm and embryo. If the pericarp is broken, fungi or bacteria can enter and reduce both germination and seedling vigor. At the end of the kernel, the pericarp merges into the pedicel tissues. If the pedicel is broken off, the thin brown or black abscission layer beneath it is exposed.

The *endosperm* occupies the bulk of the kernel. It consists mainly of starch and is the major energy source for the germinating seed and emerging seedling. The very outer cell layer of the endosperm is the aleurone, which contains enzymes important to the initiation of germination.

The *embryo* contains at its axis the initial parts of the young corn plant, including the partially developed coleoptile, plumule (containing 5-6 embryonic leaves and the apical meristem leaves), scutellar node, radicle and coleorhiza (the radicle's protective sheath). Attached to the scutellar node is the scutellum (cotyledon). Its primary function is secretion of an enzyme that helps break down the endosperm for use in the germination process. The scutellum also provides food substances directly to the embryonic axis in the early stages of germination.

Assuming the kernel has been handled correctly since maturity, it is alive and ready to germinate. It contains between 11 and 12 percent moisture and has been treated with a fungicide.

Germination

The first step in germination is the absorption of water by the corn kernel. The dormant kernel starts to swell, and the chemical changes needed for the growth process get under way. The absorbed water activates enzymes in the scutellum and aleurone layer that break down the food reserves needed to initiate growth in the embryo axis region. Endosperm starches are converted to sugars, which are readily available to the embryonic plant.

During germination, the radicle elongates and is the first structure to break through the seed coat; next comes the coleoptile, which surrounds the plumule; and then the 2-5 seminal roots emerge (Figure 2). Under optimum temperatures, this stage may be reached in 4-5 days; however, under cool soil conditions, the process will take much longer.

Initially, the radicle grows in whatever direction the kernel tip is pointed (except up). The other roots grow at varying angles from the horizontal, depending on temperature and moisture conditions. This initial root system serves to help anchor, provide moisture and eventually absorb nutrients for the young plant.

PURDUE UNIVERSITY COOPERATIVE EXTENSION SERVICE • WEST LAFAYETTE, INDIANA

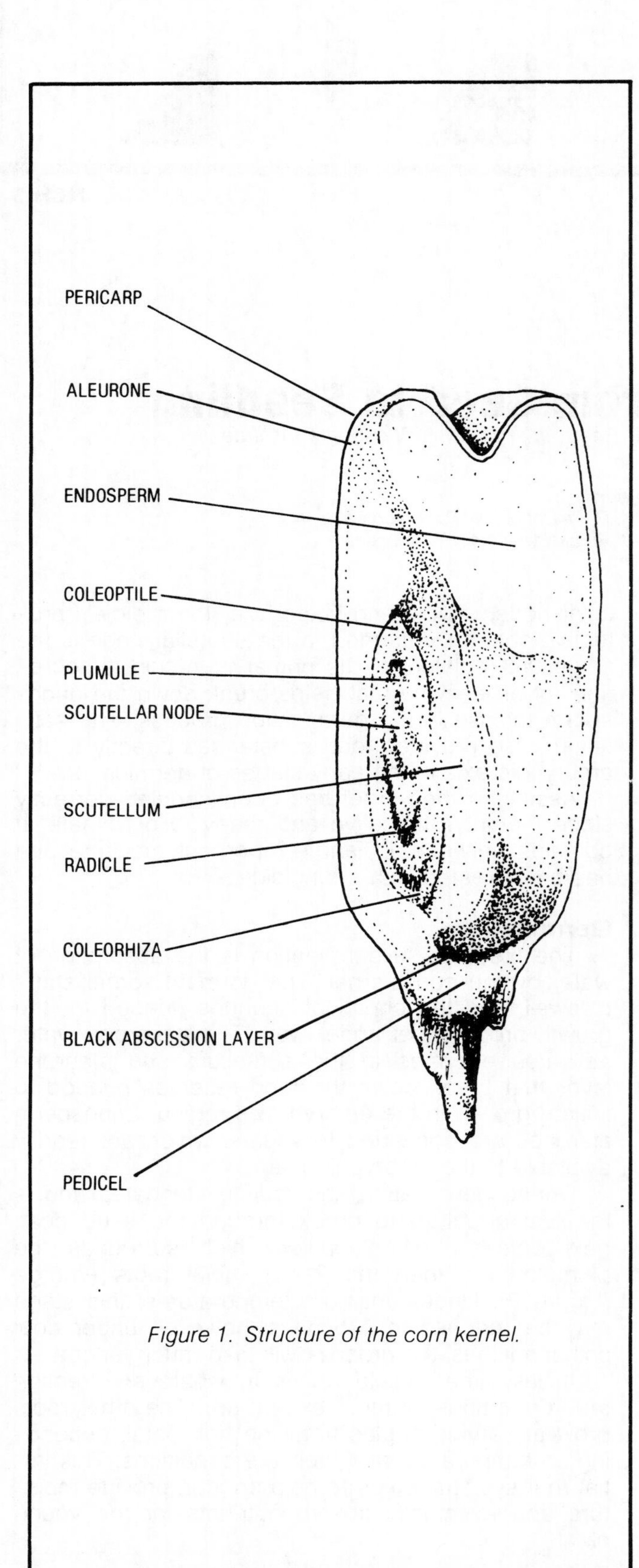

Figure 1. Structure of the corn kernel.

Emergence

Emergence is normally accomplished by a combination of coleoptile growth and mesocotyl (first internode) elongation. The mesocotyl is the structure between the scutellar and coleoptilar nodes or "crown" (Figure 3). The deeper the kernel is planted, the greater the length of the mesocotyl.

The crown area, from which the nodal or permanent roots eventually develop, will grow from 1 to 1½ inches below the soil surface with little variation due to planting depth. Thus, if a kernel is planted 1 inch deep, emergence will be accomplished entirely by growth of the coleoptile. If planting depth is greater than that, the mesocotyl becomes responsible for elevating the coleoptile close enough to the soil surface to make emergence possible.

The coleoptile pushes up through the soil until it reaches light, at which point the tip opens and the first true leaves of the plumule emerge. If the kernel is planted deeper than the mesocotyl can elongate or if the coleoptile tip ruptures below ground, emergence will not occur.

Time from planting to emergence is influenced most by temperature. Under ideal conditions, emergence can take less than a week, but typically 1-2 weeks are required. Under very cool conditions, emergence may not occur for 2-3 weeks. Soil moisture, soil compaction, crusting, planting depth, etc., also influence time to emergence.

The Seedling

The first leaf blade to emerge has a rounded tip; all others are pointed. (This has importance for leaf number identification later on.) Visible on a newly emerged seedling is all of the first leaf blade, most of the second, and parts of the third and fourth (Figure 4). Remnants of the coleoptile are seen at the soil surface.

The radicle and seminal roots of the seed root system have been taking up water and nutrients from the soil. Leaves are now carrying on photosynthesis, and the young plant is pretty much independent of the seed as a food source.

At or shortly after emergence, the nodal (crown) root system begins to develop, and soon becomes dominant as the role of the seed roots system diminishes. Some authors refer to seminal and nodal roots as adventitious roots. Establishment of the nodal roots is very important in order to provide the water and nutrients the corn plant needs for normal growth. Inability of a plant to "grow away" from many seedling problems often is associated with poor establishment of this root system.

The structure that initiates new leaves is the apical meristem or "growing point." At seedling stage, it is below the soil surface because the internodes have not yet begun to elongate. Thus, if the aboveground leaves are destroyed, additional leaves will still emerge unless the growing point has become diseased, frozen or destroyed in some other manner.

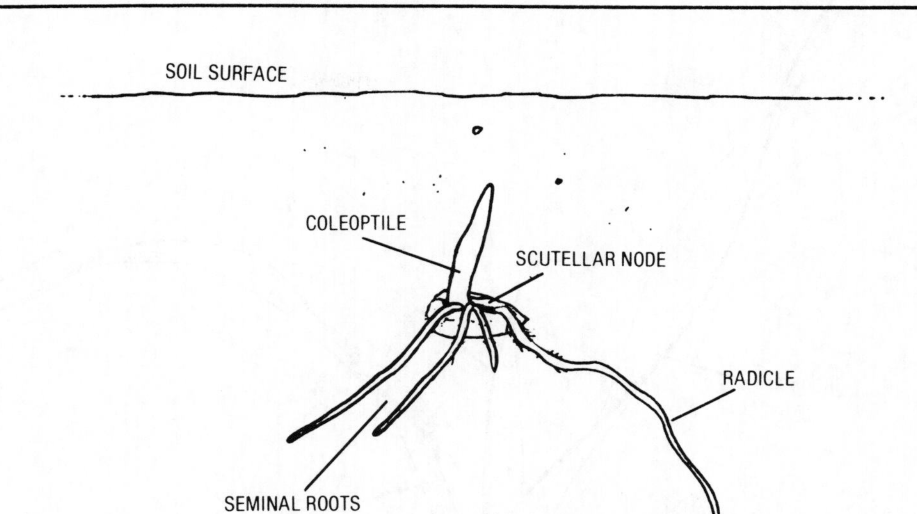

Figure 2. Germination of the corn seed.

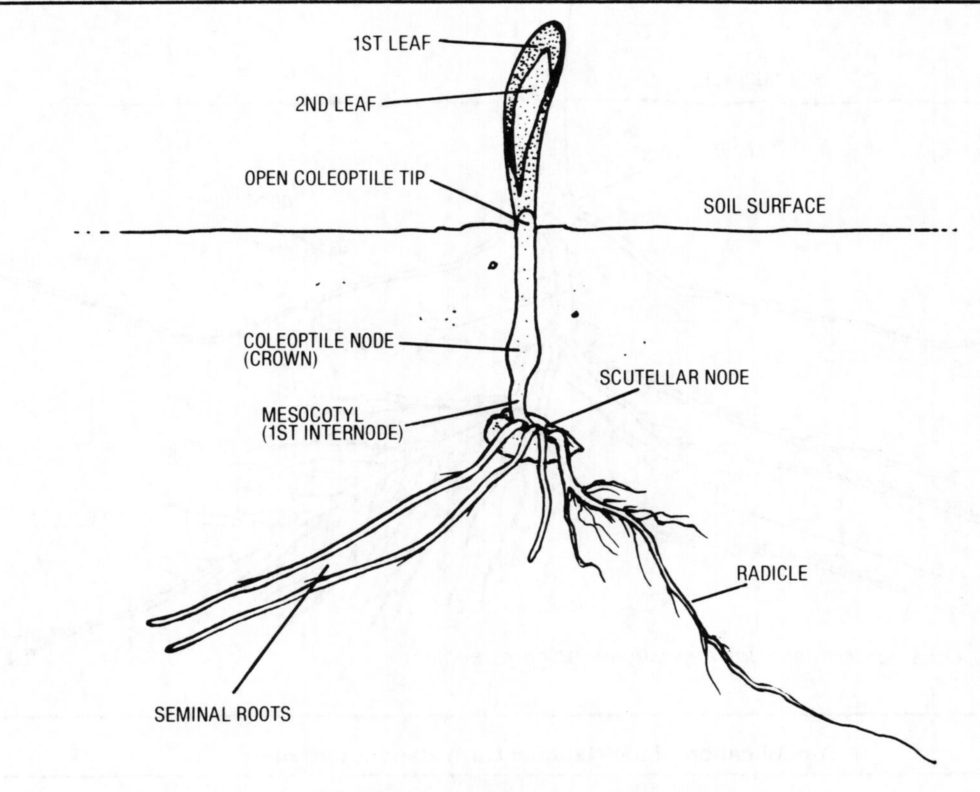

Figure 3. Emergence of the corn seedling.

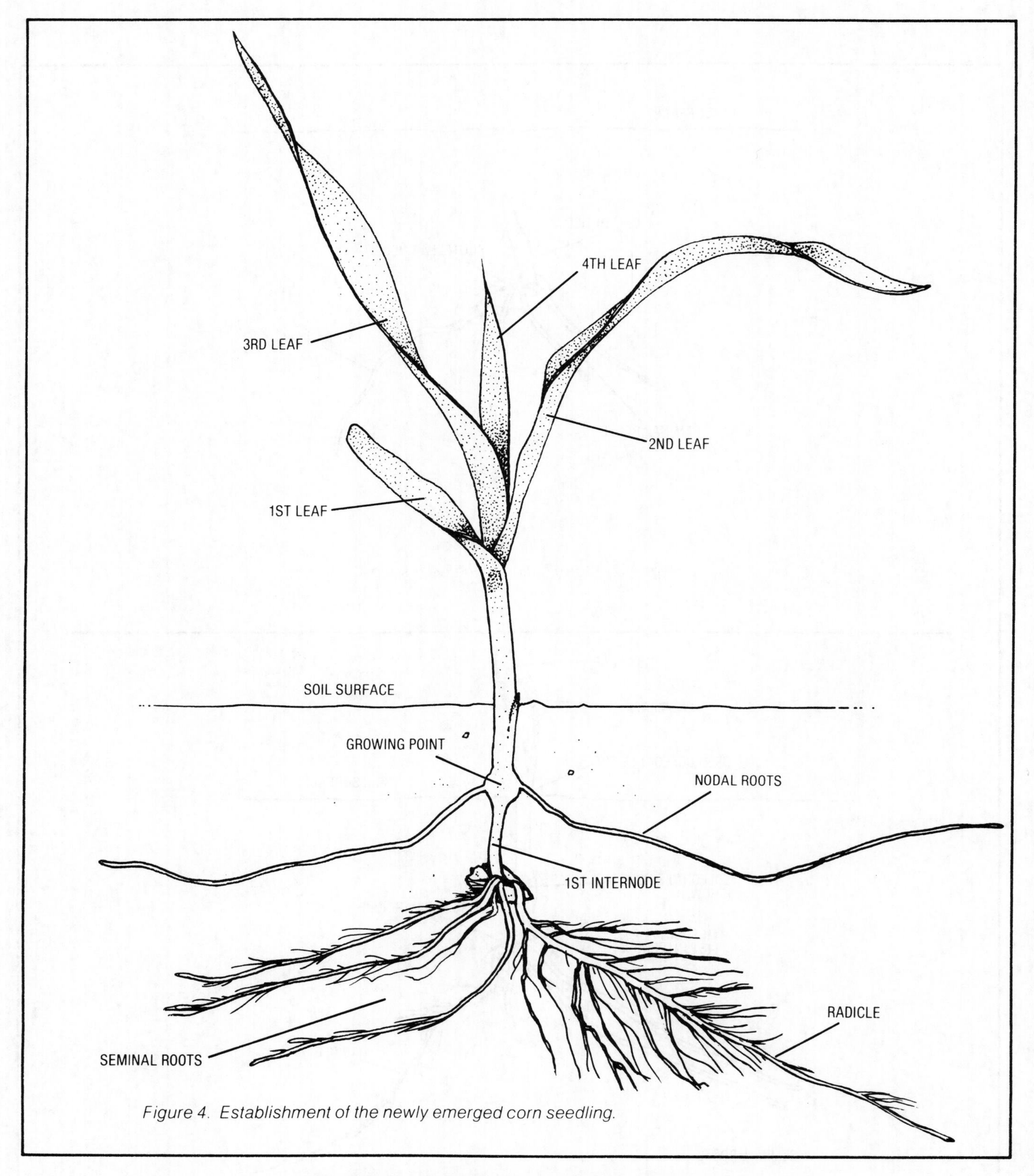

Figure 4. Establishment of the newly emerged corn seedling.

A publication of the National Corn Handbook Project

NEW 4/84

Cooperative Extension work in Agriculture and Home Economics, state of Indiana, Purdue University, and U.S. Department of Agriculture cooperating; H. A. Wadsworth, Director, West Lafayette, IN. Issued in furtherance of the acts of May 8 and June 30, 1914.
The Cooperative Extension Service of Purdue University is an affirmative action/equal opportunity institution.

D. Factors Influencing Planting Depth and Seedling Emergence

In hypogeal emergence there are nodes located below the soil surface while in epigeal emergence all nodes are pulled above the soil surface. Therefore, plants with epigeal emergence are more severely affected by an early spring "killing" frost since it has no axillary buds or nodes below the soil surface. In hypogeal emergence, nodes below the soil surface are protected from the frost and contain buds capable of initiating regrowth.

Soil crusting has a more severe effect on crops with epigeal emergence, such as soybeans, than on crops with hypogeal emergence. The breaking through the crust by the hypocotyl arch and pulling the cotyledons through the crust is difficult. Using a rotary hoe is a common tillage practice to break soil crusts. Hypogeal emerging crops are less susceptible to problems with soil crusting since the small epicotyl only has to push through the crust.

Examine the corn and soybeans seedlings which have been planted at different depths. Note that more deeply planted corn has a longer first internode and takes longer to emerge. Shallowly planted corn emerges more rapidly, but in sandy soils may not have adequate moisture for establishment. Note that corn will emerge at a deeper depth than will soybeans.

The principle difference in the emergence of grasses is the amount of first internode elongation. The first internode of corn and sorghum elongates until the coleoptilar node is just below the soil surface.

With wheat and barley the first internode does not elongate and the coleoptilar node is located near the kernel. Since the first internode of corn and sorghum elongates they can be planted deeper than crops with no first internode elongation such as wheat or barley.

DEFINITIONS

1. Adventitious roots: Those arising from tissue other than root tissue.
2. Aleurone: The outermost layer of endosperm, usually colored, and higher in protein than the rest of the endosperm.
3. Caryopsis: A dry, one-seeded, indehiscent fruit with the pericarp adhering tightly to the seed coat.
4. Cereal: A grass grown for its edible "seed", or grain.
5. Coleoptile: Sheath (second leaf) which surrounds the leaf primordia in grasses.
6. Coleorhiza: Sheath which surrounds the radicle in grasses.
7. Cotyledon: A first leaf (seed leaf) of an embryo, already formed and present in the seed.
8. Crown roots: Below-ground adventitious roots forming from or above the first internode.
9. Dicotyledon: Having two seed-leaves.
10. Embryo: A rudimentary plant, consisting of a seedling axis and one or more cotyledons.
11. Endosperm: Tissue in the grass grain in which food is stored.
12. Epicotyl: The portion of the seedling axis above the cotyledonary node, sometimes referred to as the plumule.
13. Epigeal emergence: A type of emergence where the hypocotyl elongates pulling the cotyledons above the surface of the soil.
14. Fruit: A mature ovary and its associated parts.
15. Hilum: Point on the seed where the seed was attached to the pod.
16. Hypocotyl: Zone between the cotyledonary node and just above the primary root -- grows to pull the bean cotyledons above the surface of the soil.
17. Hypogeal emergence: A type of emergence where the hypocotyl does not elongate and the cotyledons remain below the surface of the soil.
18. Monocotyledon: Having but one seed-leaf.
19. Pericarp: The mature ovary wall, plus remnants of the seedcoat (testa).
20. Radicle: Embryonic root which forms the primary root upon germination.
21. Root hair: A single-celled protrusion of a epidermal cell of a young root.
22. Seed: A mature ovule.
23. Seminal root system: All roots originating from the caryopsis.

QUESTIONS

1. What is the difference between the outer covering of the soybean seed and the corn kernel?

2. In Indiana, corn is usually planted earlier in the spring than soybeans. What seedling characteristics make corn better adapted to early planting than soybeans?

3. What parts of the legume seed are considered part of the embryo?

4. What parts of the grass caryopsis are considered part of the embryo?

5. Describe the differences between corn seedlings grown from seed planted 2.5 cm and 5 cm deep in the soil?

6. Explain how planting depth is influenced by

   a) Method of emergence

   b) Environmental conditions

7. What morphological characteristics are indications of good quality seed?

8. What are the functions of the cotyledon in:

   a. the corn kernel

   b. the soybean seed

9. What are the major differences between corn seedlings grown from seed planted one inch and three inches deep in the soil?

10. What are the three major components of the corn kernel?

11. What is the difference between the coleoptile and the coleorhiza?

12. What causes the first internode to stop elongating?

13. What is the place of food storage in the grass caryopsis?

14. What is the outer covering of the grass caryopsis?

15. What is the classification given to the members of the grass family which have only one seed-leaf per fruit?

17. What effect does the depth of planting have on the position of the origin of the seminal root system with respect to the soil surface for:

    a. corn

    b. wheat

18. What effect does the depth of planting have on the position of the origin of the crown root system for:

    a. corn

    b. wheat

HOW ROOT SYSTEMS AFFECT CROP PRODUCTION

Root systems of crop plants are very diverse, and are just as important in determining overall crop performance as are the above ground plant parts. In this exercise, root functions, structures, and types will be examined. The objective for each part is stated on the station guide beginning on page 31. The information preceding the guide will help you learn the objectives.

A. Functions of Plant Roots

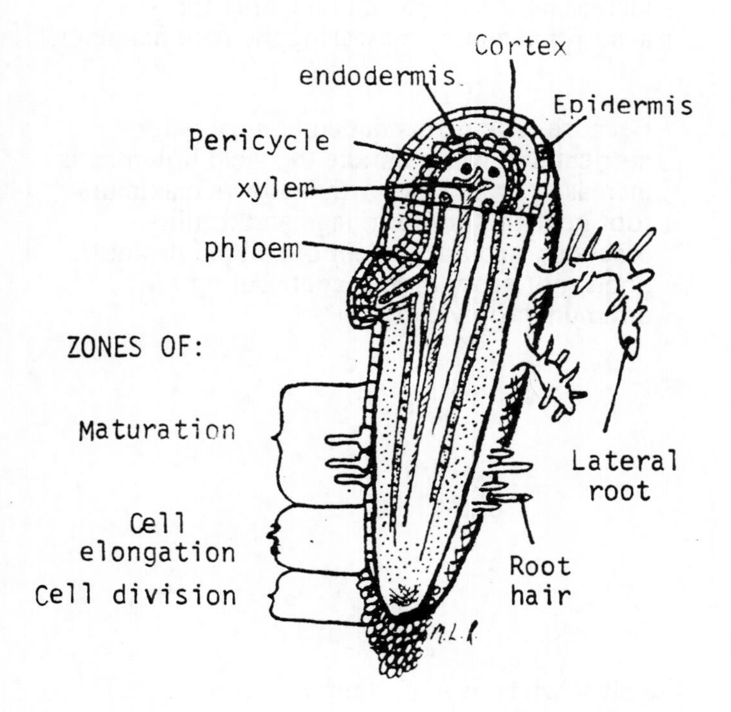

Fig. 1. Parts of a plant root

1. Water and Nutrient Uptake

The first step in nutrient uptake is establishing contact between the nutrient and the root. Roots grow into the soil, and through proliferation of secondary roots and root hairs, they intercept nutrients. This root inteception accounts for about 1% of the total uptake. Transpiration of water in the leaves creates a shortage of water in plant cells which causes a flow of soil water containing nutrients to the roots. The mass flow of water is important in the transport of nitrogen, calcium and magnesium to the root. Many other nutrients move to the root by diffusion, the movement from higher concentrations to lower concentrations.

At the root surface most nutrients are actively taken up across the epidermal cell membrane (plasma lemma) into the epidermal cell cytoplasm. They are then transported through the cytoplasm of the cortex cells, the endodermis and finally the xylem (Fig. 2). All nutrients except calcium, and water enter the plant through this path.

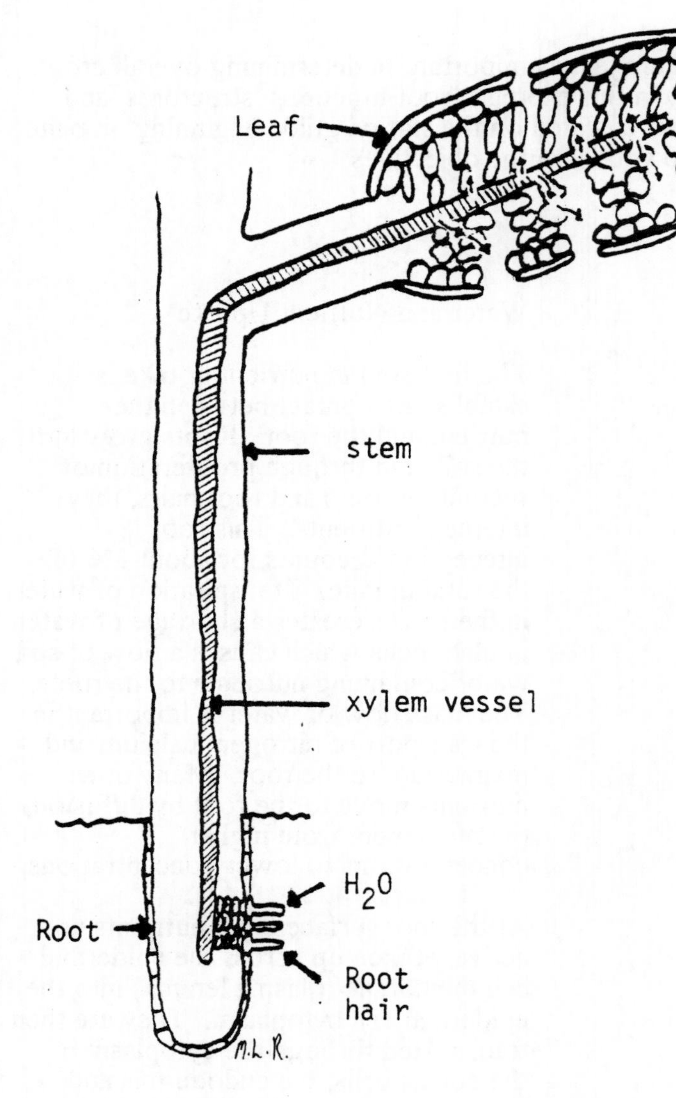

Fig. 2. Soil water movement into the plant.

The palisade and spongy cells absorb H_2O which pulls the water column upward.

Water and calcium pass through the epidermal cell walls of roots into the intercellular spaces and move through these spaces until they reach the endodermis. Endodermal cells are banded by casparian strips which blocks further movement through the intercellular spaces. Water and calcium are taken up into endodermal cells and then transported to the xylem.

Root hairs appear to improve phosphorus uptake from infertile soils by increasing the cell membrane area for absorption and by increasing the root diameter.

Because crop plants depend on roots for nutrient and water uptake the yield potential is increased by any factor that allows maximum root growth. Proper tillage and fertility practices, and adjustment of soil pH promote good root growth, thus contributing to optimum crop yields.

2. Anchorage

Roots provide physical support for the plant which is important in preventing lodging. Poor root growth, root rots and insect damage to roots increase susceptability to lodging.

3. Storage of carbohydrates

Some plants store considerable amounts of carbohydrates in roots. These are usually stored in the cortex region, although sometimes in the pith. This provides an important source of food for a large part of the world's population. Cassava (Figure 3), sweet potatoes and sugar beets are the most important root crops in the world. (Table 1)

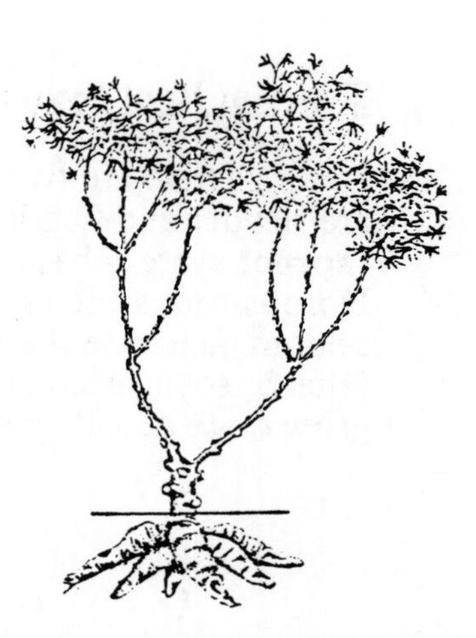

Fig. 3. Cassava is an important root crop in the lowland tropics.

Table 1. Estimated annual world production of the major root crops.

Crop	Annual World Production (million tons)
Sugar beets	200
Sweet potatoes and yams	135
Cassava	100

The world production (tons/year) of root crops (including rhizome and tuber crops) is nearly as great as that of grain crops. Because root crops usually contain large quantities of starch and sugar, and little protein and oil, they are viewed as energy sources.

Carbohydrate storage in roots is also important in the winter survival of forage legumes (especially alfalfa). Stored carbohydrates are used for regrowth after cutting until enough leaf area is present for the plant to again maintain a high photosynthetic rate. The plant then rebuilds the carbohydrate reserves in the root and crown. When the last cutting is late, within a month before the first killing fall frost, the plants start to regrow using up the carbohydrate reserves. If the frost occurs before the carbohydrate reserves are built up in the root and crown again, the plant is vulnerable to winter killing. Farmers should harvest their last hay crop for the season 30 days before the first killing frost to avoid winter kill. Figure 4 depicts what happens to root carbohydrate levels during the growing season for alfalfa.

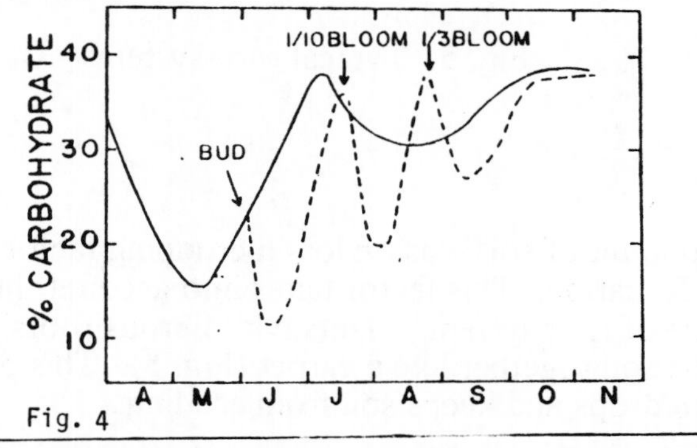

—Seasoned trends of total available carbohydrates in roots of Vernal alfalfa under a 3 cutting scheme (dashed line) and without cutting (solid line).

Source: Smith, D. R. Crop Sci. 2:75, 1962

B. Types of Root Systems

There are two types of root systems. The **fibrous root system** is one where all roots are slender and fiberlike and no one root is more prominent than any of the others. The **taproot system** has one main root that grows downward, from which branch roots arise. Fibrous root systems are common in monocotyledons (corn, wheat, and sorghum) and seldom penetrate the soil more than 2 meters. Taproot systems of dicotyledonous plants (alfalfa, soybeans, clovers) often penetrate the soil to depths of 5 meters. While roots may grow quite deeply, most root activity occurs in the upper 15 cm of soil.

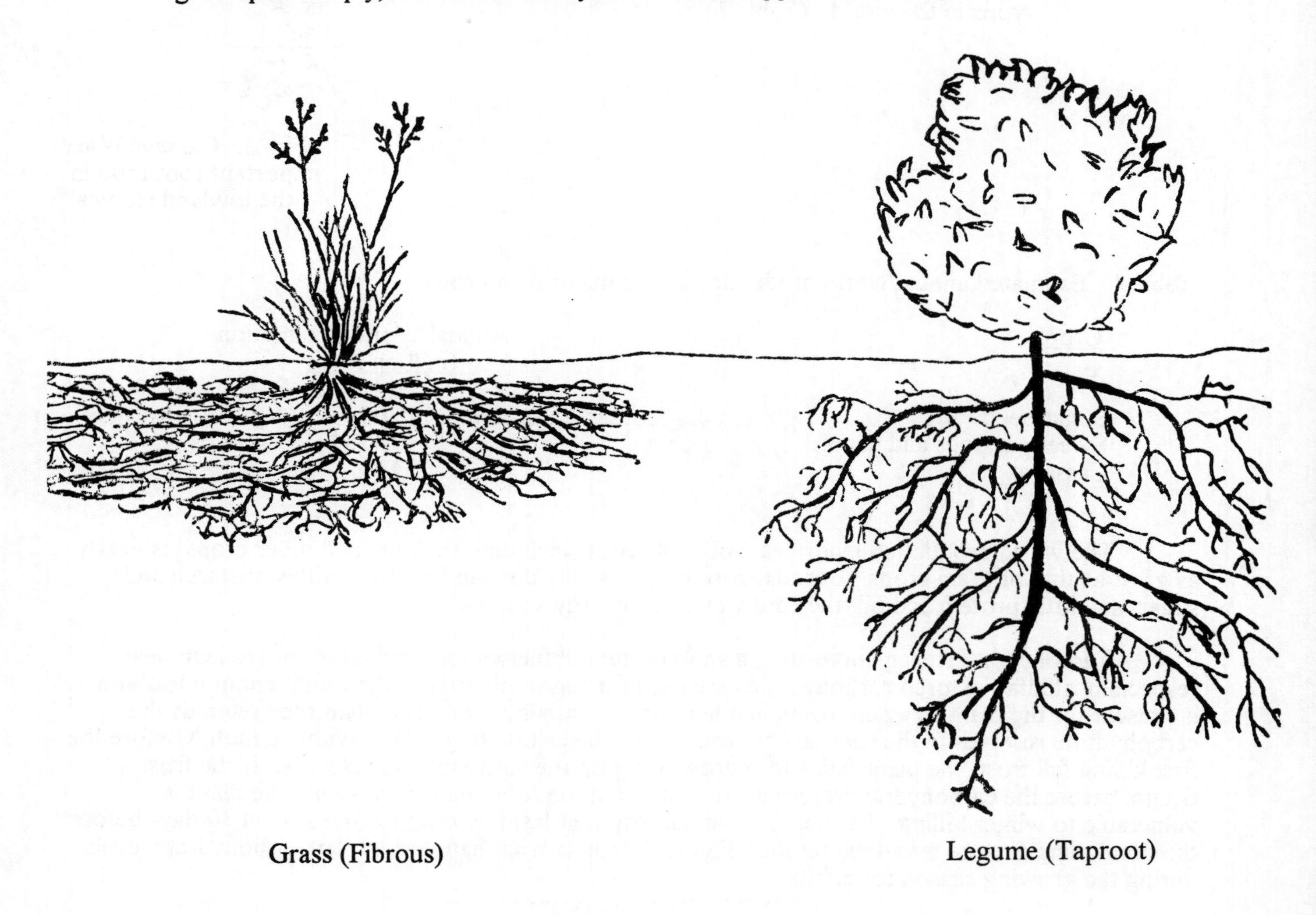

Fig. 5. Typical root systems.

1. Erosion Control

In determining the amount of soil erosion loss a cropping factor (C) is in included in the Universal Soil Loss Equation. This factor takes into account the canopy cover provided by the plant and the type of root system. Plants with fibrous roots grow so thickly in the top of soil that they bind the soil together like a carpet (Fig. 5). This protects the soil from the beating action of rain drops and keeps soil from eroding.

Another factor to consider is the life cycle of the plant (longevity). An annual crop is less effective in preventing soil loss than is a perennial crop (Table 2).

Table 2. Estimated annual C values and erosion loss for crops grown in Central Indiana on Russel silt loam, 6% slope 300 feet long, and conventional tillage without conservation practices.

Crop	Longevity	Root System	Estimated C Value	Estimated Erosion Loss (t/acre)
Corn	Annual	Fibrous	.38	27.8
Soybeans	Annual	Tap	.50	36.6
Grass	Perennial	Fibrous	.004	0.3
Alfalfa	Perennial	Tap	.02	1.5

2. Heaving

Plants with a taproot system are subject to heaving, a condition where alternate freezing and thawing forces the crown of the taproot out of the soil. This may cause root breakage, but more serious damage occurs from dessication and freezing of the exposed root. Heaving occurs chiefly during the early spring on organic soils or heavy soils which have a high water holding capacity, or on poorly drained soils.

3. Soil Compaction Effects on Root Growth

Soil conditions are constantly changing, and our modern management practices contribute to these changes. In recent years in the midwest it has become more common to find fields with slow plant emergence, variable plant sizes, shallow and malformed root systems. One of the most important causes of this is soil compaction. When a soil becomes compacted the soil pore space and pore size is decreased. Resistance to root growth then increases, (Figure 6), causing malformed roots and decreased depth of rooting (Figure 7). As the pore space of the soil is reduced, soil drainage becomes slower and oxygen level of the soil decreases. Both poor drainage and lack of oxygen can lead to reduced root growth. Soil pores must be larger than the root tip for roots to penetrate into the soil, and in compacted soils a lack of adequate pore size may greatly reduce root growth.

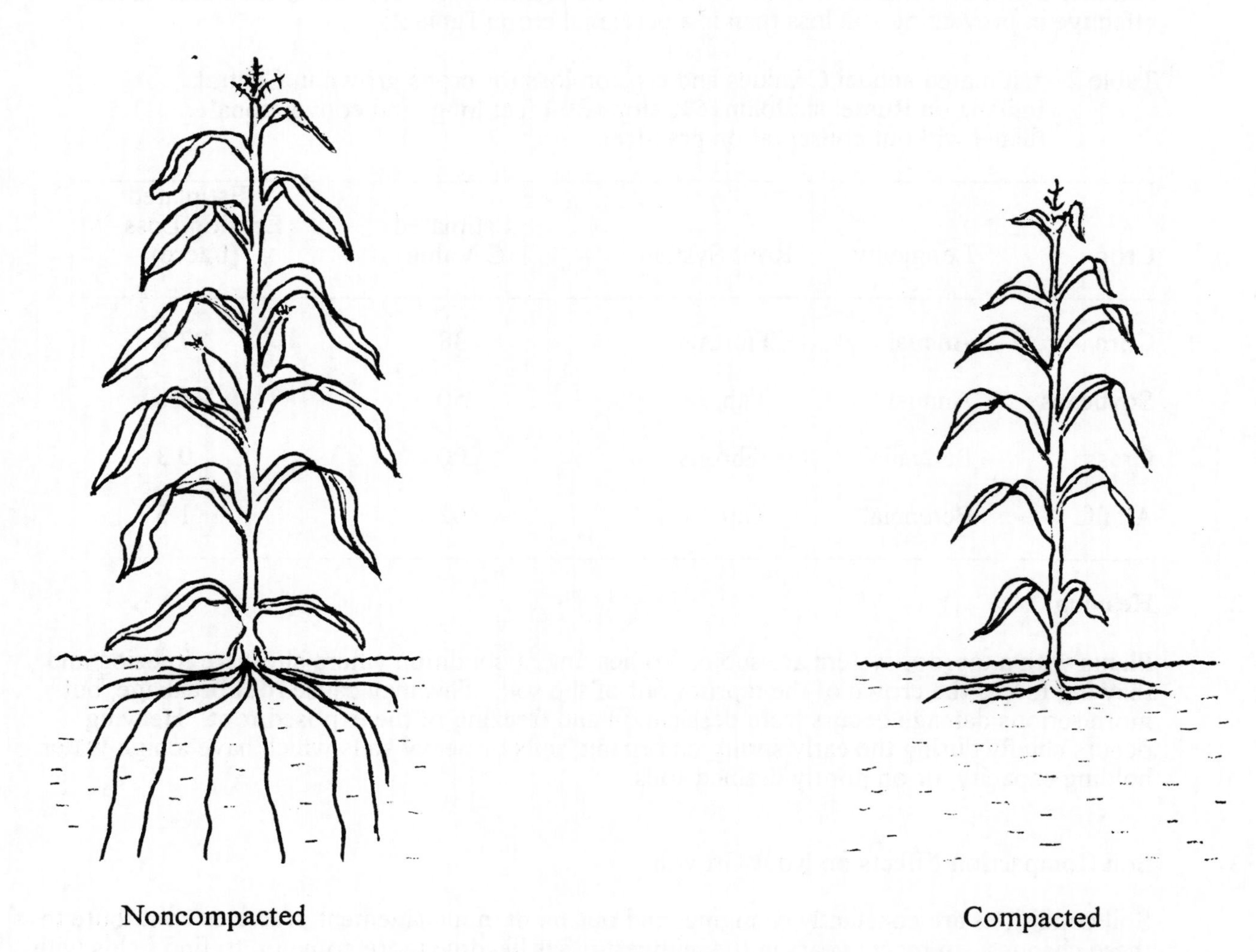

Fig. 6. Depth of rooting is influenced by soil compaction

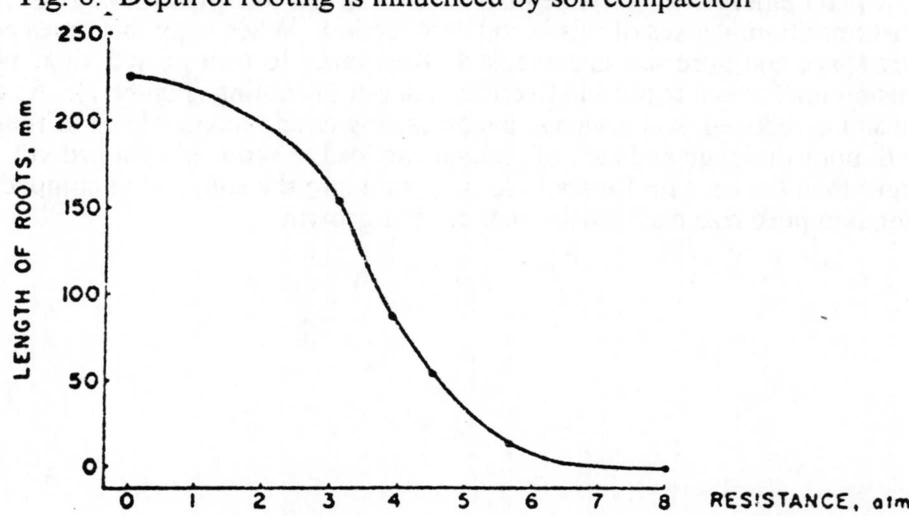

Fig. 7. Effect of soil resistance due to compaction on corn root length (adapted from Barley, 1962).

Poor root growth caused by soil compaction leads to a low ability to extract water and nutrients from the soil as demonstrated by poor plant growth, nutrient deficiency symptoms and moisture stress. Compacted soils cause shallow root systems which leads to increased harvest losses due to lodging and can cause decreased yields of root crops due to difficulty required for root expansion to occur.

C. Nitrogen Fixation and Its Management

One major difference between grasses and legumes is that legumes have the ability to form symbiotic relationship with nitrogen-fixing bacteria (**Rhizobium** spp). The bacteria invade the root and form whitish bodies on the roots called nodules. In these nodules the bacteria live on food supplied by the plant, and relationship, it is an example of **symbiosis**.

Farmers have used legumes to their advantage for many years in crop rotation to build up soil fertility. Farmers also have found that mixing legumes with grasses in pastures has reduced nitrogen fertilizer costs. A healthy legume plant, the proper **Rhizobium** species, good soil drainage, proper soil pH, and adequate phosphorus and molybdenum fertilization leads to optimum N fixation.

Definitions

1. Casparian strip - Strip around root endodermal cells that blocks water movement from the cortex to the xylem.
2. Cortex - The root region between the epidermis and endodermis whose primary function is the storage of carbohydrates.
3. Diffusion - The movement of a substance from an area of high concentration to low concentration.
4. Endodermis - The single layer of cells separating the root cortex from the stele.
5. Epidermis - The outer covering of the root.
6. Heaving - Process of winter killing of tap root crops in poorly drained soil.
7. Mass flow - Movement of nutrients in the soil to the root in water moving to the root.
8. Nodule - White growth on legume roots where nitrogen fixation occurs.
9. Rhizobium - Bacteria that fixes nitrogen in legumes.
10. Root hair - Single cell extension of root epidermal cells.
11. Root interception - Uptake of nutrients due to the roots growing to the nutrients in the soil.
13. Symbiosis - Two organisms living together and aiding each other.
14. Xylem - Vascular tissue where water and nutrient translocation from the root to the shoot occurs.

STATION GUIDE FOR HOW ROOT SYSTEMS AFFECT PRODUCTION

Station 1

Objective: Identify the major parts of the root and describe their functions. (See p. 25).

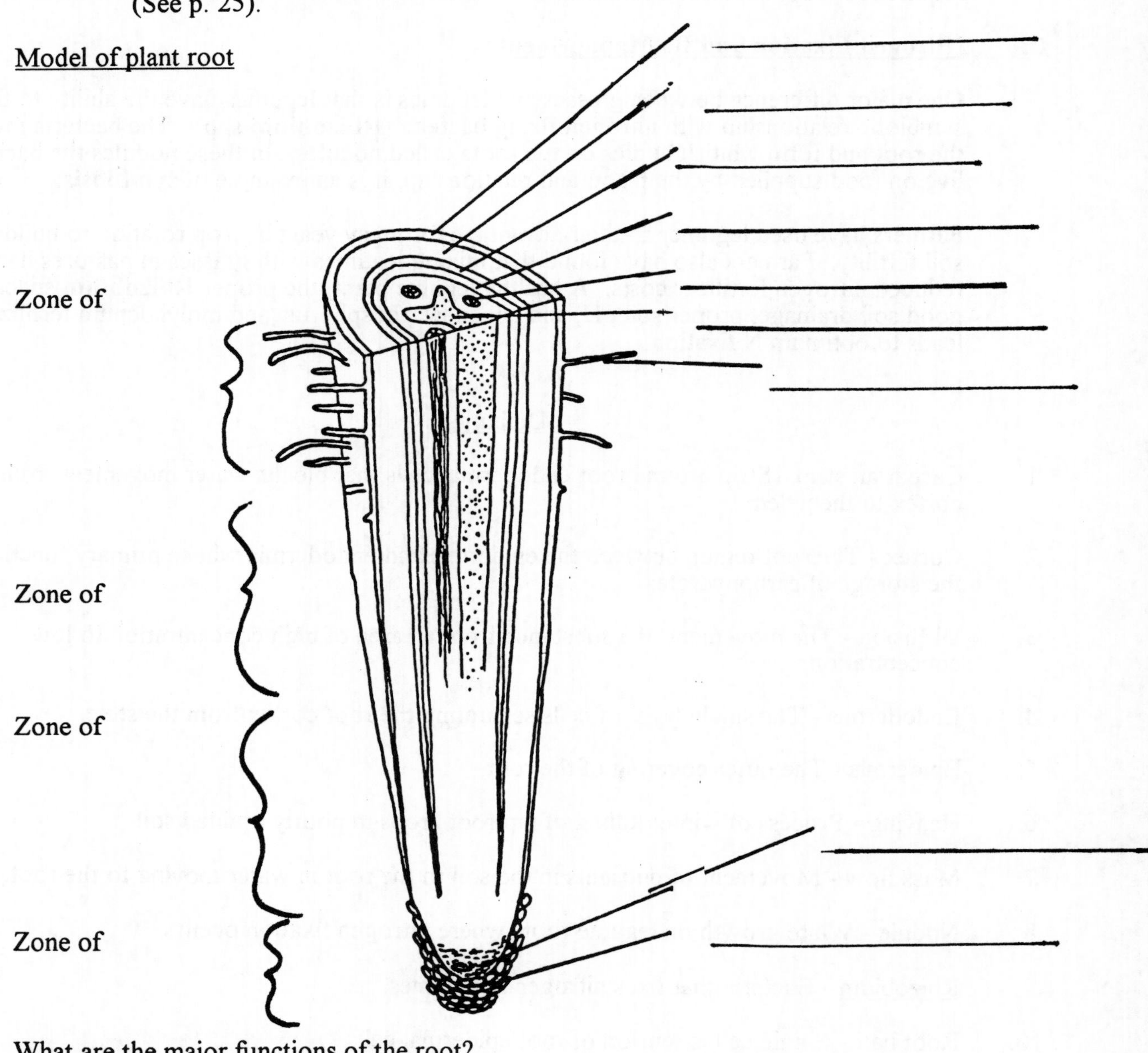

What are the major functions of the root?

1. ______________________________

2. ______________________________

3. ______________________________

Station 2

Objective: Describe the process that plant roots use to take up water and nutrients and translocate them to the shoot. (See p. 26).

Water and Nutrient Uptake Poster

What are the three ways water and nutrient uptake occur in the plant?

1.

2.

3.

Water and Nutrient Transport Poster

What are the two ways of water and nutrient transport?

1.

2.

What % of the total nutrient uptake occurs by root interception?

What root structure is important for phosphorus uptake?

What is the only barrier to water and calcium uptake by roots?

Model of plant cell

How do nutrients in the cytoplasm of one cell move to adjacent cells?

Station 3

Objective: Explain the importance of root carbohydrate reserves and how they affect winter survival of alfalfa (See p. 27).

Cassava Poster

What are three uses of cassava?

1.

2.

3.

Why are root crops viewed as energy sources?

Slide Set on Alfalfa Cutting Management

What three factors does alfalfa cutting management affect?

1.

2.

3.

What are stored carbohydrates used for:

1.

2.

3.

Does the level of carbohydrates drop after each hay cutting?

If so, explain why!

Why is it important to take your last harvest in the fall no later than four weeks before the first hard freeze?

Station 4

Objective: Explain how root systems differ and how these differences affect soil stability and erosion, and winter heaving. (pp. 28).

What type of root system does a monocot have? A dicot?

Soil Erosion Poster

Which type of root system is most effective in maintaining soil stability and thus decreasing erosion?

How do perennial and annual crops differ in ability to prevent soil erosion?

Slides on Heaving

What environmental condition causes "heaving"?

What type of root system is most susceptible to heaving? Why?

What are the harmful effects of root heaving?

What cultural practices can be used to reduce winter heaving?

Station 5

Objective: Describe the effect of soil compaction on root growth, and management practices to reduce this. (pp. 29-30).

Root Growth in a Compacted Soil Display

Why are roots unable to penetrate a compacted layer of soil?

Can a root tip penetrate a tissue with small pores? How does this relate to soil compaction?

What are principal causes of soil compaction?

1.

2.

3.

What are some management practices which help to reduce soil compaction?

1.

2.

3.

Station 6

Objective: Explain the process of symbiotic nitrogen fixation and its management (p. 31).

Nitrogen fixation

What percent of the atmosphere is nitrogen?

Is this in a plant available form?

What do we mean when we say that nitrogen is "fixed"?

Define the symbiotic relationship between nitrogen fixing bacteria and plants.

In general what types of plants form these associations with bacteria?

Do all nitrogen fixing bacteria species function on an alfalfa root?

Are the nodules on the soybean root on display active? How can you tell?

How can one introduce bacteria to a field where legumes have never been grown?

STRUCTURE AND FUNCTION OF CROP STEMS AND LEAVES

Stems and leaves of agronomic crops are quite different in structure and function. These differences require the producer to use different management principles for different crops. This exercise examines the functions, internal and external structures, types, and uses of crop stems and leaves. The objective for each part is stated in the station guide. The information preceding the guide will help you prepare for the lab.

Stem Functions

The chief functions of stems are:

1. Conduction of materials
2. Production and support of leaves and reproductive organs
3. Storage sites of food and other substances

In addition, stems containing chlorophyll are capable of photosynthesis. There are various specialized aerial and subterranean stems, many of which serve as important means of vegetative propagation.

External Structures of Stems

Upon examining an actively growing stem, a variety of structures are apparent, the most common being **leaves** and **buds**. Buds, which are meristematic tissue, are located in the axis of leaves, generally in the upper angle between the leaf and stem. A bud located at the tip of the stem or twig is a **terminal** or **apical bud**; a bud located along the side of a stem is a **lateral** or **axillary bud**. Buds which develop in other places are **adventitious buds**. The point on a stem where the leaf is attached is a **node** and the stem area between two nodes is an **internode** (Figure 1).

The first stem of a seed plant develops from the epicotyl. The epicotyl is a cylindrical structure with a mass of **meristematic cells** at its **apex**. A stem with its leaves and buds is called a shoot. Most stems grow above the soil and are erect (aerial stems), however some grow under ground (rhizomes). An aerial stem can be prostrate as with burmudagrass, or climbing as with field bindweed.

Grasses and legumes differ considerably in methods of stem growth. Legume stems increase in length by continued production of new nodes and internodes initiated from the **terminal bud**. Therefore, a soybean seedling has fewer nodes and internodes than does a mature soybean plant. Grass stems elongate in a different manner. Located at the base of each internode in the young stem is an **intercalary meristem**. Therefore, a corn seedling has all of its nodes and internodes. It then increases in height as cells divide, elongate, and mature in the intercalary meristem of each internode (Fig. 2). Because of differences in growth habit, management of grasses and legumes differs considerably.

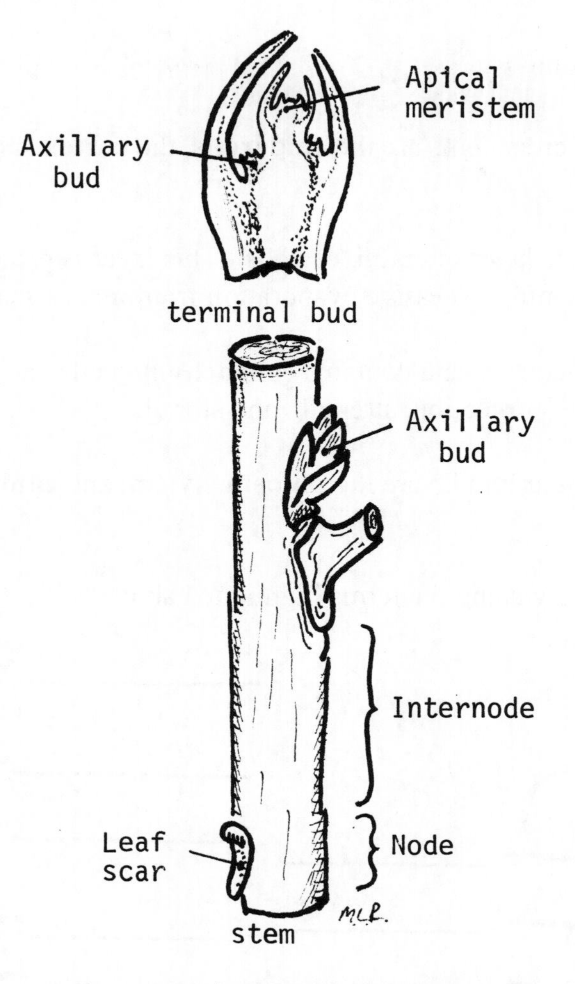

Fig. 1. Typical external stem structure

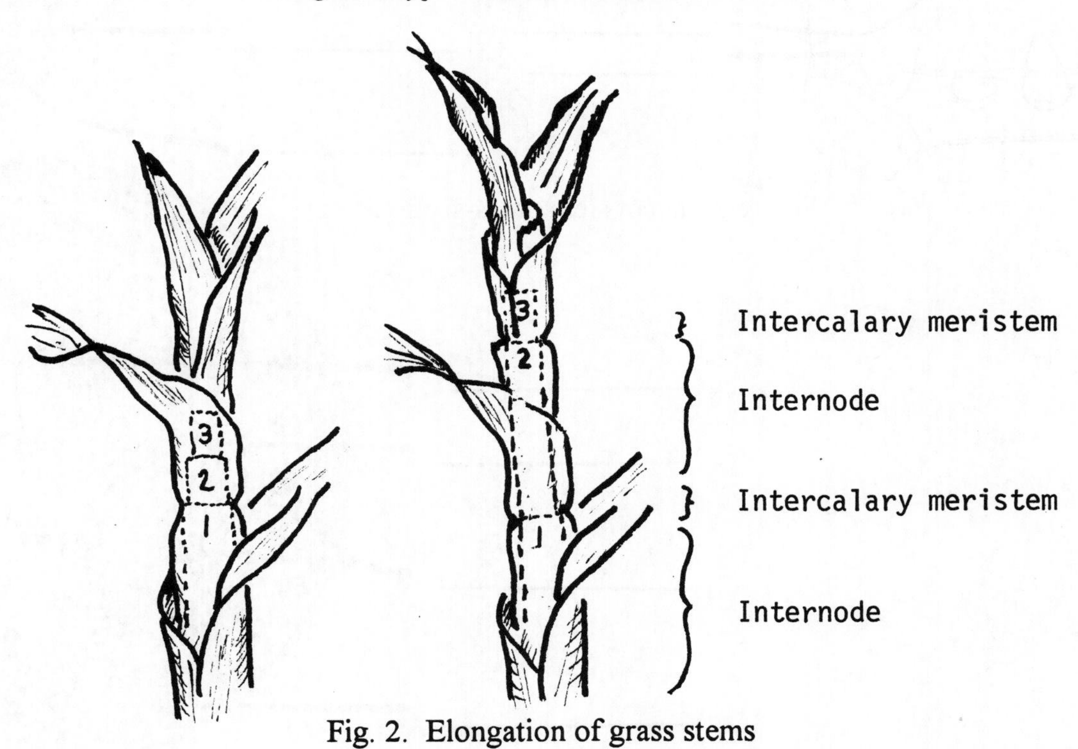

Fig. 2. Elongation of grass stems

Internal Structures of Stems

The primary tissues of crop stems are the **epidermis**, the **cortex** and the **vascular bundles**.

The epidermis is a single layer of cutinized cells. This layer is protective tissue, which is very important in preventing excessive evaporation from inner tissue.

The cortex varies in thickness, and contains **collenchyma** cells and **parenchyma** cells. The cortex is a region of protection, strength, and storage.

The tissues of the vascular bundle are the **phloem**, **xylem**, and **cambium** (found only in dicots).

Label the diagrams below using the terms highlighed above:

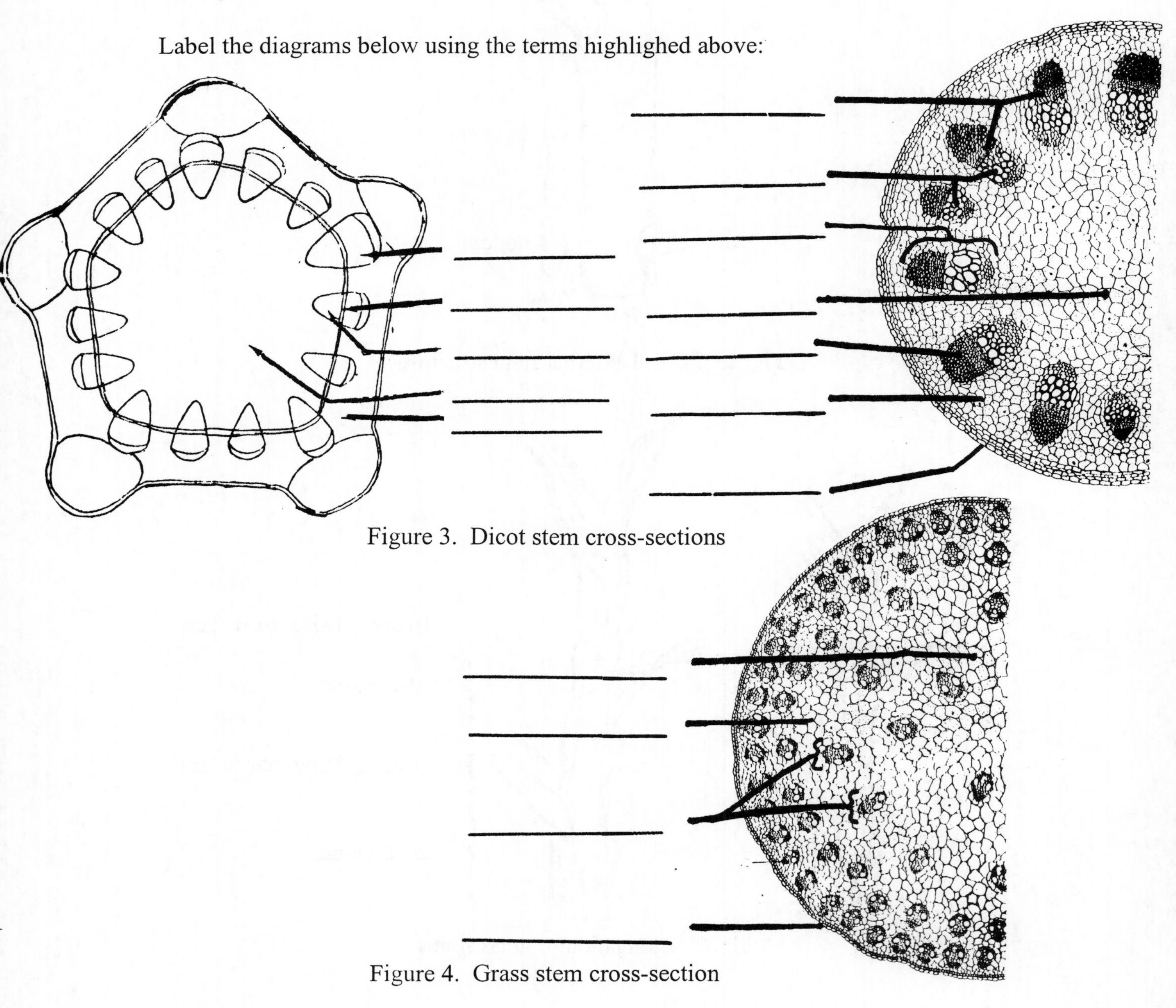

Figure 3. Dicot stem cross-sections

Figure 4. Grass stem cross-section

Types of Stems

The main aerial stem of some grass plants such as wheat and barley is a **culm**, whereas that of corn or sorghum is a **stalk**. Additional aerial stems may branch from the nodes at or below the soil surface. These stems are called **tillers**, **suckers** or **stools**. This phenomenon is a definite advantage in small grains and pasture grasses. However, in corn and sorghum these tillers may reduce grain yield and are not desirable.

Not all stems are erect. In some cases they creep on the surface of the soil, bear leaves, and often root at the nodes. Such above-ground modified stems are called **stolons**. Many grasses and some legumes have underground modified stems called **rhizomes**, which differ from roots in that the rhizomes have nodes, bear scale-like leaves, are horizontally growing, and are usually coarser than roots. A specialized, enlarged rhizome is a **tuber**. Annual plants seldom have either stolons or rhizomes.

In Fig. 5 label the following parts: rhizome, stolon, tiller, node, and internode.

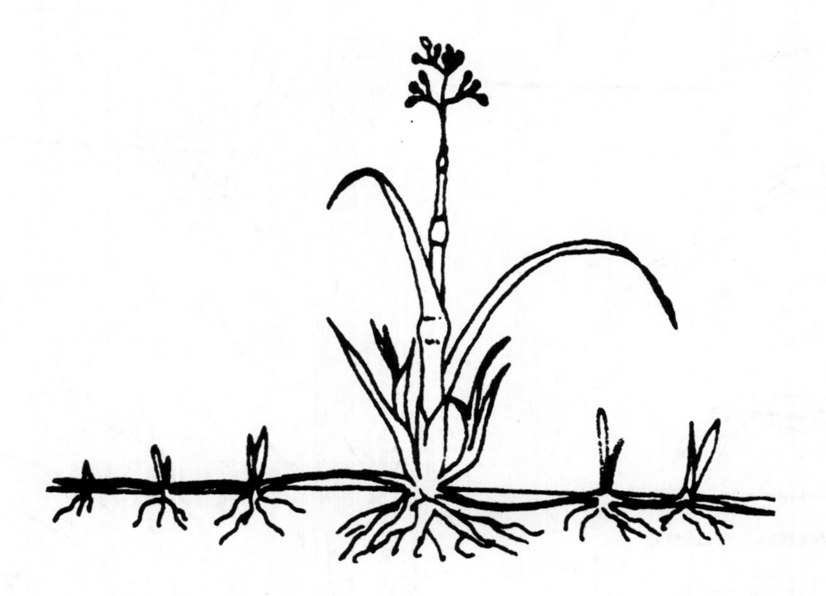

Figure 5. Grass plant

Among the common types of specialized aerial and subterranean stems are:

bulb (onion) enlarged and spherical bud. Has a small basal stem at its lower end
corm (timothy) enlarged and rounded stem base
culm (wheat) main aerial stem of some grass plants
rhizome (johnsongrass) horizontal stem that grows at or below soil surface
stalk (corn) main aerial stem of some grasses
stolon (strawberry) narrow, horizontal stem growing above soil surface
tiller (corn) additional aerial stem of some plants originating from a node near the soil surface--also called sucker or stool
tuber (potato) thickened undergound stem

Using the above information label the types of stems and their plant parts.

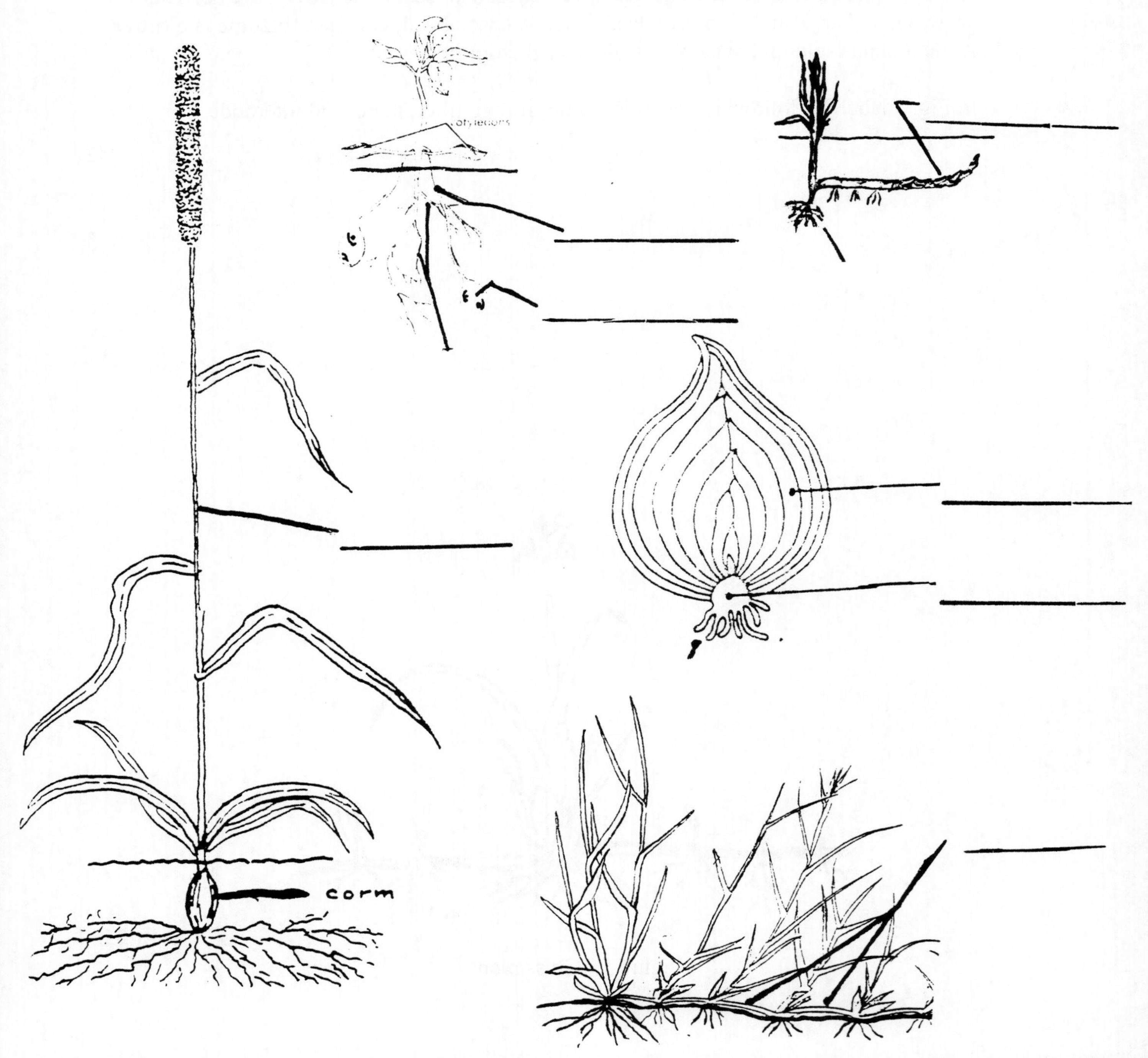

Functions of Leaves

Leaves are the major photosynthetic organs of plants. Leaf characteristics are often used to describe and identify crop plants. In addition, leaves may serve as a storage organ for photosynthates.

External Structures of Leaves

A typical legume leaf is composed of a **blade**, **petiole**, and **stipule**. The stipules are two small leaf-like outgrowths at the base of the petiole. The **vascular bundles** (phloem, xylem, and cambium) of the stem connect with the vascular bundles of a leaf or petiole at the node (Fig. 6A).

A typical grass leaf may contain four or five structures. The **blade** is the flattened part usually extending away from the stem. Nearly all the photosynthesis occurs in the blade. The **sheath** encloses the stem and attaches the blade to the node of the stem. On the outer side of the leaf where the blade and sheath joins is the **collar**. On the inner side of the leaf at this junction is the **ligule**, which is usually thin and extends upward from the sheath. Some grasses such as barley or quackgrass contain two small ear-like projections, one on either side of the sheath, called **auricles**. Both auricles and the ligule may be very useful in identification (Fig. 6B).

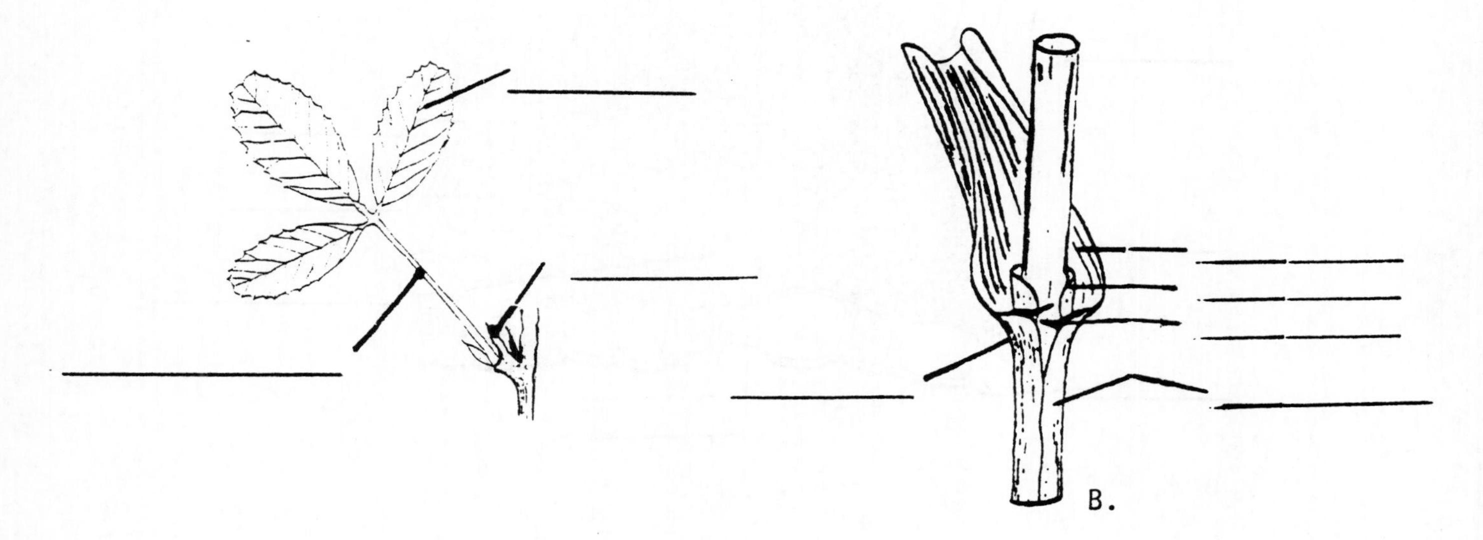

Fig. 6. Typical dicot leaf (left); typical monocot leaf (right).

Internal Structures of Leaves

The epidermis is covered with the **cuticle**, a waxy layer that retards movement of gasses and water through the epidermis. The **stomates** are lens-shaped openings in the epidermis which connect the **intercellular space** with the external atmosphere. The stomates are actually openings between the guard cells. The **guard cells** contain chloroplasts and are capable of photosynthesis.

The main photosynthesizing cells of the leaf are the **palisade** and **spongy mesophyll** cells. Generally there are 2 or 3 times as many chloroplsts in palisade cells as in spongy mesophyll cells. Sunflower has an estimated 77 chloroplasts per palisade cell.

A major difference between monocots and dicots is that dicots have a palisade layer, while monocots do not.

Xylem vessels are important water and nutrient conductors, with direction of flow generally going from root to leaf. **Phloem sieve tubes** are located nearest the lower epidermis and mostly move photosynthates and other materials away from the leaf. Label Figure 9 with the following terms:

1. Cuticle
2. Epidermis
3. Palisade mesophyll
4. Spongy mesophyll
5. Intercellular space
6. Guard cells
7. Stomatal pore
8. Xylem
9. Phloem

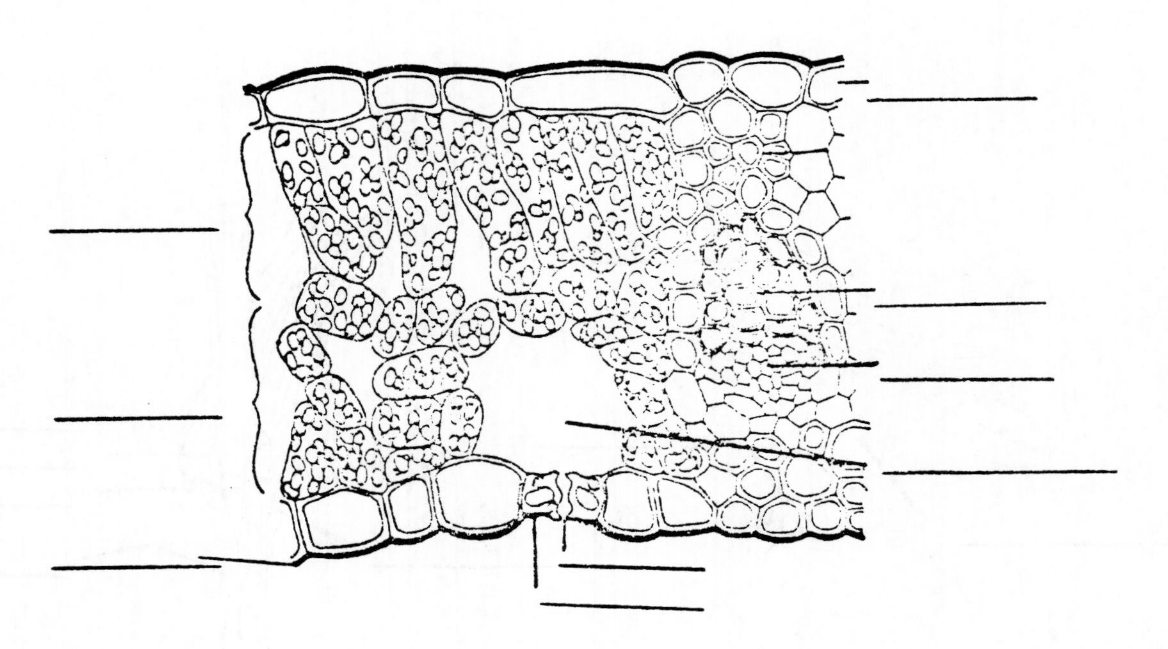

Fig. 7. Dicot leaf cross-section

Types of Leaves

There are two kinds of blade configurations: 1) simple and 2) compound. A simple leaf is one in which the blade is not divided. A compound leaf is one in which the blade is composed of two or more leaflets, such as in beans, peas, and clovers.

Leaflets can be distinguished from simple leaves by the following:

1) Buds occur in the axils of leaves, not in the axils of leaflets
2) Leaves may stand in different planes on the stem, whereas leaflets lie in a single plane.
3) Stipules, when present, are at the base of the leaf, not at the base of the leaflets.

Compound leaves commonly are either pinnately or palmately compound. Pinnately compound leaves, as in A, Fig. 7, have leaflets attached at two or more locations on the petiole. Palmately compound leaves as in B, Fig. 7, have **all** leaflets attached at one point, the end of the petiole. Some plants, such as the soybean, have a stalk or stem-like structure which attaches the leaflet to the petiole. This structure is called a **petiolule**.

The arrangement of the veins in a leaf is called **venation**. The two principal types of venation are **parallel** and **net**. In leaves containing parallel venation, common in monocots, the large veins run parallel from the base of the apex or from the midrib outward to the margin of the leaf. Leaves with **net** venation have a conspicuous network of veins throughout. In this type of venation, the larger veins branch profusely.

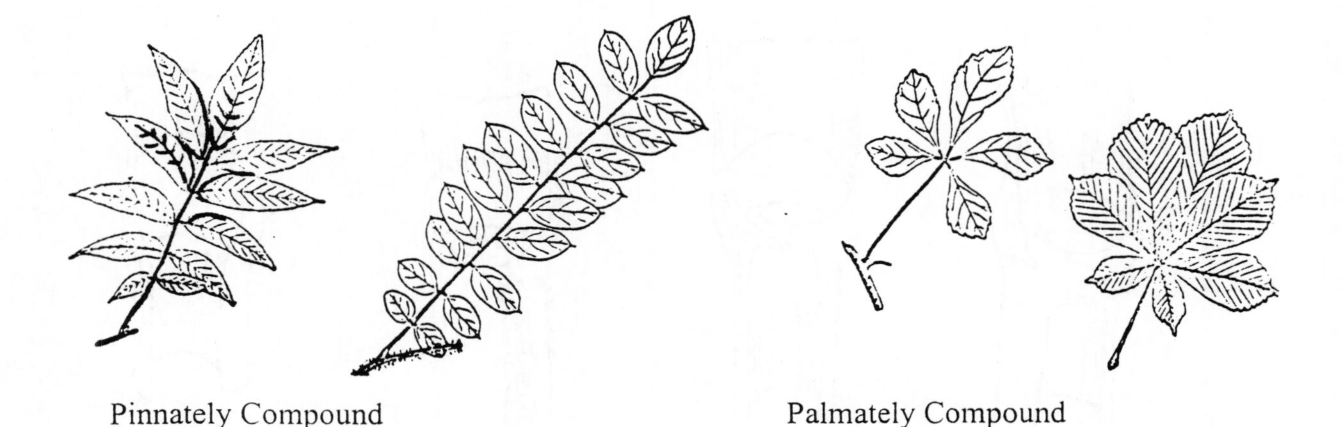

Pinnately Compound   Palmately Compound

Fig. 8. Two common types of compound leaves

Examine the leaves of young plants of wheat, oats, barley, and rye for vegetative characteristics. Identify the drawings in Figure 8.

Barley has very long and prominent auricles that clasp around the stem. The ligule is relatively short and inconspicuous. Waxy-coated leaves.

Wheat has short, slender, hairy auricles. The ligule is rounded.

Rye has very short, sharp pointed auricles with a hairy sheath and a small inconspicuous ligule.

Oat auricles are absent and both the sheath and leaf margins have hairs. The ligule is relatively large and flat topped.

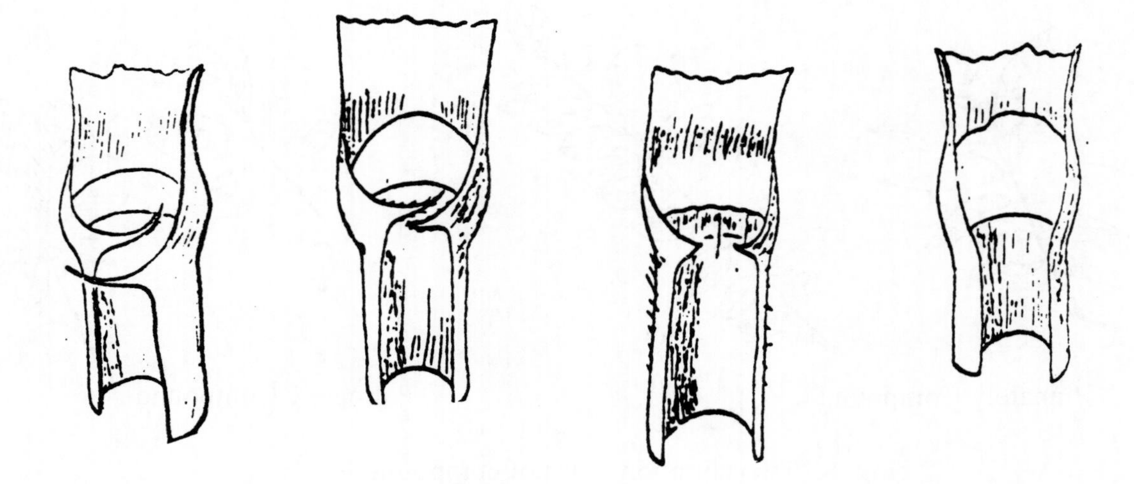

Fig. 9. Leaf structures of barley, wheat, rye, and oats.

DEFINITIONS

Auricle An ear-like lobe projecting from the top of the leaf sheath in some grasses, often clasping the stem.

Blade A portion of the leaf, usually place of greatest photosynthetic activity.

Cambium A meristematic zone of the cells between the phloem and xylem from which new tissues are developed.

Culm The jointed hollow stem of grasses.

Intercalary meristem Meristem located at the base of each internode in grasses.

Internode The region of culm between two successive nodes.

Ligule A thin fringe of membranous tissue or hairs projecting from top of the leaf sheath on the inner side.

Node Slightly enlarged portion of the culm where leaves and buds arise.

Petiole The supporting stalk of a dicot leaf.

Petiolule The supporting stalk of a leaflet.

Phloem The vascular tissue which, with xylem, makes up the vascular bundle, and mostly conducts food and material from the top to the roots of the plant.

Rhizome An underground stem, usually slender and creeping, from which new plants originate at the nodes.

Sheath The lower part of the grass leaf which encloses stem; a modified sheath.

Stipule A leaf-like appendage at the base of the petiole.

Stolon An above-ground prostrate stem tending to root and originate new plants at the nodes.

Vascular bundle A cluster or bundle of transporting vessels in the stems of plants which serve to carry dissolved food and water.

Xylem The vascular tissue which, with the phloem, makes up the vascular bundle and mostly conducts water.

STATION GUIDE FOR STRUCTURE AND FUNCTION OF CROP STEMS AND LEAVES

Upon completing this guide you should be able to

1. describe the functions of stems and leaves
2. identify external structures of stems and leaves
3. label internal structures of stems and leaves.
4. recognize specialized stems plus various leaf types

The required information to complete the guide can be found at stations 1-8 in lab or in previous pages in the lab manual.

STATION 1. FUNCTIONS OF STEMS

The three major functions of plant stems are:

-
-
-

STATION 2. EXTERNAL STRUCTURES OF STEMS

1. The first stem of a seed plant develops from the ____________ which is a cylindrical structure with a mass of ____________ at its________________.
2. Locate on a soybean plant, and define the following:
   - Node--
   - Internode--
   - Apical or terminal bud--
   - Axillary or lateral bud--
3. Examine the flat of soybeans, and describe the effect of each of the following treatments:
   - Epicotyl clipped____
   - One cotyledon removed__________________________________
   - Two cotyledons removed_________________________________
   - Hypocotyl clipped_______________________________________

4. Corn plant

   - Number of nodes on a mature corn stem

   - Length of an internode on mature corn stem

   - An intercalary meristem is

5. Examine the flat of corn seedlings, and describe how clipping affected regrowth.

STATION 3. INTERNAL STRUCTURES OF STEMS

1. The three principal stem tissues and their functions are:

   -
   -
   -

2. Vascular bundles of corn are ____________ throughout the stem and are arranged in a ____________ in alfalfa.

3. After viewing the microscope slides, label the following cross-sections of stems:

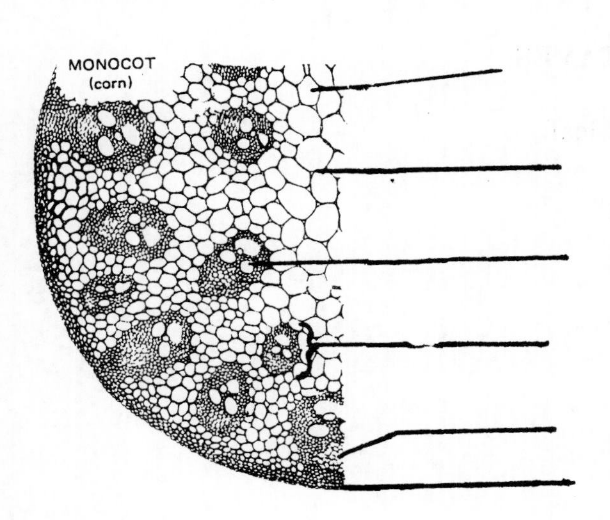

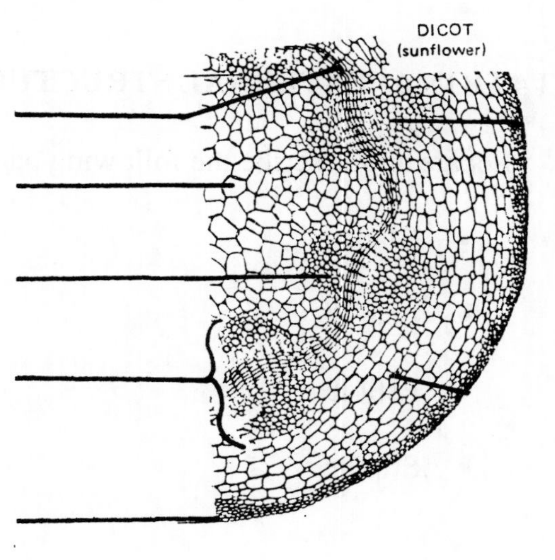

STATION 4. TYPES OF STEMS

Define, and name a plant that has each of the following types of stems:

- Bulb
- Corm
- Culm
- Rhizome
- Stalk
- Stolon
- Tiller
- Tuber

STATION 5. FUNCTIONS OF LEAVES

The three major functions of a plant leaf are:

-
-
-

STATION 6. EXTERNAL STRUCTURES OF LEAVES

1. Locate and describe the following parts of a dicot leaf:

   - Blade
   - Leaflet
   - Petiole
   - Stipule

2. Locate and describe the following parts of a monocot leaf

- Blade
- Sheath
- Auricles
- Ligules
- Collar

STATION 7. INTERNAL STRUCTURES OF LEAVES

1. Name the principal tissues in an alfalfa leaf and their functions:

Tissue	Function of Tissue
•	•
•	•
•	•
•	•

2. After viewing the microscope slides, label the following diagrams of leaf cross-sections:

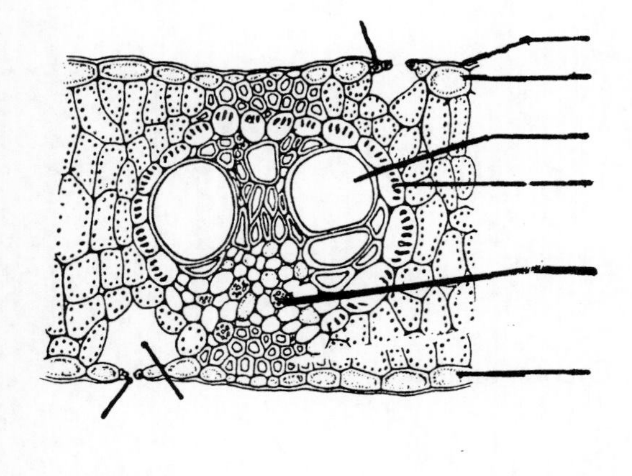
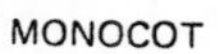

MONOCOT

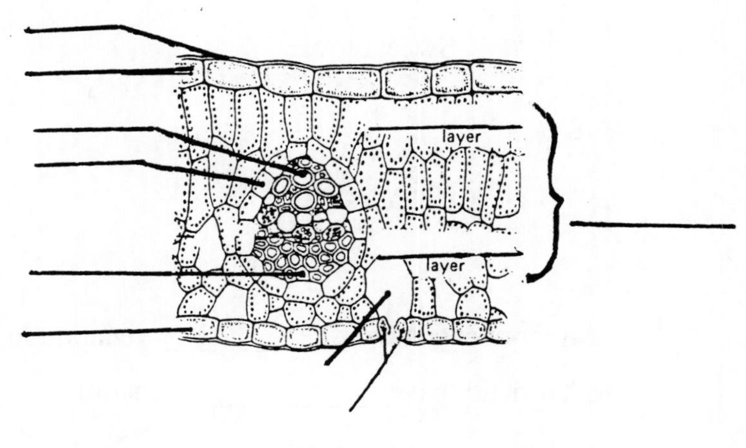

DICOT

STATION 8. TYPES OF LEAVES

1. Mount a sample of each grass and describe its auricles and ligules

<u>Oats</u> <u>Wheat</u> <u>Barley</u> <u>Rye</u>

2. Draw a typical leaf for each legume below and indicate whether it is <u>palmately</u> compound or <u>pinnately</u> compound.

Alfalfa

White clover

Red Clover

Sweet clover

Hairy vetch

3. All monocots have ____________ venation in their leaves. Dicots, on the other hand, have ____________ venation.

Z840.1

Broadleaf & Grass and Grasslike Weed Vegetative Identification Keys

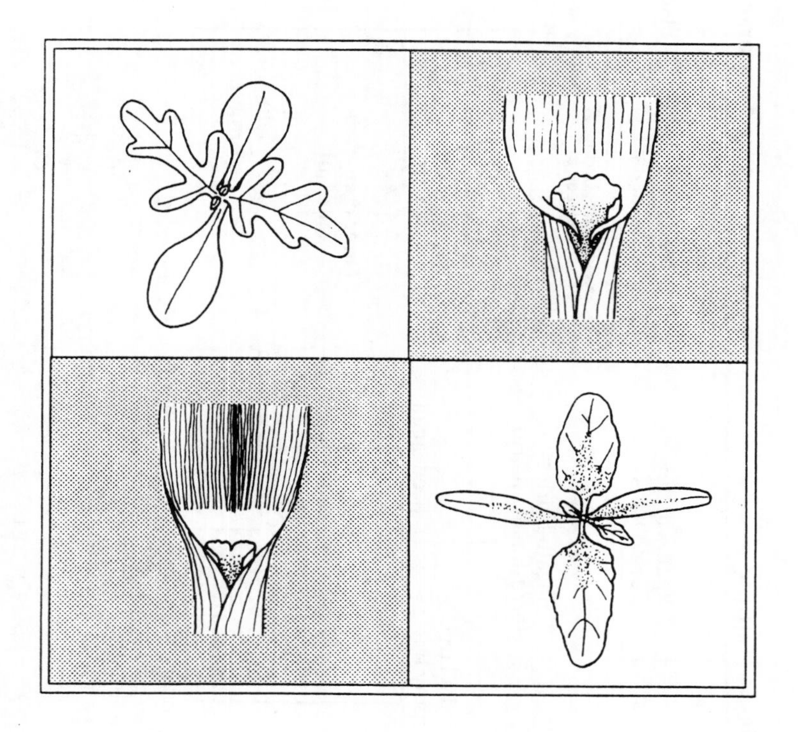

Prepared by

C. Diane Anderson, William S. Curran, and other extension specialists in weed science,
Cooperative Extension Service, University of Illinois at Urbana-Champaign.

Available from Vocational Agriculture Service, University of Illinois,
1401 S. Maryland Drive, Urbana, Illinois 61801 217/333-3871.

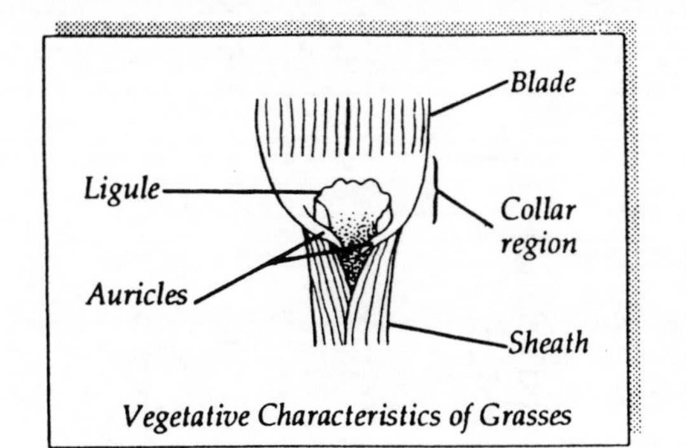

Vegetative Characteristics of Grasses

Grass and Grasslike Weed Vegetative Key

Prepared by C. Diane Anderson, and other Weed Science Extension Specialists, Cooperative Extension Service, University of Illinois at Urbana-Champaign.

Available from Vocational Agriculture Service,1401 S. Maryland Drive, Urbana, Illinois 61801 217/333-3871.

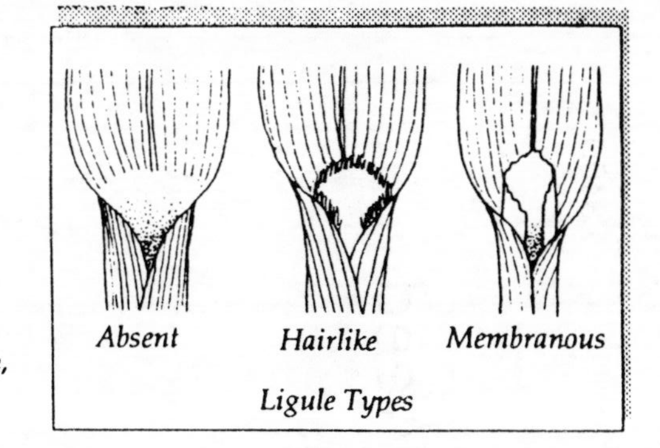

Ligule Types

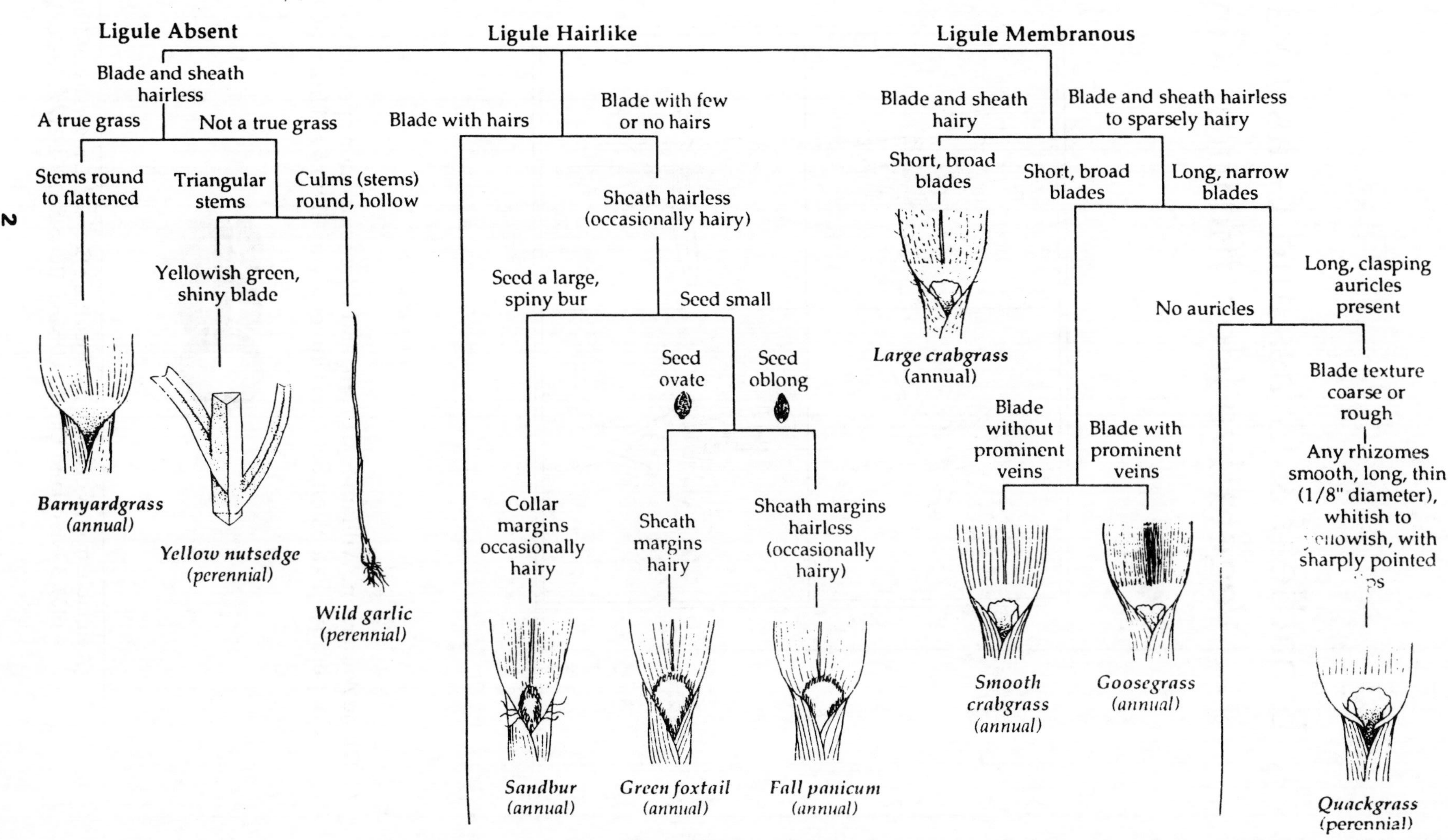

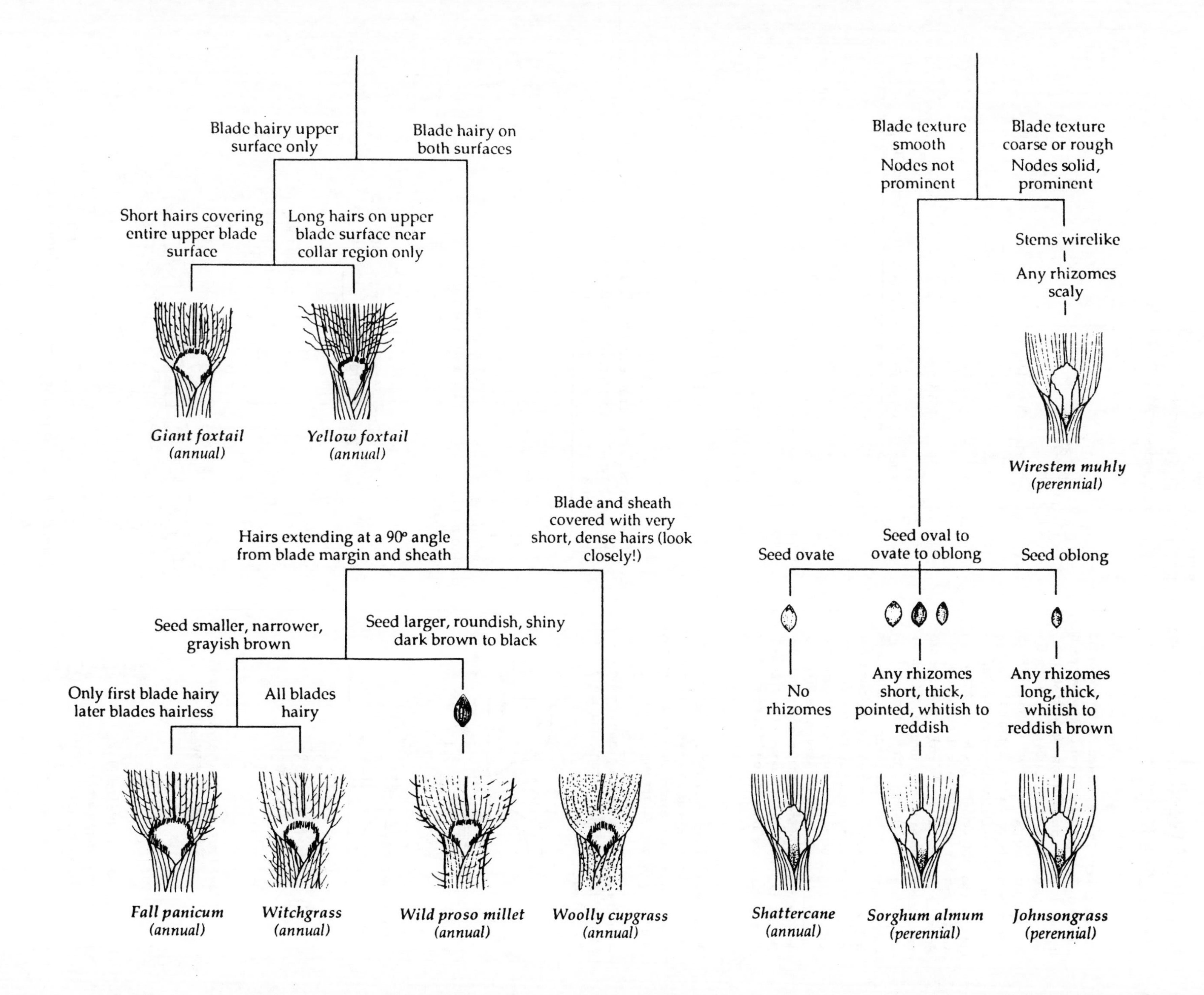
Blade hairy upper surface only
Blade hairy on both surfaces
Short hairs covering entire upper blade surface
Long hairs on upper blade surface near collar region only
Giant foxtail (annual)
Yellow foxtail (annual)
Hairs extending at a 90° angle from blade margin and sheath
Blade and sheath covered with very short, dense hairs (look closely!)
Seed smaller, narrower, grayish brown
Seed larger, roundish, shiny dark brown to black
Only first blade hairy later blades hairless
All blades hairy
Fall panicum (annual)
Witchgrass (annual)
Wild proso millet (annual)
Woolly cupgrass (annual)
Blade texture smooth
Nodes not prominent
Blade texture coarse or rough
Nodes solid, prominent
Stems wirelike
Any rhizomes scaly
Wirestem muhly (perennial)
Seed ovate
Seed oval to ovate to oblong
Seed oblong
No rhizomes
Any rhizomes short, thick, pointed, whitish to reddish
Any rhizomes long, thick, whitish to reddish brown
Shattercane (annual)
Sorghum almum (perennial)
Johnsongrass (perennial)

Broadleaf Weed Vegetative Key

Using this key. This key describes common broadleaf weed seedlings found in corn and soybeans in Illinois. It focuses primarily on characteristics of the true leaves, but in some cases the cotyledons are important (see figure below). Options A, B, and C describe weeds that have an alternate leaf arrangement. Option D contains weeds with an opposite leaf arrangement. The leaves of most weeds are either all alternate or all opposite. However, in some weeds, the early true leaves are opposite but later leaves are alternate. Note these exceptions given in the key. (Cotyledons are always opposite.) Once leaf arrangement has been determined, other characteristics of the leaves and cotyledons are needed to follow the key. It's very common when using a key to try more than one route before reaching the correct species. The sketches of many of the weeds are approximately life size. Others are roughly one-half as large as actual size and are indicated by 1/2X beside the sketch.

Prepared by C. Diane Anderson, William S. Curran, and other extension specialists in weed science, Cooperative Extension Service, University of Illinois at Urbana-Champaign.

Available from Vocational Agriculture Service, 1401 S. Maryland Drive, Urbana, Illinois 61801 217/333-3871.

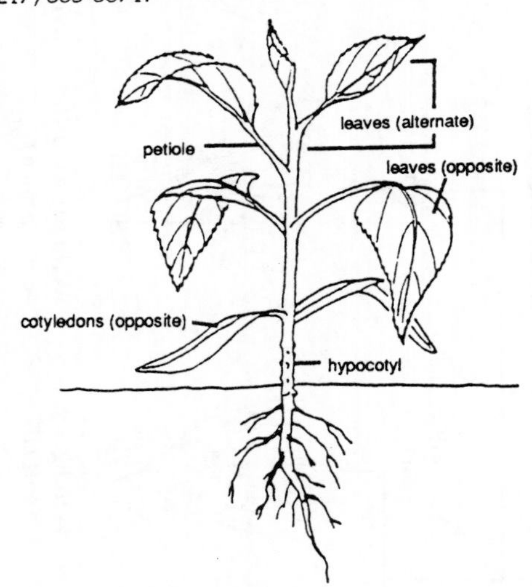

Plant Parts and Leaf Arrangement of Broadleaf Weeds

Ochrea of plants in the family Polygonaceae (Buckwheat or Smartweed family)

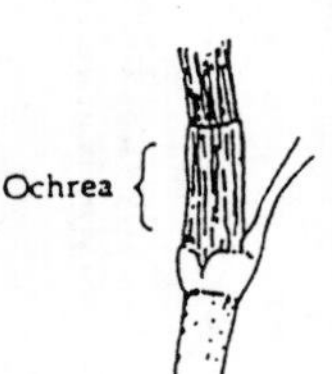

OPTION A.
FIRST TRUE LEAVES
Arrangement: ALTERNATE

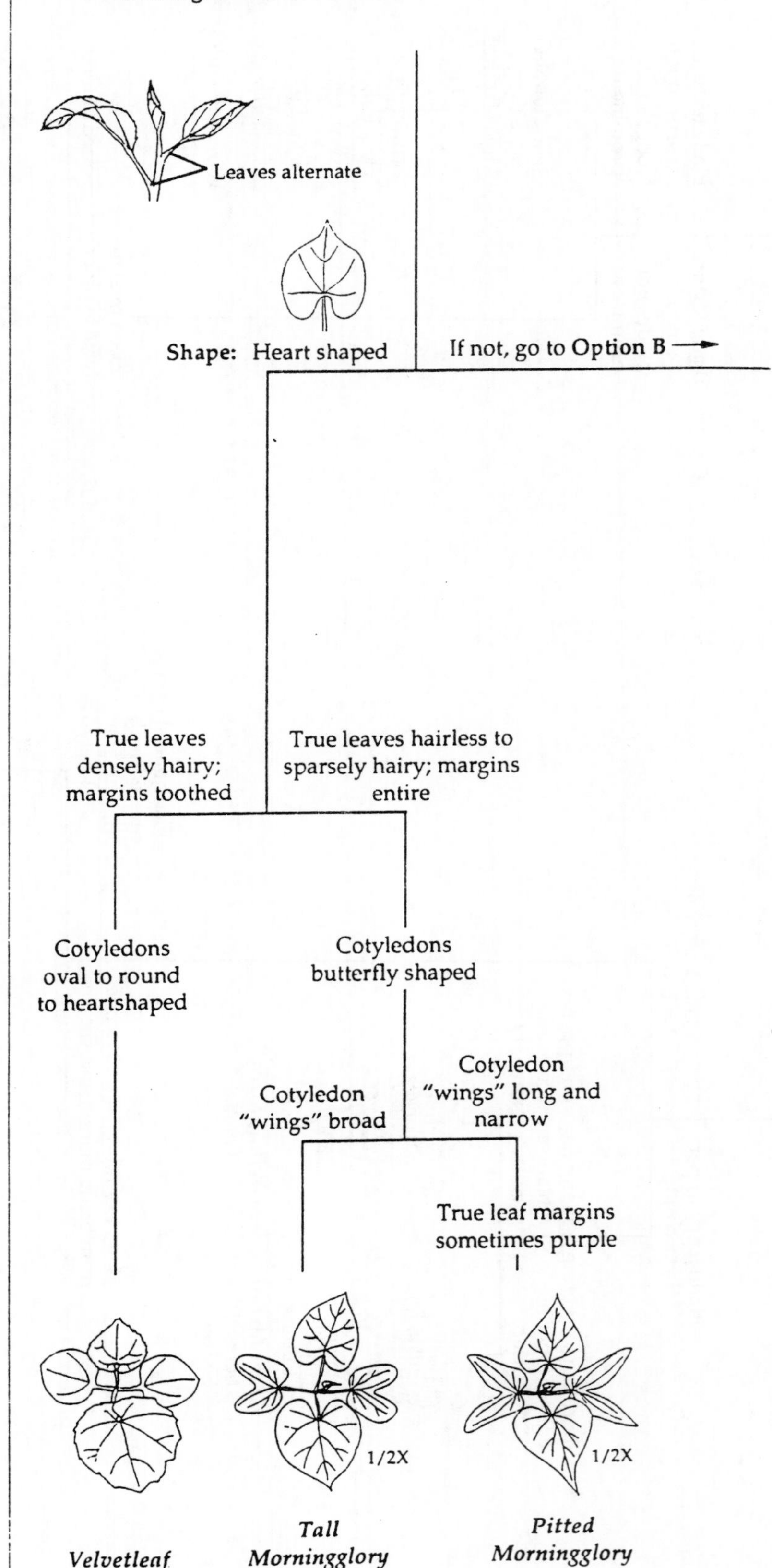

OPTION B.
FIRST TRUE LEAVES
Arrangement: ALTERNATE

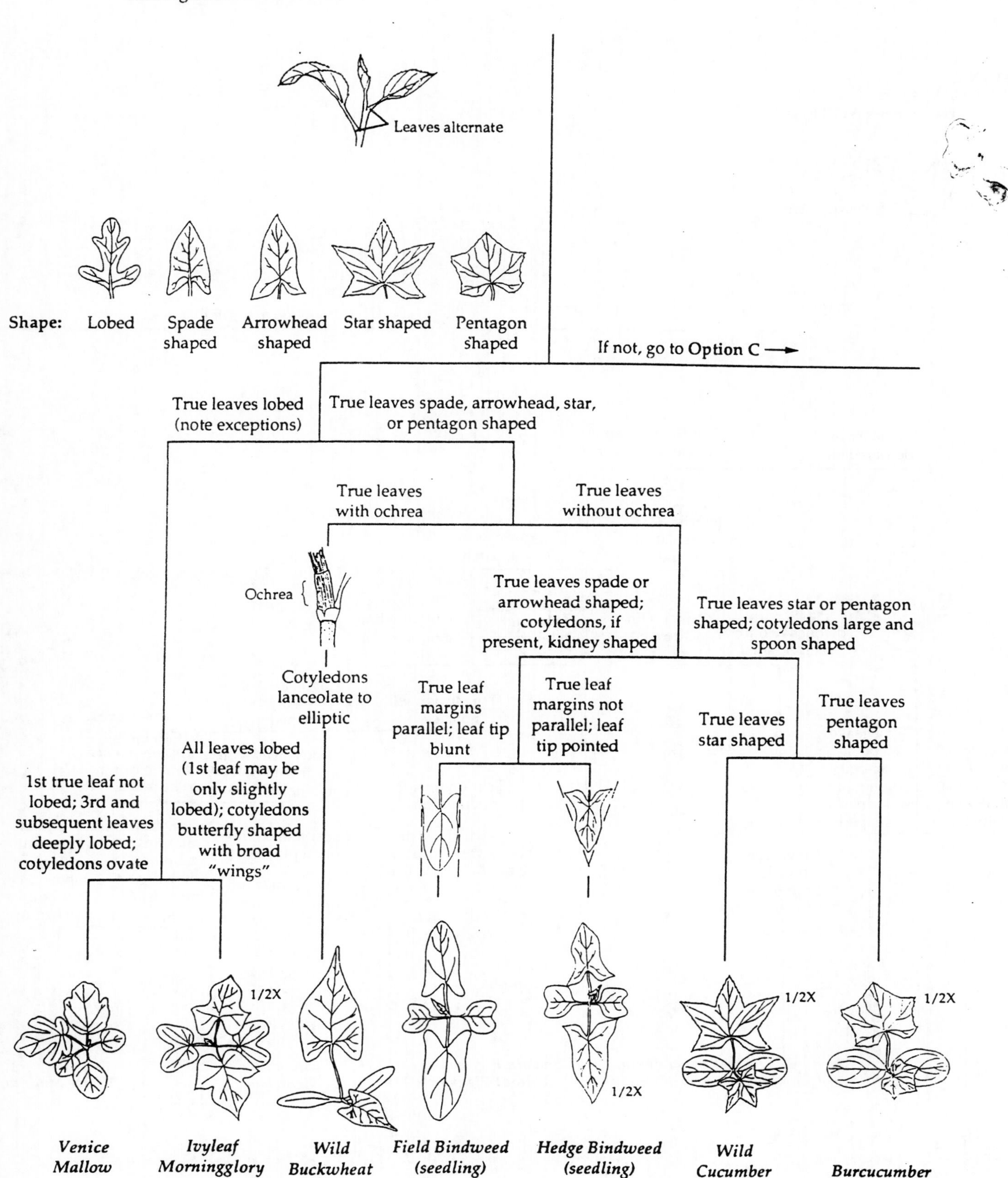

Leaves alternate
Shape:
Lobed
Spade shaped
Arrowhead shaped
Star shaped
Pentagon shaped
If not, go to Option C
True leaves lobed (note exceptions)
True leaves spade, arrowhead, star, or pentagon shaped
True leaves with ochrea
True leaves without ochrea
Ochrea
True leaves spade or arrowhead shaped; cotyledons, if present, kidney shaped
True leaves star or pentagon shaped; cotyledons large and spoon shaped
Cotyledons lanceolate to elliptic
True leaf margins parallel; leaf tip blunt
True leaf margins not parallel; leaf tip pointed
True leaves star shaped
True leaves pentagon shaped
1st true leaf not lobed; 3rd and subsequent leaves deeply lobed; cotyledons ovate
All leaves lobed (1st leaf may be only slightly lobed); cotyledons butterfly shaped with broad "wings"
1/2X
1/2X
1/2X
1/2X
Venice Mallow
Ivyleaf Morningglory
Wild Buckwheat
Field Bindweed (seedling)
Hedge Bindweed (seedling)
Wild Cucumber
Burcucumber

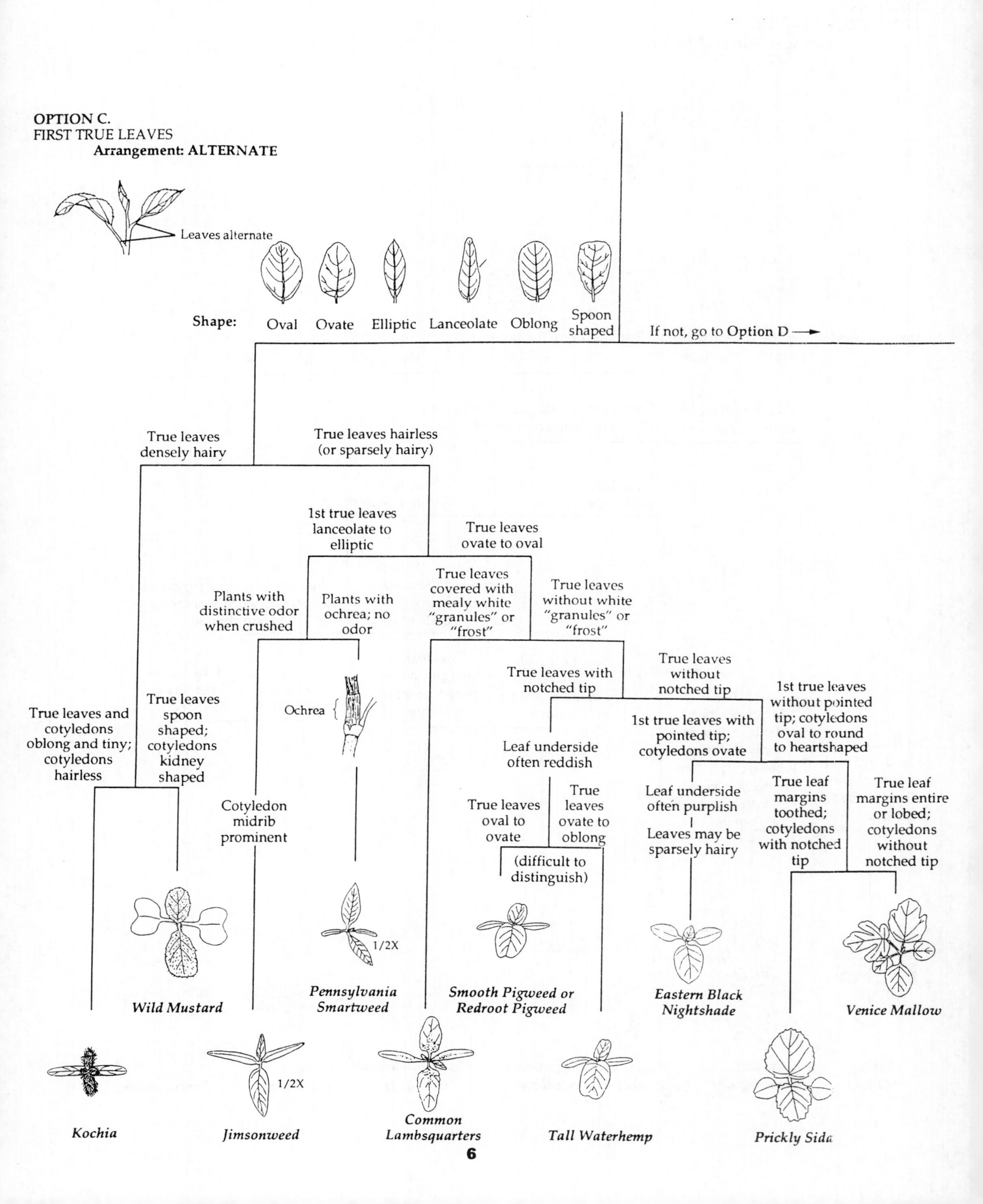

OPTION C.
FIRST TRUE LEAVES
Arrangement: ALTERNATE
Leaves alternate
Shape:
Oval
Ovate
Elliptic
Lanceolate
Oblong
Spoon shaped
If not, go to Option D →
True leaves densely hairy
True leaves hairless (or sparsely hairy)
1st true leaves lanceolate to elliptic
True leaves ovate to oval
Plants with distinctive odor when crushed
Plants with ochrea; no odor
True leaves covered with mealy white "granules" or "frost"
True leaves without white "granules" or "frost"
True leaves with notched tip
True leaves without notched tip
1st true leaves with pointed tip; cotyledons ovate
1st true leaves without pointed tip; cotyledons oval to round to heartshaped
True leaves and cotyledons oblong and tiny; cotyledons hairless
True leaves spoon shaped; cotyledons kidney shaped
Ochrea
Cotyledon midrib prominent
Leaf underside often reddish
True leaves oval to ovate
True leaves ovate to oblong
(difficult to distinguish)
Leaf underside often purplish
Leaves may be sparsely hairy
True leaf margins toothed; cotyledons with notched tip
True leaf margins entire or lobed; cotyledons without notched tip
Wild Mustard
1/2X
Pennsylvania Smartweed
Smooth Pigweed or Redroot Pigweed
Eastern Black Nightshade
Venice Mallow
Kochia
1/2X
Jimsonweed
Common Lambsquarters
Tall Waterhemp
Prickly Sida

OPTION D.
FIRST TRUE LEAVES
Arrangement: OPPOSITE

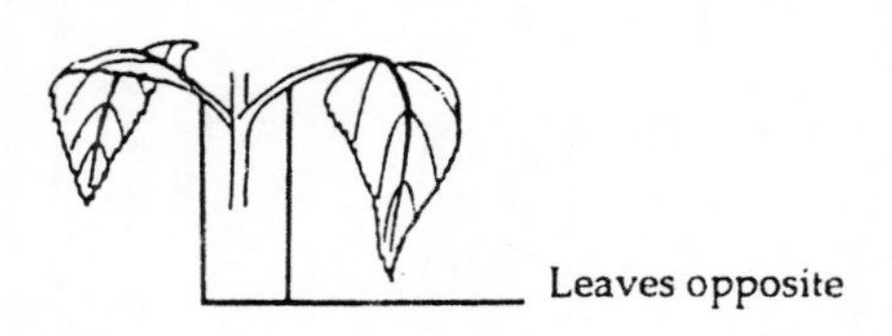

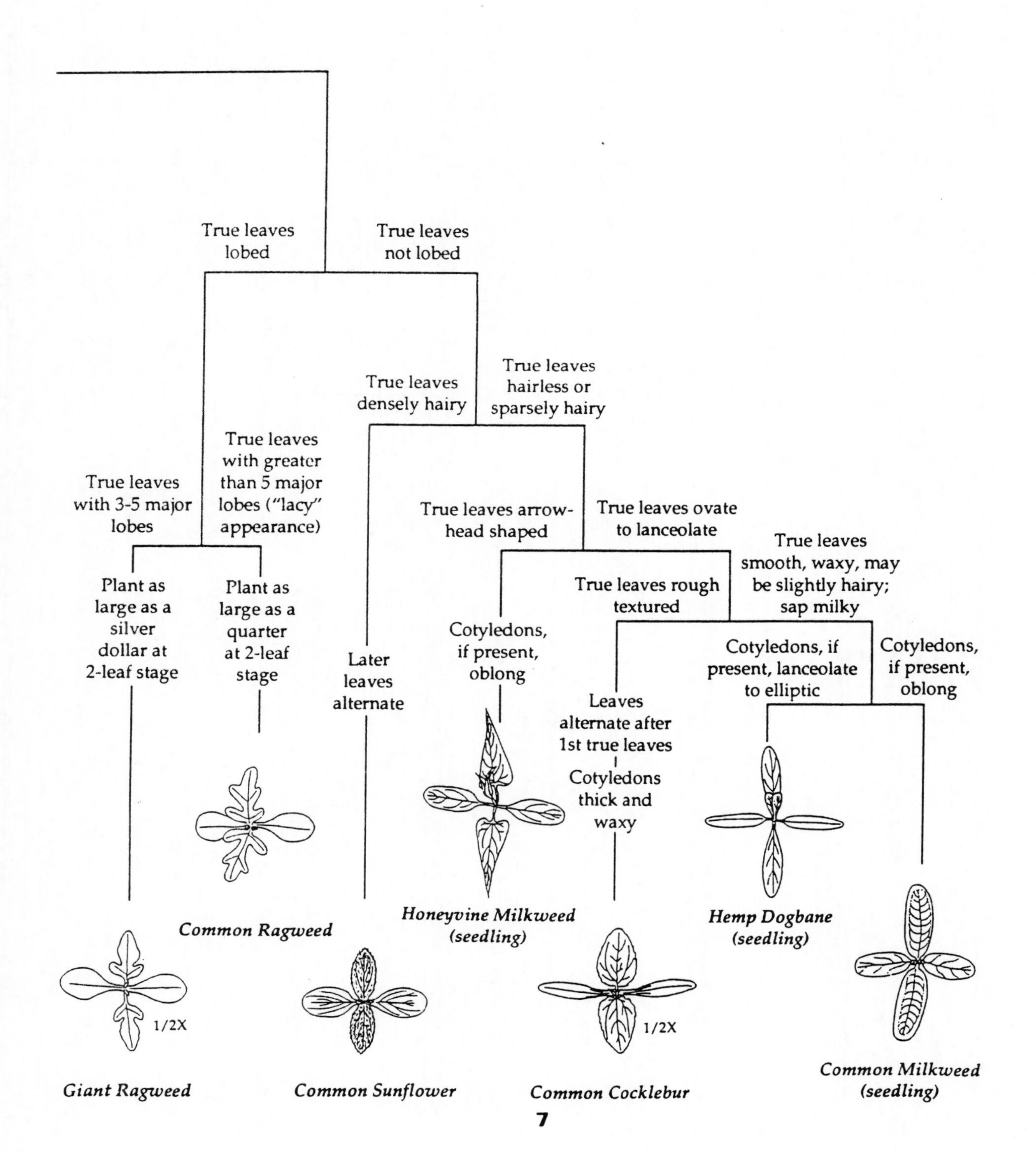

Common Ragweed

Honeyvine Milkweed (seedling)

Hemp Dogbane (seedling)

Giant Ragweed

Common Sunflower

Common Cocklebur

Common Milkweed (seedling)

Assessing Hail Damage to Corn

James V. Vorst, Purdue University

Reviewers
George Bender, National Crop Insurance Adjusters
Dale Hicks, University of Minnesota
Gerry Posler, Kansas State University
Jim Rink, Farm Bureau Crop Insurance

In the U.S., approximately half of all hailstorms occur between March and May. These early storms are responsible for only minor corn yield losses, however, because the corn either has not yet been planted or is too small to be damaged significantly. Even when fields are severely damaged early in the growing season, they can often be replanted.

On the other hand, about a third of all hailstorms occur between June and September. These have resulted in yield losses to corn estimated at $52 million annually.

Hail affects yields primarily by reducing stands and defoliating the plant. Defoliation causes most of the losses. Thus, knowing how to recognize hail damage and assess probable loss is a very valuable decision-making aid.

For instance, proper assessment of yield loss after an early-season storm can help you determine whether or not to replant or understand how an insurance adjuster determines yield loss. An accurate estimate of loss from a late storm is important for making correct harvesting and marketing decisions.

In this publication, we will examine how hail damages the corn plant, how the degree of damage can be determined and how the extent of yield loss is estimated.

When Corn is Most Susceptible to Hail Damage

Prior to, and for some time after emergence, the corn plant is affected very little by hail damage. At emergence, the plant's growing point is below the soil surface and remains there for about 3 weeks (until five to seven leaves have fully emerged). Because the growing point is below the soil surface and in the leaf whorl, plant damage due to hail at these early stages rarely results in any significant stand or yield loss.

Approximately 3 weeks after emergence, all nodes and internodes have developed, and the growing point is elevated above the soil surface due to internode elongation. For the next 4-5 weeks, the plant grows rapidly and becomes increasingly vulnerable to hail damage up through the tasseling stage, which is the most critical period. Once past tasseling, hail would cause progressively less yield loss as the plant approaches maturity.

Determining Yield Loss Due to Stand Reduction

When a hailstorm occurs early in the growing season, an accurate stand reduction assessment is important if replanting is still a management option. Because it is difficult to distinguish living from dead tissue immediately after a storm, the assessment should be delayed for a week to 10 days. By that time, regrowth of living plants will have begun and discolored dead tissue will be apparent. (Another reason for assessment delay is that some plants initially surviving a storm may soon die because of disease infection entering at the sites of plant damage.)

To get an accurate estimate of the extent of damage, observe and sample plants from at least three parts of affected fields, totaling about 1/100 acre. Use Table 1 to determine how many feet of row are required to make 1/100 acre at various row spacings. Then divide that figure by the number of sampling locations to determine how many feet to sample at each location.

Table 1. Total Feet of Row Required to Make 1/100 Acre at Various Row Spacings.

Row Spacing	Row Length	Row Spacing	Row Length
20″ Rows	261 Ft.	32″ Rows	163 Ft.
22″ Rows	238 Ft.	34″ Rows	154 Ft.
24″ Rows	218 Ft.	36″ Rows	145 Ft.
26″ Rows	201 Ft.	38″ Rows	138 Ft.
28″ Rows	187 Ft.	40″ Rows	131 Ft.
30″ Rows	174 Ft.		

To sample pre-tasseled corn, split the stems of several obviously damaged plants with a knife to observe the growing point. If it is whitish-yellow in color, the plant is alive and should survive; if discolored and soft, the plant is dead or dying.

Some plants may be "tied" or "crippled"; i.e., the leaves fail to expand in a normal manner from the whorl. Since it cannot be determined until much later whether or not these crippled plants will develop normally, they should be classified as non-living if replanting is being considered.

Percent yield loss due to stand reduction is estimated by comparing yield potential of the field at its original plant population with yield potential at its now-reduced population. Table 2 has made these estimates for 25 different original and remaining population levels in 1/100 acre. When determining the advisability of replanting, be sure to consider availability and cost of adapted hybrid seed, replanting costs, calendar date and the possible need for additional weed control. (For more information on replant decisions, see NCH-30, "Guidelines for Making Corn Replanting Decisions.")

Determining Yield Loss Due to Defoliation

Most yield reduction in corn due to hail damage is a result of the loss of photosynthetically active leaf area. How severe that reduction is likely to be depends on not only the amount of leaf area removed, but also the corn's growth stage when hail occurs (Table 3). When leaf area is removed, the plant loses some of its capability to produce dry matter, resulting in reduced grain yields. However, grain yield reductions are not directly proportional to leaf area reductions, because of increased dry matter production in the remaining leaf area and movement of dry matter from other plant parts into the developing ear.

1. The first step in assessing yield loss due to defoliation is to establish the stage of plant growth at the time of the storm.

Growth stages BEFORE tasseling are defined in terms of the number of leaves exposed, e.g., 7-leaf stage, 13-leaf stage, etc. Counting of leaves starts with the lowermost leaf, which has a rounded tip (Leaf 1 in Figure 1), and continues up to the "indicator leaf," which is the uppermost leaf that is 40-50

Table 2. Estimated Percent Corn Yield Loss Due to Stand Reduction Occurring Through the Tenth-Leaf Stage of Growth (1/100 Acre Area).*

READ TOP LINE FOR REMAINING PLANTS

ORIGINAL STAND	320	310	300	290	280	270	260	250	240	230	220	210	200	190	180	170	160	150	140	130	120	110	100	90	80	ORIGINAL STAND
320	0	1	2	3	4	5	6	7	8	9	11	13	16	18	21	23	26	29	32	35	38	41	45	49	53	320
310		0	1	2	3	4	5	6	7	8	10	12	14	16	19	21	24	27	30	33	36	39	43	47	52	310
300			0	1	2	3	4	5	6	7	9	11	12	14	17	20	23	25	28	31	34	37	41	45	50	300
290				0	1	2	3	4	5	6	8	10	11	13	15	18	21	23	26	29	32	35	39	43	48	290
280					0	1	2	3	5	6	7	9	10	12	14	16	19	21	24	27	30	34	37	41	46	280
270						0	1	3	4	5	6	7	9	10	12	14	16	18	21	24	28	31	35	40	45	270
260							0	1	3	4	5	6	7	9	10	12	14	16	19	22	25	29	33	38	43	260
250								0	1	2	3	4	6	7	8	10	12	14	17	20	23	27	31	36	41	250
240									0	1	2	3	4	5	6	9	10	12	15	18	22	26	29	34	40	240
230										0	1	2	3	4	5	8	9	11	14	17	21	25	29	33	39	230
220											0	1	2	3	4	7	8	10	13	16	20	24	28	33	38	220
210												0	1	2	4	6	7	9	12	16	20	24	27	32	37	210
200													0	1	3	5	6	8	11	15	19	23	27	31	36	200
190														0	2	4	5	7	10	14	17	21	25	30	35	190
180															0	2	4	6	9	12	15	19	23	28	33	180
170																0	2	4	7	10	13	17	21	26	31	170
160																	0	2	5	8	11	15	19	24	29	160
150																		0	3	5	8	12	16	21	26	150
140																			0	3	6	10	14	18	23	140
130																				0	3	6	10	15	20	130
120																					0	3	7	12	17	120
110																						0	3	8	12	110
100																							0	4	8	100
90																								0	4	90
80																									0	80

* Reprinted by permission from the National Crop Insurance Service's "Corn Loss Instructions" (Rev. 1984)

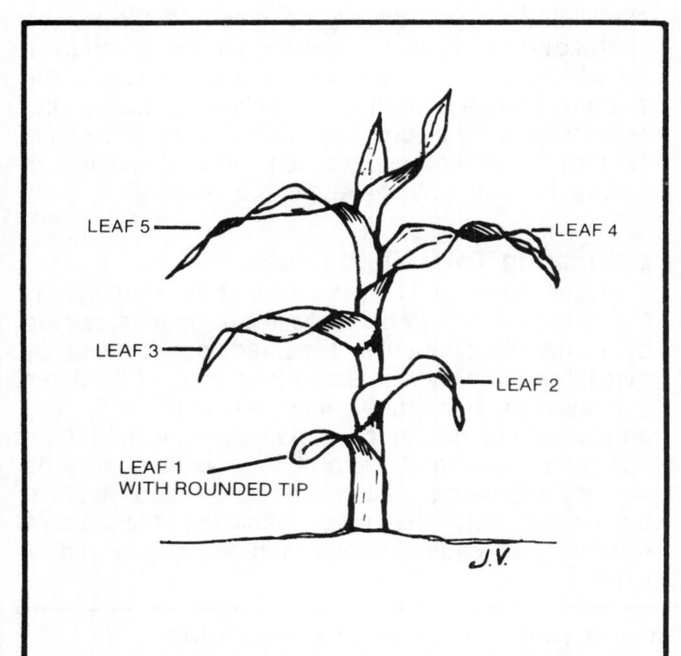

Figure 1. Corn plant in the 5-leaf growth stage. (Tip of Leaf 5 no longer points upward, so it is the "indicator leaf.")

Figure 2. Longitudinal section through base of corn plant, showing fifth leaf attachment (at top of first noticeable elongated internode).

percent exposed from the whorl and whose tip points below a horizontal line (Leaf 5 in Figure 1).

In some situations (particularly as the plant nears tasseling), Leaf 1 may have been removed, in which case the lowermost leaf will not have a rounded tip. By splitting the stalk, you can positively identify leaf location, since Leaf 5 is attached to the top of the first noticeably elongated internode (Figure 2.) Growth stage can then be determined by counting upward from the fifth leaf to the indicator leaf. If, for example, this indicator leaf is the eleventh above Leaf 5 (the leaf attached to the top of the first elongated internode), the corn is considered to be in the 16-leaf stage of growth (i.e., 11 + 5).

Growth stages AFTER tasseling are identified according to development of the ear shoot and kernels as follows:

- Silked stage: silks have emerged and tassel is shedding pollen.
- Silks brown stage: 75 percent of silks on ear shoot show purple to brown color but are not dry to the touch.
- Pre-blister stage: silks all brown but not dry; no fluid in kernels, which look like pimples.
- Blister stage: kernels look like white, water blisters; fluid is colorless.
- Milk stage: roasting ear stage with cob at its maximum length and kernels yellow in color and containing only milky fluid (no solid substance).
- Late milk stage: milky fluid thickening and solids forming in base of kernels.
- Soft dough stage: kernels contain semi-solid substance, but still produce thick milky material when squeezed; kernels near butt end of ear beginning to dent.
- Early dent stage: all kernels beginning to dent and containing thick gummy substance; but many still squirt "milk" when mashed.
- Dent stage: kernels denting or dented and can be cut easily with fingernail.
- Nearly mature stage: kernel hull on opposite side of embryo has shiny appearance halfway to cob.
- Mature stage: black layer formed at base of kernels; kernel moisture 35-40 percent.

2. The next step in assessing yield loss due to defoliation is to estimate percent of leaf area destroyed per plant. In making this estimate, consider both leaf area removed and leaf area still attached to the plant but no longer green. Live green tissue remaining on the plant, even though mutilated, should not be considered as leaf area destroyed. Examine plants in each of at least three areas of a damaged field to be assured of an accurate estimate.

3. With the corn growth stage established and amount of leaf area loss estimated, use Table 3 to determine likely yield reduction from defoliation.

Determining Direct Ear Damage

Ear damage due to hail may result from hailstorms occurring late in the season. To determine the extent of crop loss due to ear damage, select ears from 10 consecutive plants and count the number of damaged kernels on all of the ears. Next, determine the total number of kernels on all 10 ears, and calculate the percent of total kernels damaged. This figure is the percent of loss due to direct ear damage. If direct ear damage occurs in association with stand reduction, the percent ear damage is adjusted to account for plants lost. This is done by multiplying the percent ear damage times the percent of plants remaining.

Bruising and Stalk Damage

After the corn has reached the 10-leaf stage, stem bruising may occur. To calculate the damage due to bruising, determine the number of totally destroyed plants out of 100 consecutive plants. Bruised plants that do not actually go down should not be counted. Bruising may allow an avenue of infection for stalk rots, which may increase lodging later in the season. Since weather conditions during the remainder of the growing season affect disease severity, it may not be possible to evaluate fields with severe bruising until the end of the season.

Estimating Total Yield Loss

Total corn yield loss from hail damage is estimated by adding the expected yield loss caused by stand reduction, the expected loss caused by defoliation, and the expected loss caused by direct ear damage. Remember, however, that this is only an estimate of the percent yield loss. As with undamaged corn, extremely favorable weather during the rest of the growing season can cause actual yields to be higher than expected. Similarly, unfavorable weather can cause greater-than-anticipated reductions.

Table 3. Estimated Percent Corn Yield Loss Due to Defoliation Occurring at Various Stages of Growth.*

Stage	------Percent Leaf Area Destroyed------																		
of Growth	10	15	20	25	30	35	40	45	50	55	60	65	70	75	80	85	90	95	100
7 Leaf	0	0	0	0	0	0	1	1	2	3	4	4	5	5	6	7	8	9	9
8 Leaf	0	0	0	0	0	1	1	2	3	4	5	5	6	6	7	8	9	10	11
9 Leaf	0	0	0	1	1	2	2	3	4	5	6	6	7	7	9	10	11	12	13
10 Leaf	0	0	0	1	2	3	4	5	6	7	8	8	9	9	11	13	14	15	16
11 Leaf	0	0	1	1	2	3	5	6	7	8	9	10	11	12	14	16	18	20	22
12 Leaf	0	0	1	2	3	4	5	7	9	10	11	13	15	16	18	20	23	26	28
13 Leaf	0	1	1	2	3	4	6	8	10	11	13	15	17	19	22	25	28	31	34
14 Leaf	0	1	2	3	4	6	8	10	13	15	17	20	22	25	28	32	36	40	44
15 Leaf	1	1	2	3	5	7	9	12	15	17	20	23	26	30	34	38	42	46	51
16 Leaf	1	2	3	4	6	8	11	14	18	20	23	27	31	36	40	44	49	55	61
17 Leaf	2	3	4	5	7	9	13	17	21	24	28	32	37	43	48	53	59	65	72
18 Leaf	2	3	5	7	9	11	15	19	24	28	33	38	44	50	56	62	69	76	84
19-21 Leaf	3	4	6	8	11	14	18	22	27	32	38	43	51	57	64	71	79	87	96
Tassel	3	5	7	9	13	17	21	26	31	36	42	48	55	62	68	75	83	91	100
Silked	3	5	7	9	12	16	20	24	29	34	39	45	51	58	65	72	80	88	97
Silks Brown	2	4	6	8	11	15	18	22	27	31	36	41	47	54	60	66	74	81	90
Pre-Blister	2	3	5	7	10	13	16	20	24	28	32	37	43	49	54	60	66	73	81
Blister	2	3	5	7	10	13	16	19	22	26	30	34	39	45	50	55	60	66	73
Early Milk	2	3	4	6	8	11	14	17	20	24	28	32	36	41	45	50	55	60	66
Milk	1	2	3	5	7	9	12	15	18	21	24	28	32	37	41	45	49	54	59
Late Milk	1	2	3	4	6	8	10	12	15	18	21	24	28	32	35	38	42	46	50
Soft Dough	1	1	2	2	4	6	8	10	12	14	17	20	23	26	29	32	35	38	41
Early Dent	0	0	1	1	2	3	5	7	9	11	13	15	18	21	23	25	27	29	32
Dent	0	0	0	1	2	3	4	6	7	8	10	12	14	15	17	19	20	21	23
Late Dent	0	0	0	0	1	2	3	4	5	6	7	8	9	10	11	12	13	14	15
Nearly Mature	0	0	0	0	0	0	0	0	1	2	3	4	5	5	6	6	7	7	8
Mature	0	0	0	0	0	0	0	0	0	0	0	0	0	0	0	0	0	0	0

* Reprinted by permission from the National Crop Insurance Service's "Corn Loss Instructions" (Rev. 1984)

A publication of the National Corn Handbook Project

REV 11/91 (5M)

Cooperative Extension work in Agriculture and Home Economics, state of Indiana, Purdue University and U.S. Department of Agriculture cooperating. H.A. Wadsworth, Director, West Lafayette, IN. Issued in furtherance of the acts of May 8 and June 30, 1914. The Cooperative Extension Service of Purdue University is an affirmative action/equal opportunity institution.

REPRODUCTIVE MORPHOLOGY AND SEED DEVELOPMENT

I. REPRODUCTIVE MORPHOLOGY

Objectives

1. Identify legume and grass flowering structures.
2. Describe the function of legume and grass flowering structures.
3. Recognize the different types of inflorescences.
4. Describe the importance of the processes of seed development, gamete formation, pollination and fertilization.

A. Basic Floral Structures

Flowers are stem structures which arise from buds. The parts of a flower are therefore modified stems and leaves which differ greatly from the green vegetative leaves.

The **perianth** of a flower is composed of two whorls of modified leaves which are attached to the receptacle which is enlarged end of the **pedicel** or **peduncle**. One whorl is collectively called the **calyx** and is made up of the **sepals**. The second whorl is collectively called the **corolla** and is made up of the **petals**. Both the calyx and corolla function as protective coverings of the reproductive portions of the flower and in some cases as a colorful attraction to insects which aid in pollination.

The floral structures directly associated with seed formation are the **pistil** and **stamens**. These structures are also modified leaves although their appearance is very different from the vegetative leaves. The **pistil** represents the female portion of the flower and consists of the **stigma**, **style**, and **ovary**. Within the ovary, an **ovule** is formed which contains the egg cell. The male reproductive parts are called **stamens**. Each stamen consists of a **filament** which supports the pollen producing **anther**.

Complete flowers contain petals, sepals, pistil, and stamens. If one or more of these structures are absent, the flower is termed incomplete. A perfect flower contains both the pistil and stamens. Absence of either of these structures results in an imperfect flower which may be either staminate (containing only stamens) or pistillate (containing only the pistil).

A great deal of floral variability exists among plant species. Most plants produce perfect flowers which are termed **monoclinous**. Other species may produce pistillate or staminate flowers (**diclinous**) either on the same plant (**monoecious**) or different plants (**dioecious**). Corn is a good example of a monoecious plant species, producing pistillate flowers (ear) and staminate flowers (tassel). Dioecious species differ in that plants contain either staminate flowers or pistillate flowers resulting in male and female plants. Canada thistle and Buffalograss are excellent examples of such a condition.

Label the following flower parts in the diagram below:

perianth	petals	stamens	style
corolla	sepals	ovary	anther
calyx	pistil	stigma	filament

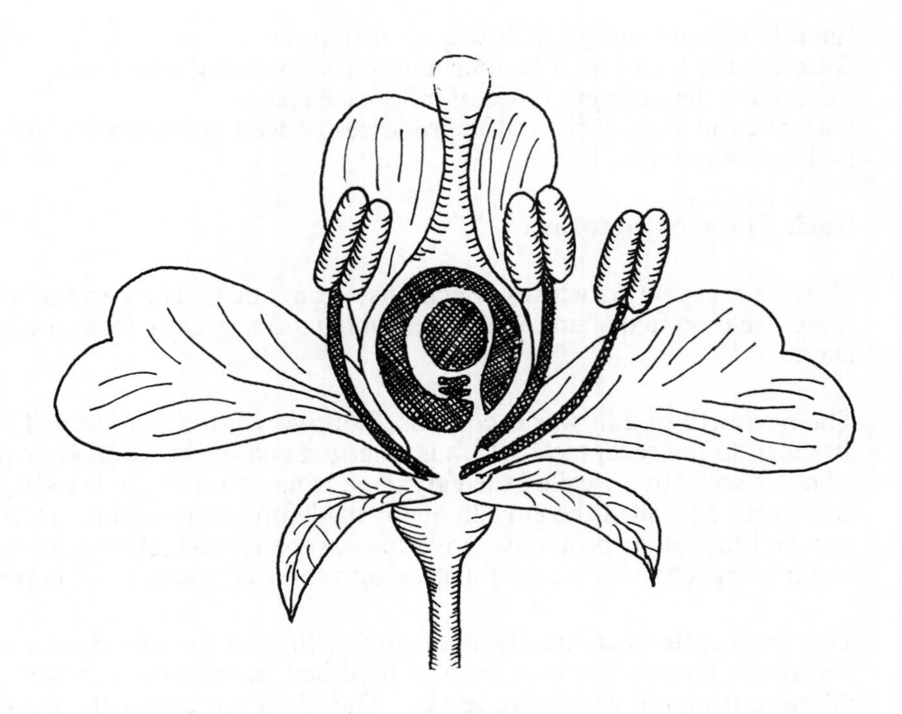

Fig. 1. Basic flower

B. The Legume Flower

A legume flower consists of a **calyx**, composed of five **sepals** united at the base, and a **corolla** with five **petals**. The largest is the standard, the two laterals are called **wings**, and the two petals which are united at the base form the **keel**. The **pistil** and **stamens** are located inside the keel. They are held under tension within the keel and will break through the upper junction of the keel with slight pressure, which is generally applied by insects. This process, called **tripping**, distributes pollen onto the body of the insect entering the flower and brings the stigma of the flower in contact with pollen from other plants which is already present on the body of the insect.

The legume flower contains ten stamens. Nine of the filaments of these stamens are partially fused forming a cylinder around the pistil, with the one remaining stamen remaining free. Enclosed within the cylinder of stamens is the pistil which consists of the **ovary**, **style**, and **stigma**. The ovary develops into the <u>pod</u>. The stigma and style function in pollination and then dry up. Inside the ovary are the **ovules** which form the seed of the legume. The pistil and stamens are attached to the enlarged base called the **receptacle**.

Label the diagram of the legume flower and its parts.

Standard	Ovary	Anther
Wings	Stigma	Filament
Keel	Style	Calyx
Sepals	Receptacle	Corolla

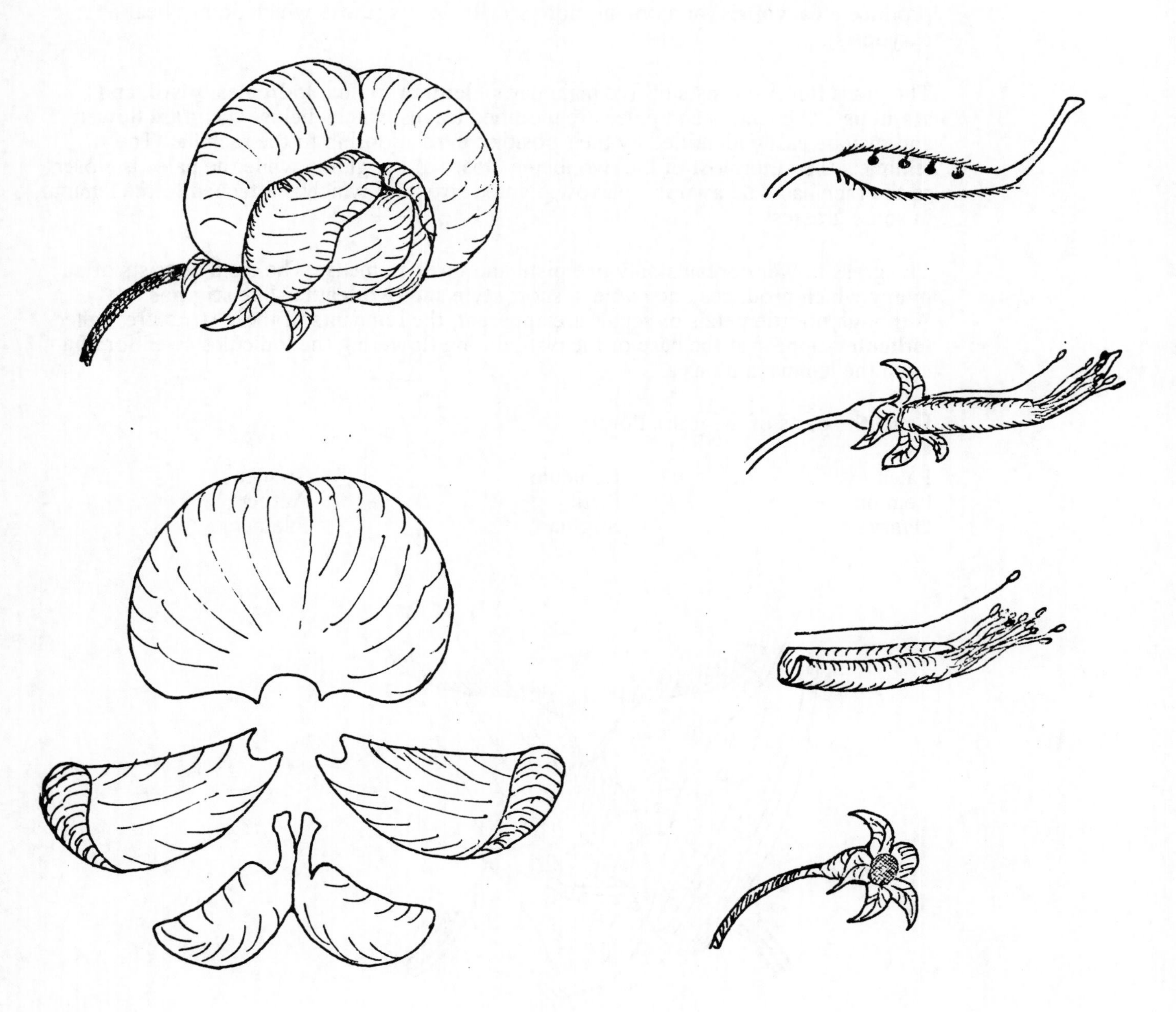

Fig. 2. The legume flower.

C. The Grass Flower

The flowers of grasses have no distinct perianth and are borne singularly or in groups of two or more in compact units called **spikelets**. These spikelets are the basic units of a grass **inflorescence**. A spikelet consists of a pair of modified leaves called **glumes** which enclose one or more **florets**. In grasses that have more than one floret per spikelet, the florets are arranged alternately on a small central axis called the **rachilla**. Spikelets generally contain one or more **fertile** florets (those which produce a **caryopsis**) and one or more **sterile** florets (those which do not bear a caryopsis).

The grass **floret** consists of five main parts: **lemma**, **palea**, **lodicules**, **pistil**, and **stamens**. The lemma and palea are modified leaves which enclose the grass flower, and may be easily identified by their position in relationship to the rachilla. The lemma is the outermost of the two, being closer to the glume while the palea is closer to the rachilla. The **awn** is a narrow pointed structure which is attached to the lemma in some grasses.

The grass flower contains only one pistil and three stamens. The pistil consists of an **ovary** which produces one ovule, a short **style** and two feather-like **stigmas**. Although no true petals or sepals are apparent, the remnants of these structure, called **lodicules**, appear at the base of the pistil during flowering the lodicules swell forcing open the lemma and palea.

Label the parts of the grass floret.

Palea	Lodicule	Stamens
Lemma	Pistil	Anther
Ovary	Stigma	Filament

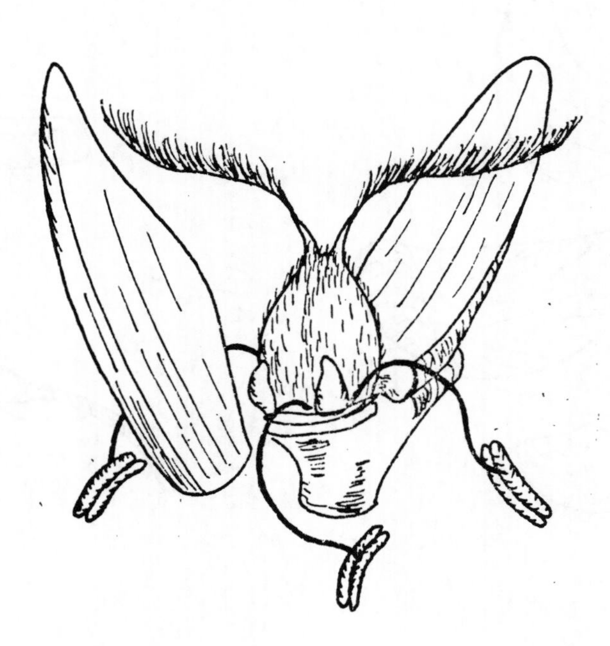

Fig. 3. The grass floret.

Dissect the oat spikelet given to you and label the parts on the diagram below. Note that the harvested grain of oats is actually the entire flower.

Glume
Lemma
Sterile floret
Awn
Palea
Rachilla
Fertile floret

Fig. 4. The oat spikelet.

Dissect the wheat spikelet given to you and label the parts on the following diagrams. Note the different appearance of the glumes and other structures of wheat as compared to oats.

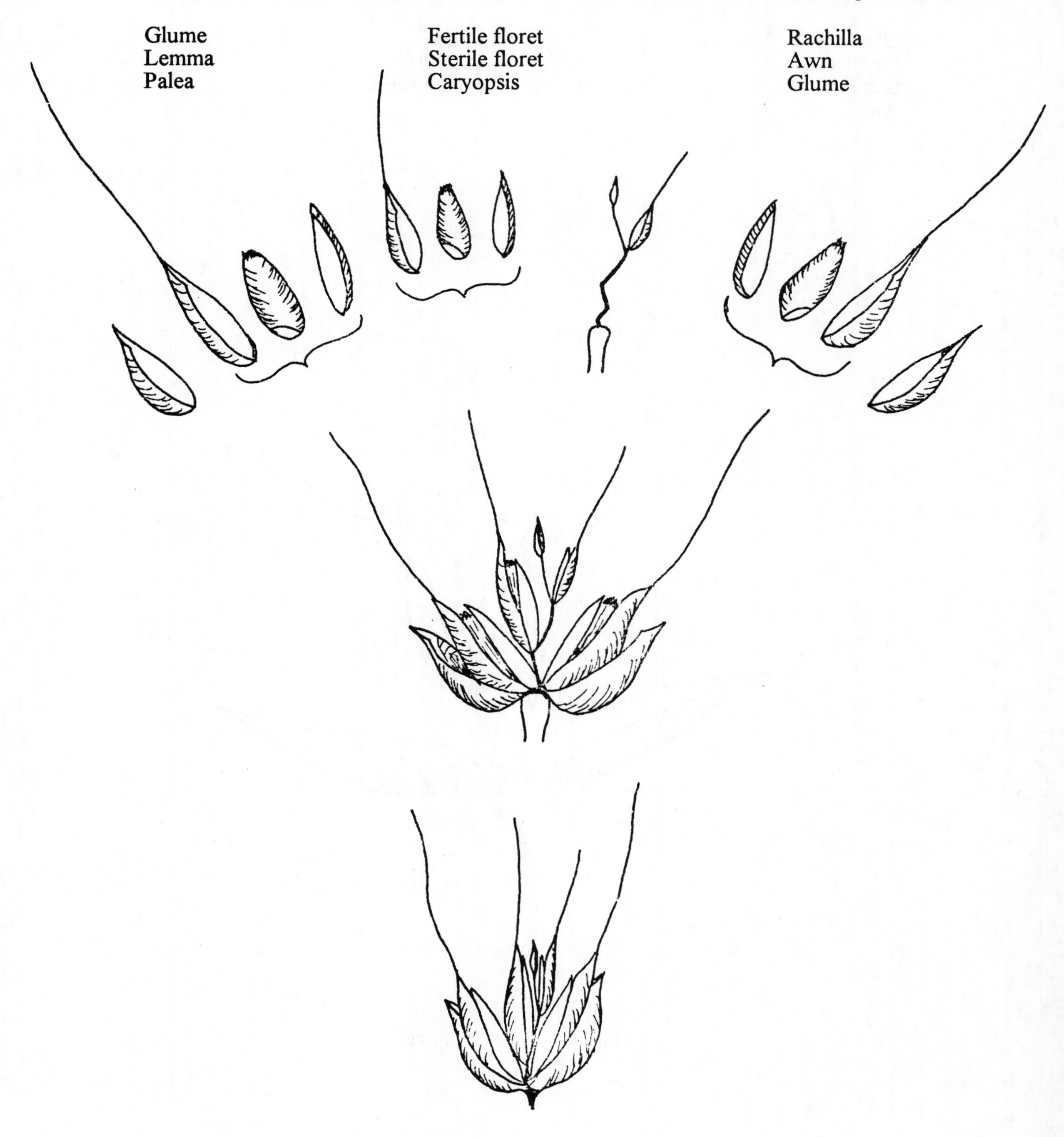

Fig. 5. The wheat spikelet.

D. Inflorescence Types

The flowers of legumes and spikelets of grasses appear on groupings called **inflorescences** which are highly variable and bear special names.

Inflorescence of most legumes is generally either a head or a receme. The **head** inflorescence has a very short floral axis which is called the rachis. Flowers are borne on very short compacted branches called **pedicels**, providing a spherical type of inflorescence. Red clover and white clover are good examples of such legumes. The **raceme** inflorescence is characterized by a long rachis to which flowers are attached by means of short branches called pedicels. Alfalfa and sweetclover are good examples of legumes exhibiting a raceme inflorescence.

Grass species may have one of three types of inflorescence, which is supported by a stalk called the **peduncle**. The **spike** inflorescence consists of a central rachis with the spikelets attached directly or sessils to it. Wheat, rye and barley are good examples of the spike inflorescence. The **panicle**, the most common type of grass inflorescence, consists of spikelets held by pedicels which are in turn attached to branches from the rachis. Panicles are usually more open and irregularly branched than are other inflorescence types. Examples of the panicle inflorescence include smooth bromegrass, timothy, and oats. The least common grass inflorescence type is the **raceme** which is characterized by spikelets attached to the rachis by pedicels. Examples of this inflorescence type include crabgrass and goosegrass.

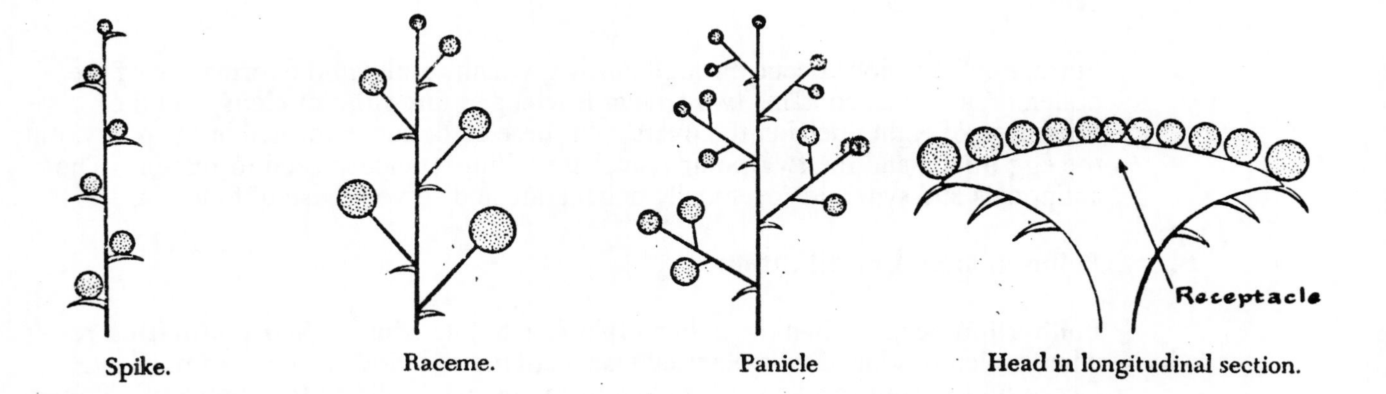

Fig. 6. Types of inflorescence.

E. Identification of Plants by Floral Characteristics

Variability of floral characteristics among legume and grass species makes this a reliable tool for the identification of an individual species. In legumes, characteristics such as flower size, flower color, and inflorescence type show wide differences among species. Similarly, grasses can be compared for such traits as number of florets per spikelet, relative size and texture of glumes, location of sterile florets relative to fertile florets, and inflorescence type. General characteristics such as production of perfect of imperfect flowers can be used in both legumes and grasses. If the species produces imperfect flowers, we can look further to see if the species is monoecious or dioecious.

II. SEED DEVELOPMENT

Objectives

1. Define self pollination, cross pollination, cross fertilization, self fertilization and double fertilization.
2. Describe the steps in the reproduction of seed plant.

Seed formation and development is critical to grain crop production. An adequate understanding of the principles of seed formation will aid in relating plant growth and development to the environment.

A. Gamete Formation

Seed development begins with the formation of the male gametes (microspores) in the anther and female gametes (megaspores) in the ovary. The steps involved in this process are briefly outlined in Figure 7. Formation of both male and female gametes starts with a plant cell which contains the regular chromosome number. **Chromosomes** occur in the nuclei of cells and carry the genetic material which influences the development of the different plant characteristics. Each cell contains two identical copies of each chromosome (2N). Cells resulting from the divisions of meiosis contain only 1 copy of each chromosome, and are therfore designated (1N) cells.

Further cell divisions occur through mitosis, which results in the formation of the pollen grain which contains **two sperm nuclei** and **one tube nucleus**, and the formation of eight nuclei in the ovary. Of these eight nuclei formed in the ovary, only the **egg nuclei** and the **two polar** nuclei are of importance in seed formation. The antipodals and synergids eventually degenerate and serve no useful function.

B. Pollination and Fertilization

Pollination occurs when the pollen grain falls on the stigma. **Self-pollination** results when pollen produced on a plant falls on a stigma of a flower on the same plant. **Cross-pollination** results when pollen from one plant falls on the stigma of a flower on a different plant.

Shortly after pollination, the pollen grain germinates and pollen tube grows through the style and enters the ovary. Tube growth is controlled by the tube nucleus. The sperm nuclei then enter the ovary and one of the nuclei unites with the egg cell to form the zygote. The other sperm nucleus unites with the two polar nuclei to form the endosperm. This process which involves two nuclear fusions is called **double fertilization**.

Since the endosperm formation involves the fusion of 1 nucleus from the male and 2 nuclei from the female, each containing a 1N number of chromosomes, the resulting endosperm contains three copies of each chromosome (3N) instead of the normal two copies (2N).

Only one ovule can be fertilized by the nuclei of a given pollen tube. Since there is more than one ovule per ovary in many legumes this means that separate pollen tubes must enter the ovary to fertilize each ovule. After fertilization the ovules grow and develop into seeds. In legumes, the ovary wall expands and matures into the pod.

Self-fertilization occurs when male and female gametes from the same plant unite. **Cross fertilization** results when male and female gametes from different plants unite.

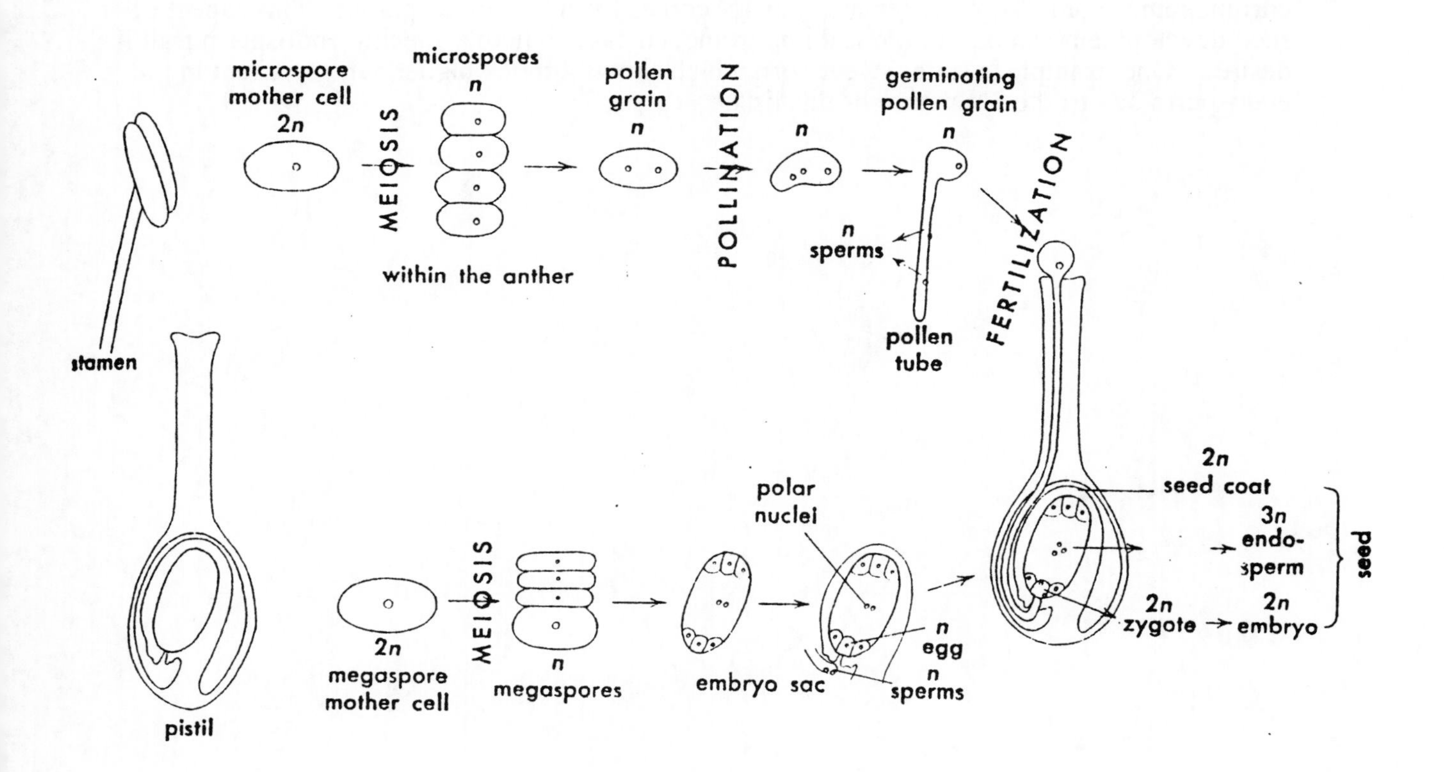

Fig. 7. Steps in the reproduction of a seed plant. Starting with the spore mother cells in the anthers and in the ovules, a succession of events takes place which leads to fertilization and eventual formation of a seed.

III. PRACTICAL APPLICATION

The study of floral morphology and seed development may seem to be a fruitless area in terms of application to crop production, however, an understanding of floral morphology and reproduction can very quickly become important when grain production and seed production are considered.

One area of agronomy which deals directly with the effect of the environment on seed formation and plant flowering habit is plant breeding. Hybridization techniques depend on the floral characteristics of individual plant species and the environment. For example if a breeder chooses to work with buffalo grass, which is dioecious, he must have both male and female plants present to make plant hybridization or crossing possible. Similarly a seed producer of this crop will not make much money from a field of male plants.

Changes in the environment can also strongly affect both the plant breeder and producer. Low temperature and lack of rainfall can seriously influence when pollen is shed by the anthers and the receptiveness of the stigma and style. If environmental conditions are not ideal, changes in the timing of pollen shed (**anthesis**) in relationship to stigma receptiveness may be affected. If these two events are sufficiently offset, the breeder will not be able to make his cross, and the farmer will suffer reductions in seed production.

Certain aspects of double fertilization are also of concern in some crop production situations. Since the endosperm of a seed has **three** copies of each chromosome (3N), traits obtained from the female parent are expressed to a greater degree, since 2N chromosomes come from the female, and 1N comes from the pollen parent. This aspect of seed development can be of practical importance in cases where a specific endosperm trait is desired. One example is super sweet corn which shows a much higher sugar content in the endosperm due to the 3N nature of this tissue.

DEFINITIONS

1. Anther - Pollen-bearing portion of the stamen.

2. Anthesis - The process of dehiscence (splitting open) of the anthers; period of pollen distribution.

3. Awn - Long, slender extension of the lemma or glume, sometimes called the "beard".

4. Calyx - Whorl of modified leaves called sepals associated with a flower.

5. Chromosomes - Structures occuring in the nucleus of the cell which carry the genetic information.

6. Complete flower - A flower which has sepals, petals, pistil and stamens present.

7. Corolla - Whorl of modified leaves called petals associated with a flower.

8. Dioecious - Plants having the stamens and the pistils in different flowers on different plants, resulting in male and female plants.

9. Diclinous - Plants having the stamens and the pistils in separate flowers, either on the same plant or different plants.

10. Endosperm - Tissue which arises from the triple fusion of a sperm nucleus with the polar nuclei, contains three copies of each chromosome (3N).

11. Fertilization - Union of an egg and a sperm (gametes) to form a zygote. Self fertilization is the union of an egg with a sperm from the same plant. Cross fertilization is the union of an egg with a sperm from a different plant or a different clone.

12. Filament - Thread-like stalk which supports an anther.

13. Floret - Flowering unit of the grass plant consisting of the lemma, palea, pistil, stamens, and lodicules. Florets which develop a caryopsis are called fertile florets, while those which do not develop a caryopsis are called sterile florets.

14. Head - Inflorescence type with sessile or nearly sessile flowers on a very short rachis or receptacle.

15. Imperfect flower - A flower that does not contain both pistil and stamens.

16. Incomplete flower - A flower which is missing one or more parts (sepals, petals, pistil, stamens.).

17. Lemma - Floral modified leaf adjacent to glume side of caryopsis.

18. Lodicule - Structures at the base of the grass flower that swell and open the palea and lemma at flowering. Remnants of the calyx and corolla.

19. Meiosis - Two successive nuclear divisions in the course of which the 2N chromosome number is reduced to 1N in the nucleus of the cell.

20. Mitosis - Nuclear division in which the chromosomes are duplicated longitudinally, each having a 2N chromosome number.

21. Monoclinous - Plant having both the stamens and the pistils in the same flower.

22. Monoecious - Plants having the stamens and the pistils in separate flowers on the same plant.

23. Ovary - The enlarged basal portion of the pistil, in which seed is borne.

24. Ovule - Structure which bears the female gamete and becomes the seed after fertilization.

25. Palea - Thin paper-like structure adjacent to rachilla side of caryopsis.

26. Panicle - Inflorescence type in which spikelets are borne on pedicel from branches on the rachis.

27. Pedicel - Branch supporting a spikelet or flower in an inflorescence.

28. Peduncle - Section of grass stem supporting the grass inflorescence.

29. Perfect flower - A flower which contains both pistil and stamens.

30. Perianth - Floral structure consisting of the calyx and corolla.

31. Pistil - Flower part that produces female gametes, composed of stigma, style, and ovary.

32. Pistilate flower - Flower which contains only a pistil and no stamens.

33. Pollination - Transfer of pollen from the anther to a stigma.

34. Raceme - Inflorescence type in which spikelets are born on pedicels from the rachis.

35. Rachilla - Central axis of a spikelet.

36. Rachis - Central axis of a grass inflorescence.

37. Sessile - Attached directly to the rachis or main stem.

38. Spike - Inflorescence type in which spikelets are sessile to the rachis.

39. Spikelet - Basic unit of the grass inflorescence, consists of two glumes and all that lies between.

40. Stamen - Part of a flower that produces male gametes. Composed of anther and filament.

41. Staminate flower - Flower which contains stamens and no pistil.

42. Stigma - Expanded, fuzzy or sticky tip of the style to which the pollen adheres.

43. Style - Slender column of tissue which arises from the top of the ovary through which the pollen tube grows.

44. Zygote - Cell resulting from the fusion of the gametes.

QUESTIONS

1. What composes the perianth of a soybean flower? How does a wheat flower differ?

2. What floral characteristics of grasses can be used to distinguish grass species? Can these traits be considered reliable?

3. Why would a seed producer want to know if the species he is working with is mono-clinous or diclinous? Monoecious or dioecious?

4. Can a flower be both incomplete and perfect? Explain.

5. How does pollination differ from fertilization?

6. What is double fertilization?

7. What is the difference between a spikelet and a floret?

8. When producing certified seed, why should a farmer be aware of whether his crop is self or cross-pollinated?

SEED QUALITY AND CERTIFICATION

Objectives

1. List factors that determine seed quality.
2. Describe the three commonly used tests which measure "vigor" of seeds.
3. Conduct and interpret a standard warm germination test
4. List information found on seed tags.
5. Use information on seed tags to do cost analysis and comparison of seed.
6. Determine planting rate required to get a desired population.
7. Analyze a seed sample for its contents.
8. Calculate percent purity and pure live seed.
9. Construct a seed tag which reflects the characteristics of a seed lot.

A. Seed Quality

Seed quality is determined by (1) the kind and variety of seed, (2) weed seed content, (3) purity, (4) germination percentage, and (5) vigor of the seed.

1) Kind and variety of seed

Only crop varieties that are well adapted to the soil and environmental conditions of given area should be considered for commercial production. Cotton and hard red winter wheat are important crops, but should not be produced in Indiana since they are not well adapted to our climate. Jupiter soybeans, a variety produced in the southern United States, are not adapted to Indiana because day length during the growing season is too long for them to flower and set seed. Also, a variety resistant to certain diseases or insects may be a necessity when grown where these pests are prevalent.

2) Weed seed content

The kind and amount of weed seed present is very important when purchasing seed. Most farmers have enough trouble controlling weeds already in the soil without sowing more weeds with their crop. The Indiana Seed Law prohibits the sale of seed containing any prohibited noxious weed seed, more than 1/4 of 1 percent restricted weed seed, or more than 2-1/2 percent total weed seed.

3) Purity

The percent purity of the seed indicates how much of a seed lot contains material other than the desired seed. This may include other crop seed, other variety seed, dirt, pieces of stems, or other undesirable material.

4) Germination percentage

The germination percentage is an indication of the amount of viable or living seed in the seed lot. It also indicated the ability of the seed lot to produce a stand under ideal environmental conditions. The germination percentage is determined in the laboratory by use of the warm ("rag-doll") germination test. For soybeans this consists of placing the seed in rolled paper germination towels and placing them in a germinator at 77°F and 100% relative humidity for 8 days. After this period, each seed is classified as dead, abnormal, or germinable. A germination percentage is then calculated.

5) Vigor

Vigor is defined as the ability of seed to germinate under adverse conditions. Many seeds in the germination percentage are "weak" seeds due to the presence of disease, mechanical damage, or improper storage. These "weak" seeds may be unable to produce a healthy plant when planted under adverse field conditions. Vigor of seeds is commonly measured by the cold test, tetrazolium test, or accelerated aging test.

In the cold test, soybean seeds are stored in moistened muck or soil-sand mix at 50°F for 7 days. Then, they are germinated for 5 days at 77°F and the germination percentage is determined. Results of this test are consistently lower than those of the warm test. This test is a better indication of a seed's performance under cold, wet conditions common in late April or early May.

The tetrazolium test is a chemical stain test in which seeds are placed in a 1% tetrazolium solution for 5-7 hours. Healthy tissue is indicated by a bright pink color, damaged tissue by a dark red to black color, and dead tissue by a white color. The stained soybean is split with a razor blade and a vigor rating is determined by location and severity of damaged and dead tissue.

The accelerated aging test is an indication of the storability of seed. The seed is subjected to high temperature and high humidity for 2 days. A warm germination test is conducted on the seed. Seed with high vigor should store better and have a higher percentage germination than low vigor seed.

B. Pure Live Seed

After selecting the desired variety and checking the weed seed content, the purity and germination percentages can be used to determine the percentage of pure live seed.

The percentage of pure live seed can serve as a simple index of seed quality. It is the number of pounds out of every 100 pounds of bulk seed capable of producing seedlings.

Pure live seed is determined by the following formula:

percent purity x percent warm germination = percent pure live seed

Consider the following example of two seed lots:

Lot 1: 95% pure x 90% germination = 85.5% pure live seed.
Lot 2: 80% pure x 60% germination = 48% pure live seed.

It would take nearly twice as much seed from Lot 2 to supply 100 pounds of pure live seed. When the price of bulk seed is known, a simple calculation will reveal which seed lot is actually cheaper. This formula is:

$$\frac{\text{bulk seed price}}{\text{percent pure live seed}} = \text{price on a pure live seed basis}$$

Suppose in the above example Seed Lot 1 costs $50/100 pounds bulk seed and Lot 2 costs only $30/100 pounds bulk seed. It would at first appear that Lot 2 would be the best buy. Seed should be bought on a pure live seed basis as bulk seed prices may be misleading.

On a pure live seed basis the calculations actually show that Lot 1 is the best buy!

$$\text{Lot 1 costs } \frac{\$50}{.855} = \$58.48/100 \text{ pounds pure live seed}$$

$$\text{Lot 2 cost } \frac{\$30}{.48} = \$62.52/100 \text{ pounds pure live seed}$$

C. Seed Tag Information

Much information accompanies the seed bought by the farmer. This information can be found on the seed tag. Items that may be found on the tag are:

1) Lot number
2) Seedsman and his address
3) Kind and variety of seed
4) Minimum pure seed percentage
5) Crop seeds percentage
6) Minimum germination percentage
7) Date of germination test
8) Inert matter percentage
9) Weed seed percentage
10) Where the seed crop was grown
11) Hard seed percentage -- viable legume seed that is physically dormant because of a seed coat impermeable to water; **scarification** is the process that removes the physical dormancy.

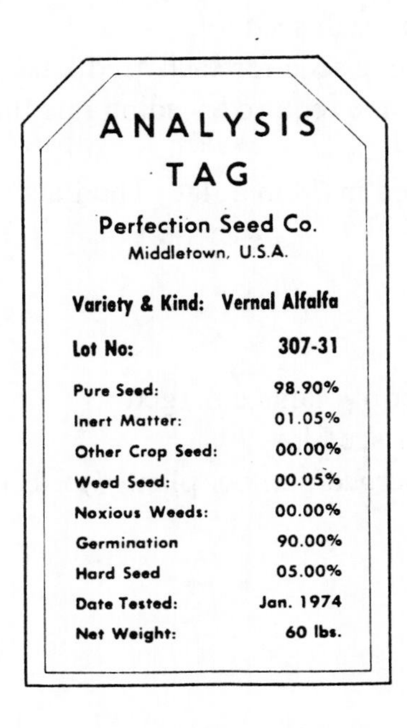

ANALYSIS TAG

Perfection Seed Co.
Middletown, U.S.A.

Variety & Kind:	Vernal Alfalfa
Lot No:	307-31
Pure Seed:	98.90%
Inert Matter:	01.05%
Other Crop Seed:	00.00%
Weed Seed:	00.05%
Noxious Weeds:	00.00%
Germination	90.00%
Hard Seed	05.00%
Date Tested:	Jan. 1974
Net Weight:	60 lbs.

Fig. 1. Seed Tag Example

SEED QUALITY EXERCISE

Work in groups of three or four. Each group obtains a sample from one seed lot:
CORN 1, CORN 2, SOYBEAN 1, OR SOYBEAN 2.

WEEK 1

INDIVIDUALLY

1. Obtain a sample from your group's seed lot.
2. Determine total weight of sample, excluding bag weight.
3. Separate sample into the following components, weigh, and record.
   - desired crop seed
   - other crop seed (may not be present)
   - prohibited noxious weed seed (may not be present)
   - inert matter
4. Calculate the percent purity of your sample.
5. Prepare a ragdoll to determine warm germination from your seed sample. See instructions in lab.

WEEK 2

INDIVIDUALLY

1. Count germinated seeds in ragdoll.
2. Calculate percent germination. Record.

AS A GROUP

3. Average your individual scores to obtain results for your lot of seed.
4. Obtain results of percent purity and percent germination for each seed lot from other groups (areas shaded on chart).
5. Calculate seeding rate per acre for each lot of seed to obtain a corn stand of 28,000 plants/A or a soybean stand of 180,000 plants/A. Base your seeding rate on germination percent.
6. Calculate percent pure live seed for each seed lot.
7. Calculate price per pound of pure live seed for each seed lot.
8. Use your group results to calculate the required seeding rate that will result in 30 plants in a flat.
9. Plant the number of seeds calculated in #8 in a flat. The flats will be taken to the greenhouse for growout.

WEEK 3

INDIVIDUALLY

1. Check your group's flat for number of plants emerged
2. Calculate percent emerged for your seed lot.

   Percent emerged=(number emerged/number planted) x 100.

Name______________________ID#______
Group Members ____

Seed Lot	YOUR SAMPLE	GROUP'S SAMPLE	CLASS RESULTS Corn 1	Corn 2	Soyb. 1	Soyb. 2
		________				
WEEK 1						
Total sample weight, grams	________	________				
Weight of crop seed, grams	________	________				
Weight of other crop seed, grams	________	________				
Weight of noxious weed seed, grams	________	________				
Weight of inert matter, grams	________	________				
Percent Purity	________	________	________	________	________	________
WEEK 2						
Percent germination (from ragdoll))	________	________	________	________	________	________
Seeding rate for 30 plants/flat (calculate)		________				
WEEK 3						
Number of plants emerged (from growout)		________				
Percent Emergence (calculate)		________				
Seeding rate per acre (calculate)			________	________	________	________
Percent pure live seed (calculate)			________	________	________	________
Price per pound of pure live seed (calculate)			________	________	________	________

EACH PERSON HANDS IN THEIR OWN WORKSHEET. BE SURE TO COMPLETE THE FOLLOWING:

1. Fill in all blanks on seed quality worksheet.
2. Make a seed tag for your group's lot of seed using your group's results.
3. Please answer questions A, B, C, and D of this worksheet. Information may be found in the Crops Resource Center.

QUESTIONS

A. Compare and contrast the four seed lots. Which would you buy if you were farming?

B. What other factors should you consider when purchasing seed?

C. What does the tetrazolium test tell you?

D. What is an advantage and a limitation of the tetrazolium test?

E. What colors are the following seeds in a tetrazolium test?

Dead
Damaged
Healthy

GRAIN GRADING

Introduction

The primary function of the USDA Grain Grading System is to provide a uniform standard of quality in the grain marketing system. Grading standards have been established for wheat, barley, oats, rye, soybeans, corn, sorghum, flax, canola, and mixed grains.

The Grading Procedure

Each step in the grain grading procedure is presented at a station. These stations are:

1. Obtaining the Sample
2. Moisture Determination
3. Test Weight Determination
4. Grain Grading Factors
5. Grade Determination
6. Marketing and Pricing

To complete this exercise, proceed through the stations and place a grade on your sample of grain. At the marketing and pricing station, you will use the information you obtained in the first five stations to arrive at a monetary value for your sample.

As you proceed through the stations fill out the data sheet for your grain. The data sheet can be obtained in the lab review area of the Crops Resource Center.

When finished, place the data sheet in the Assignment Box in the Crops Resource Center.

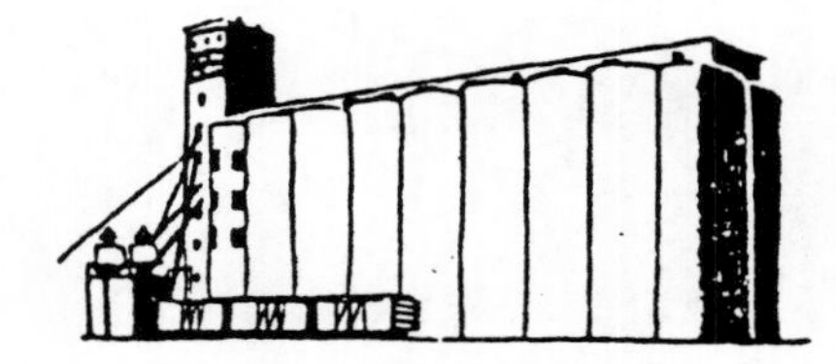

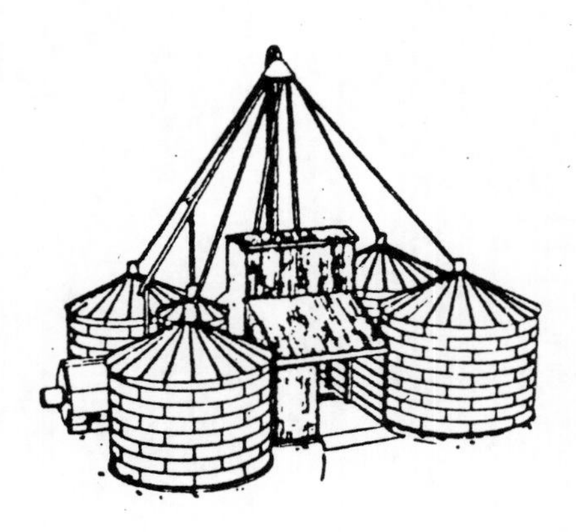

Station 1. OBTAINING THE SAMPLE

Objective: To become familiar with different types of sampling equipment and learn how to obtain a representative sample from a lot of grain.

Obtaining a sample that is truly representative of the lot of grain is an essential part of grain grading. If the sample obtained is not representative, it will be very difficult to determine a true grade for the lot involved. It is essential that samples are randomly taken from several areas in the lot of grain. After each sample is taken, the grain is visually examined for quality and uniformity. The samples are then bulked together to form the representative sample of approximately 1 1/2 to 2 quarts. There are several types of sampling equipment available that can be used to obtain a representative sample. A brief description of some of these are listed below.

1. Pelican sampler: used for obtaining a representative sample from a free-falling stream of grain.

2. Probe or trier: used for sampling bulk grain in trucks or boxcars. These can be of various sizes, but a common type is a 12-compartment apparatus made of brass.

3. Automatic sampler: works on the same principle as the probe or trier, except that the sample is taken automatically. An operator inside the elevator has control over the sampling process.

Procedure

Read the information, and follow the instructions to obtain a sample of grain. You will use the sample at the other stations to determine its grade.

Station 2. MOISTURE DETERMINATION

Objectives:

1. To state the moisture content needed for safe storage of grain.
2. To determine the moisture content of a grain sample.

The moisture content of grain is important for 2 reasons:

1. The moisture content largely determines how well the grain can be stored without spoiling.

   Numerous species of fungi are always present on grain. At moistures below 14% their development is inhibited and molding is not usually a problem.

2. Moisture content affects the volume of grain being marketed.

   Consider the following: Water has weight, and must be paid for when grain is bought or sold. For example: A freight car will hold approximately 200,000 lb. of wheat. With 14% moisture, 200,000 lb. of wheat contains 28,000 lb. of water. With 8% moisture, the same carload contains only 16,000 lb. of water--a difference of 12,000 lb. of water. Since a gallon of water weighs 8.345 lb., the 14% moisture wheat contains 3,355.3 gal. of water, and the 8% moisture wheat contains only 1,917.3 gal. of water. Assuming 60 lb. of wheat per bu., the quantity of water present would be equivalent to 466.7 and 266.7 bu. of wheat, respectively.

There are several types of moisture meters available, most work on electrical conductivity, and wetter grain produces a greater electrical flow, giving a higher meter reading.

Procedure

Determine the % moisture of your grain sample using both the hand-held moisture meter, and the Burrows moisture meter. Follow the operating instructions carefully.

Did you obtain similar readings? If not, which do you think is most reliable?

Station 3. TEST WEIGHT DETERMINATION

Objective: To determine the test weight of a grain sample.

Test weight is commonly expressed as pounds per bushel (lb/bu). Since the amount of grain sold is determined by weight and the price paid for the grain is quoted by volume (1 bushel = 2150.4 cubic inches) a conversion factor for weight per bushel is necessary.

Several instruments may be used to determine test weight. Traditionally this method involves using a balance scale, however, we will use a more recently developed electronic scale.

Procedure

1. Fill the one quart container with the grain sample (strict guidelines must be followed). The bottom of the hopper is exactly two inches above the top of the one quart container and the opening of the hopper is 1 1/4 inches in diameter. Any deviations from this may cause variations in the weight-per-bushel test.

2. Using the striker level the grain with 3 zig-zag motions so the quart container is level full.

3. Place the container on the scale. The reading on the scale will tell you the pounds per bushel of the grain.

Station 4. GRAIN GRADING FACTORS

Objective: To identify grain damage factors in a sample of grain.

Determining the grade

In addition to test weight and moisture, the grade and price received for the grain is determined by several other factors.

CORN GRADING FACTORS

A. Cracked corn and foreign material.

Includes cracked corn and all matter other than corn including other grains that passes through a 12/64 sieve, and all matter other than corn that remains on the sieve after screening. Corn of another class is a form of foreign material.

B. Heat damage

Severely discolored (brown to black) from external heating, or from excess moisture and spoilage.

C. Total damage

Includes heat-damaged, sprouted, frosted, badly ground damaged, badly weather damaged, some forms of insect damage, and kernels that have become slightly discolored from heat. Note that the percent heat damage is added to other types of damage for obtaining the percent of total damage.

D. Special grades

Corn that contains 2 or more live weevils, one live weevil and 5 or more other injurious insects, or 15 or more other injurious insects has the word "weevily" added to the grade designation.

E. Factors causing grain to be automatically placed in SAMPLE GRADE are:

1. Stones and Cinders

   Sample may not contain more than 7 stones, or 0.2% of sample weight.

2. Musty, sour, hot, or objectionable odor

   Any of these conditions render the grain unfit for normal use.

3. <u>Distinctly low quality</u>

Grain containing rocks, glass, etc., or over 0.2% rodent excreta.

SOYBEAN GRADING FACTORS

A. Foreign material

Anything that passes through the 8/64 sieve, including soybeans and pieces of soybeans plus all matter other than soybeans that remain on the sieve after sieving.

B. Splits

Soybeans that remain on the sieve and are missing more than one-fourth of the seed. Expressed as a whole percent.

C. Heat damage

Soybeans that are materially discolored by heat.

D. Total damage

Includes all types of damage, such as: Heat-damaged, frost and immaturity, sprout, insect, slight discoloration by heat, ground damage and moldy.

E. Brown, black or other colored soybeans

Soybeans of other seed and seed coat colors (not due to heat damage).

F. Other grading factors

"Garlicky" contains 5 or more garlic bulbets per 1000 g. "Weevily" contains live weevils or other injurious insects. Beans which are garlicky or weevily have the word "garlicky" or "weevily" added to the grade designation.

Station 5. GRADE DETERMINATION

Objective: To place a grade on a sample of grain.

Minimum and maximum requirements for the various grading factors have been established for each possible grain grade. Minimum limits restrict test weight and maximum limits restrict foreign material and amount of damage that is permitted in each of the grades. If a sample of grain fails to meet the requirements established for the numerical grades it is graded U.S. Sample Grade.

For example, U.S. No. 1 yellow corn must have a test weight of at least 56.0 lbs/bu. If a sample only weights 55.9 lbs/bu then it cannot be graded as U.S. No. 1 because of low test weight. The factor or factors that reduce the sample to the lowest grade (regardless of what the other factors indicate) determines the grade of the sample of grain.

A sample of yellow corn that tests as follows:

Test weight	56.2 lb/bu
Cracked/foreign	1.8%
Total damage	5.4%
Heat damage	0.0%

is graded as U.S. No. 3 yellow corn, because, even though test weight, cracked/foreign, and heat damage fulfill the requirements for U.S. No. 1 yellow corn, total damage meets the requirements for U.S. No. 3 yellow corn, causing the sample to be graded in U.S. No. 3 yellow corn. See page 90 for the official grading standards.

In addition to determining the proper grade you must write the grade correctly. The following procedure, used in commercial channels is suggested for writing the grade.

1. Write the abbreviation "U.S. No."
2. Write the numerical grade you have given the sample
3. After the grade write the class of grain

The following are examples of correctly written grades:

U.S. No. 2 yellow corn
U.S. No. 3 yellow soybeans, weevily
U.S. Sample Grade

Station 6. MARKETING AND PRICING

Objectives:

1. To recognize the different types of information found on a trade sheet report from the Chicago Board of Trade.

2. To determine the value of your sample of grain.

Marketing the crop is the final step in a growing season. There are two markets for agricultural commodities--cash and futures. The markets are separate but will influence each other.

Placing a grade on the commodity offers both the buyer and the seller the opportunity to pay and receive an appropriate price. It seems logical that U.S. Grade No. 1 grain should be worth more than grain that is, for example, only sample grade. But how is the price determined?

The principles behind supply and demand are the major factors that influence the value of a commodity. Typically, prices for corn and soybeans are based on the value of U.S. Grade No. 2. Grain of lower quality (Grades 3 through sample) are discounted and are of lower value.

Rather than formally grading grain as it comes in to an elevator (too time consuming), a system of price discounting is established based on the market price of a standard grade. Examples of typical discount schedules are found in the display area. Using these discount schedules you can determine the price a farmer would receive at the elevator when he delivers a truckload of grain.

Grades and Grade Requirements for Corn

Grade	Minimum Test Weight Per Bushel	Maximum limits of -- Damaged Kernels Total	Heat-damaged Kernels	Broken Corn and Foreign Material
	lbs.	%	%	%
U.S. No. 1	56.0	3.0	0.1	2.0
U.S. No. 2	54.0	5.0	0.2	3.0
U.S. No. 3	52.0	7.0	0.5	4.0
U.S. No. 4	49.0	10.0	1.0	5.0
U.S. No. 5	46.0	15.0	3.0	7.0
U.S. Sample Grade:				

U.S. Sample Grade shall be corn which:
- a) Does not meet the requirements for the grades U.S. Nos. 1, 2, 3, 4, or 5; or
- b) In a 1,000 gram sample, contains 8 or more stones which have an aggregated weight in excess of 0.20 percent of the sample weight, 2 or more pieces of glass, 3 or more crotalaria seeds (Crotalaria ssp.), 2 or more castor beans (Ricinus communis), 8 or more cocklebures, 4 or more particles of an unknown foreign substance(s), or a commonly recognized harmful or toxic substance(s), or animal filth in excess of 0.20 percent; or
- c) Has a musty, sour, or commercially objectionable foreign odor; or
- d) Is heating or otherwise of distinctly low quality.

Grades and Grade Requirements for Soybeans

Grade	Minimum Test Weight Per Bushel	Maximum limits of -- Splits	Damaged Kernels (Total)	Heat-damaged Kernels	Foreign Material	Soybeans of Other Colors
	lbs.	%	%	%	%	%
U.S. No. 1	56.0	10.0	2.0	0.2	1.0	1.0
U.S. No. 2	54.0	20.0	3.0	0.5	2.0	2.0
U.S. No. 3 1/	52.0	30.0	5.0	1.0	3.0	5.0
U.S. No. 4 2/	49.0	40.0	8.0	3.0	5.0	10.0
U.S. Sample Grade:						

U.S. Sample Grade shall be soybeans which:
- a) Do not meet the requirements for U.S. No. 1, 2, 3, 4; or
- b) Contain 8 or more stones which have an aggregate weight in excess of 0.2 percent of the sample weight, 2 or more pieces of broken glass, 3 or more crotalaria seeds (Crotalaria ssp.), 2 or more castor beans (Ricinus communis), 4 or more pieces of an unknown foreign substance(s) or a commonly recognized harmful or toxic substances, 10 or more rodent pellets, bird droppings, or an equivalent quantity of other animal filth in 1,000 grams of soybeans; or
- c) Have a musty, sour, or commercially objectionable foreign odor (except garlic odor); or
- d) Are heating or otherwise of distinctly low quality.

1/ Soybeans which are Purple Mottled or Stained shall be graded not higher than U.S. No. 3.

2/ Soybeans which are Materially Weathered shall be graded not higher than U.S. No. 4.

WEED CONTROL

Objectives:

1. Define a "weed", and describe possible harmful and beneficial effects of weeds.
2. Describe the four main methods of weed control.
3. Describe methods of applying chemical herbicides.
4. Devine herbicide "mode of action."
5. List ways that spray drift can be reduced.
6. Explain why use of a wetting agent is desirable for post emergent herbicide spraying.
7. Recognize the ten parts of a herbicide label.

A. Weeds

A weed is a plant growing where it is not desired, or a plant out of place. Rye in a wheat field is a weed. Tall fescue in a Kentucky bluegrass lawn can be considered a weed if it is not desired. Weeds include all types of plants - trees, grasses, broadleaved plants, sedges, rushes, aquatic plants, and parasitic flowering plants. The losses from these plants to agriculture alone total about $4 billion yearly.

In our efforts to produce food, we often think of only the harmful effects weeds have upon our environment. Think for a moment and list what the harmful aspects of weeds are. Do weeds have any beneficial aspects to society? You should be able to list at least three harmful and three beneficial effects of weeds.

Effects of weeds

A. Harmful effects

B. Beneficial effects

B. Methods of Weed Control

Weed control is one of the most expensive steps in crop production. Methods that control certain species will not even slow the growth of other species. Factors to be considered in selecting the proper method of control include longevity of the weed, growth habit, morphology, physiology, age, area where it's growing, methods and equipment available, and timeliness.

Improper weed control for one year can make serious problems many years later. Dr. Beal's seed-viability test started in 1879 at Michigan Agricultural College points out this fact. Pint bottles of various seed were planted and excavated every ten years. In 1970 three of the original 21 species planted still contained viable seed. Yearly weed control is important to keep weed populations limited through the years.

1. Mechanical

The oldest methods of weed control include mechanical methods such as hand weeding, using the garden hoe, and the weed-hook. More commonly used mechanical methods include cultivation, mowing, and flaming.

a. Cultivation

Although cultivation may have additional benefits, the main function is to control weeds. Most weeds are easily killed by cultivation just as they emerge from the soil. Some of the primary tools for use after planting are the spike-tooth harrow, spring-tooth harrow, rotary hoe, and row cultivator. The proper use of a plow, disk-harrow, or field cultivator can also eliminate a large number of weeds prior to seeding.

After a corn or soybean crop has been planted, a rotary hoe can be an economical weed killer. The rotary hoe should be pulled about ten to thirteen miles per hour and possibly weighted so that it stirs the soil. Best results are attained if the field is rotary hoed when the weeds are in the "white stage". This stage is when the weeds are usually less than one-half inch tall.

The spring-tooth harrow has been used quite effectively to control Johnsongrass and quackgrass as well as other weeds. The spring-tooth harrow pulls soil-covered rhizomes onto the surface, where they then dry out and die.

The row cultivator is effective in controlling weeds if properly used. However, if the shovels are set too deep or too close to the row, this will prune crop roots. Shallow cultivation is preferred for soybeans. This keeps ridging to a minimum and will make harvesting easier.

b. Mowing

Mowing may effectively control tall annual weeds and certain perennials. The main control it offers is in prevention of seed production. It may also starve underground parts. In order to be effective on perennial species, mowing should be repeated several times at intervals when the leaves have recovered from the previous mowing just enough to again start storing food reserves in the roots. Weeds that form mats, rosettes, or low-growing sod usually cannot be controlled by mowing.

c. Flaming

Flaming gas has been used to control weeds in cotton, in fence rows, and flaming has been partly successful in controlling weeds in corn fields. When corn is 10-12 inches tall the burners of the flamer can be dropped close to the ground to kill weed seedlings, with little damage to the corn.

One draw-back to flame weed control is that it is quite expensive. Flaming costs are about 4-5 dollars per acre for fuel alone. This does not include the expense of the flamer.

2. Cultural

a. Crop Rotations

Proper rotation of crops is an efficient way to control weeds. Certain weeds are more prominent in some crops than in other crops. Smartweed, pigweed, ragweed, velvetleaf, and jimsonweed are more common in row crops than in small grains. Thistles, curly dock, and downy brome are common in pastures in row crops. Wild mustard, wild garlic, and thistles are common in small grains. A rotation including summer row crops and winter small grains generally aids in controlling weeds, and is environmentally sound.

b. Crop Competition

Weed control by crop competition involves using the most favorable crop production methods available. For example, using the correct seed bed preparation methods, planting dates, row spacings, and seeding rates generally helps to control weeds by encouraging growth of the crop.

3. Biological

Biological weed control involves using a natural enemies, such as insects and diseases, of the weedy plant which will not harm desired plants. In the Western United States, St. Johnsonwort is being controlled by leaf-eating beetles. In the cotton fields of the South, domesticated geese have been used to rid cotton fields of foxtail seedlings. Prickly pear cactus has been controlled by the use of a moth borer.

4. Chemical

Weed-killing chemicals are called herbicides. Herbicides are generally described as preplant herbicides, pre-emergence herbicides, or post-emergence herbicides according to the time they are applied. A **preplant herbicide** is any herbicide that is applied before the crop is planted. A **pre-emergence herbicide** is any herbicide which is applied prior to emergence of the crop. A **post-emergence herbicide** is a herbicide applied after the crop has emerged.

Herbicides are applied **broadcast**, as a **band**, or as a **direct spray**. The broadcast application refers to a uniform application to an entire area. Band application refers to treatment of a narrow band directly over a crop row. Direct sprays are applied to a specfic part of the plant, usually the lower leaves.

A **translocated herbicide** is one where the herbicide moves though the phloem tissue of the plant. It may cause necrosis to occur where the herbicide may not actually have been sprayed. A **contact herbicide**, like Paraquat, will kill plant parts covered by the herbicide and not affect parts of the plant untouched by the chemical.

Herbicides can be classified as being **selective** or **non-selective**. A selective herbicide will control some weed species and leave your crop intact. A non-selective herbicide will destroy all vegetation it contacts.

C. Classification of Herbicides by Mode of Action

Most herbicides can be grouped according to their mode of action. Herbicides within each group react similarly on plants, and if you know the characteristics of each group you will be able to make better decisions when you decide which herbicide to use. Read WS-13, Diagnosing Herbicide Injury, to learn the modes of action groups of some common herbicides.

DEPARTMENT OF

B·O·T·A·N·Y

PLANT·PATHOLOGY

PURDUE UNIVERSITY • WEST LAFAYETTE IN

Weed Science

Purdue University
Cooperative Extension Service

Diagnosing Herbicide Injury

Thomas N. Jordan, Daniel J. Childs
Extension Weed Specialists

Consider all possible causes when diagnosing herbicide injury, including environmental and mechanical factors, as well as damage from disease and insects. Observe the condition of the plant. Also, look for the effectiveness of the herbicide on weeds in the treated area. Herbicide injury will usually occur in definite patterns within a field. Look for:

- drift patterns across a field
- overlapped rates at the end of rows
- uniformly injured strips caused by application equipment
- differential injury across soil types (injury on light soils such as sandy knolls and/or lack of weed control on heavy soils such as in low spots in the field).

Primary Sites of Action and Injury of Some Common Herbicides

Most herbicides can be grouped according to their mode of action (the type of injury symptoms they express). By knowing a few characteristics of each group of herbicides, the injury in a particular field can be identified through a process of elimination. The specific herbicide may not be readily determined, but all herbicides that do *not* produce the characteristic injury symptoms can be eliminated. This leaves a much smaller group of herbicides to consider. Many new trade names of herbicides on the market today are simply prepackaged mixtures of two or more herbicides (i.e. Bicep, a mixture of Dual and atrazine). Several of these mixtures combine herbicides from different mode-of-action groups. Using the example above, Bicep is composed of Dual, a cell-growth inhibitor, and atrazine, a photosynthetic inhibitor. While injury to a given crop could be caused by each representative herbicide of the mixture, giving a variety of symptoms, it is most often noted that one class of herbicidal injury will be prominent. By checking records of the fields in question, or by determining what was applied in adjacent fields, the herbicide which caused the injury can usually be identified.

Note: The premixes listed for each mode-of-action group contain at least one representative herbicide from that group. The other herbicides in the premix may be from another group. Therefore, if symptoms do not fit the particular herbicide from that group, then refer to characteristics of other groups.

1. Photosynthetic Inhibitors

Action: These herbicides are most often applied to the soil and are absorbed by plant roots. The herbicide moves with the flow of water (systemic) into the foliage by the xylem system but does not return to the roots by the phloem system. When photosynthetic-inhibitor herbicides are used postemergence, their action is contact and requires thorough wetting of the foliage, usually with the aid of an adjuvant. Injury occurs on leaves and above-ground stems. No damage is caused to the root system. Plants must germinate and turn green before they die.

Injury: Plants turn yellow then die back from the bottom to the top. Leaves turn yellow between the veins, then begin dying from the tip toward the base and from the outer edges toward the center. Leaves will fall off the plant, leaving only a stem with an apical bud.

Specific Herbicides: Atrazine, Bladex, Sinbar, Lorox, Princep, Sencor/Lexone

Premixes: Bicep, Extrazine II, Marksman, Preview, Salute, Laddock, Lariat, Bullet, Buctril/Atrazine, Lorox Plus, Sutazine+, Turbo.

2. Cell-growth Inhibitors

Action: Herbicides work on the germinating seedlings and stop growth of the roots and/or shoots before they emerge.

Injury: Symptoms may be expressed in the roots and/or the shoots, but mostly in one or the other. Those herbicides that affect the root system usually do not cause any visible damage to the above-ground system other than stunting or discoloration of foliage, while those that affect the shoot system usually cause no visible damage to the below-ground system.

Root injury symptoms: Injury may appear as root pruning and the inhibition of secondary roots. Roots may be swelled or club-shaped. The underground portion of the stem will be thickened and shortened. Stems of soybean plants may become callus and brittle at the soil surface.

Specific Herbicides: Treflan, Balan, Sonalan, Prowl

Premixes: Salute, Squadron, Commence, Tri-Scept, Pursuit Plus, Passport

Shoot injury symptoms: Corn shoots appear twisted and leaves tightly rolled. Stems can become ruptured, with new growth protruding out of the ruptured tissue. Soybean leaves are dark green, crinkled, and/or leaf tips flattened. Leaves may fail to unfold from buds. Root damage may also occur, but not very often.

Specific Herbicides: Lasso, Dual, Eptam, Sutan+, Vernam, Genep, Genate, Eradicane, Eradicane Extra

Premixes: Bicep, Sutazine+, Turbo, Lariat, Bullet

3. Growth Regulators

Action: Growth-regulator herbicides can be taken up from the soil by the root system. However, most growth-regulator herbicides are used as postemergence treatments; thus, these cause more damage to the shoot system than to the already established root system. When growth-regulator herbicides are used as pre-plant treatments in reduced or no-tillage systems, root damage can occur.

Injury: Soil-absorbed herbicides can cause both root and shoot injury. Usually Banvel injury to soybean shoots is more severe than 2,4-D or 2,4-DB at equal rates. Both 2,4-D and Banvel cause plants to turn darker green than normal. In corn, the brace roots will fuse together and the shoot will "buggy whip" (onion leaf).

Root injury symptoms: A proliferation or clusters of short secondary roots are visible along the tap root of soybeans. Growth of secondary roots (brace roots) on corn is inhibited and usually the brace roots are fused together into a leaf-like structure. With Alanap, roots may turn upward as if growing out of the ground.

Specific Herbicides: 2,4-D, Banvel, Amiben, Alanap

Premixes: Rescue, Marksman

Shoot injury systems: 2,4-D, Banvel and 2,4-DB cause bending and twisting of soybean stems. 2,4-D causes leaf strapping (feathering) of soybean leaves and brittle stems of corn. Banvel additionally causes new growth of soybean leaves to cup upward. Plant tissue may also become dark green in color.

Specific Herbicides: 2,4-D, 2,4-DB, Banvel

Premixes: Rescue, Marksman

4. Cell-membrane Disruptors

Action: This herbicide class consists mostly of postemergence contact herbicides. Thorough coverage of foliage is needed for complete weed control. Toxicity increases with high temperatures and direct sunlight. Incomplete coverage or spray drift can cause spots of dead tissue which may be confused with plant diseases.

Injury: Injury symptoms include desiccation of leaf tissue (leaf burn) caused by disruption of cell membranes (Gramoxone Extra, Blazer, Cobra, Reflex, Basagran). Applications of Cobra to soybeans may cause crinkling of new growth similar to Lasso or Dual injury, besides the characteristic leaf speckling or burning symptoms.

Specific Herbicides: Gramoxone Extra, Basagran, Blazer/Tackle, Cobra, Reflex (**Note:** Basagran is actually a photosynthetic inhibitor; however herbicidal action occurs so quickly that it is sometimes classified under this mode-of-action group)

5. Growing-point Disintegrators (Post Grass Materials)

Action: This herbicide class is active on only grass species, while broadleaf plants are tolerant to these herbicides. Although some of these herbicides have shown minimum soil activity, the major activity is from postemergence applications. Most of the herbicides have both annual and perennial grass activity; however, sensitivity will vary with species. Movement of the herbicide within the plant occurs in both the xylem and phloem.

Injury: Root and shoot growth ceases very rapidly and plants begin to turn color (usually red to purple). In about 7 to 14 days the growing points decay. The outward appearance of the plant will indicate the grass is still alive, but removing the top leaves from the whorl will show that the growing point is dead and rotten. Cutting through each node will reveal that the meristem tissue is dead and disintegrating. In perennial grasses the same decay will be found in the nodes of the underground stems (rhizomes).

Specific Herbicides: Poast, Fusilade, Option, Assure

Premixes: Tornado

6. Amino-acid Inhibitors (Slow Death)

Action: The herbicides in this group are effective on both broadleaf and grass weed species. Roundup is nonselective on most all species; Scepter, Classic and Pursuit are selective in soybeans and Harmony Extra is selective in wheat and barley. The herbicides prevent the production of essential amino acids, which cause the plants, in essence, to slowly starve to death.

Injury: The affected plants stop growing and stay green for several days. Roundup and related compounds may show a small amount of bleaching of the foliage of grasses and sedges around the new growth. Most plants die slowly and turn a uniform harvest death color (golden in grasses and flat green in broadleaves). The meristematic tissues (nodes) do not deteriorate like those treated with the postemergence grass herbicides. Canopy, Lorox Plus, Preview, Pursuit or Scepter injury symptoms on soybeans may include shortened internodes and possibly a reduced root system. With a postemergence application of Classic or Pinnacle herbicides, temporary leaf yellowing and/or retardation of growth may occur on soybeans under certain conditions with an occasional reddening of the leaf veins. Injury to corn from Canopy, Lorox Plus, Preview, Pursuit or Scepter carryover may include one or all of the following injury symptoms: stunted plant, interveinal chlorosis, purpling, red-purple midrib of lower leaves, reduced root system including bottle-brush appearance of lateral roots. **Note:** Both the imidazolinone (Scepter, Pursuit) and the sulfonylurea (Classic) chemistry inhibit the same enzyme within the plant; therefore, these herbicides will display the same injury symptoms.

Accent and Beacon, two other sulfonylureas, may cause stunting, chlorosis and possible malformation of corn leaves under certain conditions.

Specific Herbicides: Roundup, Arsenal, Scepter, Glean, Classic, Oust, Harmony Extra, Pursuit, Pinnacle, Accent, Beacon

Premixes: Squadron, Tri-Scept, Passport, Canopy, Preview, Lorox Plus, Pursuit Plus

7. Pigment Inhibitors

Action: These herbicides are mostly applied as preplant or preemergence treatments. These herbicides inhibit carotenoid synthesis, but do not affect preexisting carotenoids. Therefore, new growth turns white while growth prior to treatment does not. Carotenoids protect chlorophyll from light (photo-oxidation); therefore, destruction of chlorophyll occurs, since chlorophyll is not protected by the carotenoid pigments. The herbicides used in row crops may, under certain conditions, persist long enough to affect rotational crops the following year after application.

Injury: The primary injury to crops is the bleached white appearance of leaves. In soybeans this appearance may be interveinal. In some cases the affected areas of the soybean leaves will be highlighted with a pink to red margin.

Specific Herbicides: Command, Sonar

Premixes: Commence

Reference to products in this publication is not intended to be an endorsement to the exclusion of others which may be similar. Persons using such products assume responsibility for their use in accordance with current label directions of the manufacturer.

RR 8/92 (4M)

Cooperative Extension work in Agriculture and Home Economics, state of Indiana, Purdue University, and U.S. Department of Agriculture cooperating; H. A. Wadsworth, Director, West Lafayette, IN. Issued in furtherance of the acts of May 8 and June 30, 1914. The Cooperative Extension Service of Purdue University is an affirmative action/equal opportunity institution.

Herbicide Mode-of-Action Categories

Merrill A. Ross and Thomas N. Jordan
Botany and Plant Pathology - Purdue University

I. Herbicides causing injury to new growth and with the potential to move from leaves to roots

Auxin Growth Regulators

Trade Name	Common Name	Trade Name	Common Name	Trade Name	Common Name
Phenoxy acids		**Benzoic acids**		**Pyridine acids**	
Butyrac	2,4-DB	Banvel	dicamba	Garlon	triclopyr
Butoxone	2,4-DB	Clarity	dicamba	Stinger	clopyralid
Rhonox	MCPA			Tordon	picloram
Rhomene	MCPA				
various	2,4-D				
various	2,4-DP				

Premixes containing one of the above:
Crossbow, Hornet, Marksman, Resolve, Scorpion III, Shotgun, Tiller, Accent Gold, Optill

Amino Acid Synthesis Inhibitors

Branch Chain Amino Acid Inhibitors

Trade Name	Common Name	Trade Name	Common Name	Trade Name	Common Name
Imidazolinones (ALS)					
Contain	imazapyr	Plateau	imazapic		
Pursuit	imazethapyr				
Raptor	imazamox				
Scepter	imazaquin				

Premixes containing one of the above:
Detail, Contour, Lightning, Pursuit Plus, Resolve, Squadron, Steel, Tri-Scept, Passport

Trade Name	Common Name	Trade Name	Common Name	Trade Name	Common Name
Sulfonylureas (ALS)					
Accent	nicosulfuron	Express	tribenuron	Permit	halosulfuron
Ally	metsulfuron	Glean	chlorsulfuron	Pinnacle	thifensulfuron
Beacon	primisulfuron	Telar	chlorsulfuron	Titus	rimsulfuron
Classic	chlorimuron	Oust	sulfometuron	Matrix	rimsulfuron
Skirmish	chlorimuron	Peak	prosulfuron	Expert	oxasulfuron

Premixes containing one of the above:
Accent Gold, Authority Broadleaf, Basis, Basis Gold, Canopy, Canopy XL, Concert, Exceed, Harmony Extra, Reliance, Synchrony, Spirit

Trade Name	Common Name
Sulfonanilides (ALS)	
Python	flumetsulam
FirstRate	cloransulam

Premixes containing one of the above:
Accent Gold, Broadstrike+Dual, Broadstrike+Treflan, Hornet, Scorpion III

Trade Name	Common Name
Aromatic Amino Acid Inhibitors (EPSPS)	
Roundup Ultra	glyphosate
Touchdown	sulfosate

Pigment Inhibitors

Trade Name	Common Name
Command	clomazone
Balance	isoxaflutole

Premixes containing one of the above:
Commence, Authority One-Pass

Lipid Synthesis (ACCase) Inhibitors (Grass Growing Point Disintegrators)

Trade Name	Common Name	Trade Name	Common Name
Aryloxyphenoxyproprionates		**Cyclohexanediones**	
Assure II	quizalofop	Poast Plus	sethoxydim
Fusilade DX	fluazifop	Prestige	sethoxydim
Hoelon	diclofop	Select	clethodim
Horizon	fenoxaprop	Prism	clethodim
Option	fenoxaprop		

Premixes containing one of the above:
Conclude, Fusion, Manifest, Rezult, Tiller, Tornado, Headline

II. Herbicides causing injury to old growth and with the potential to move only upward

Photosynthesis Inhibitors

Trade Name	Common Name	Trade Name	Common Name	Trade Name	Common Name
Triazines		**Substituted ureas**		**Others (contact)**	
Aatrex	atrazine	Karmex	diuron	Basagran	bentazon
Princep	simazine	Lorox	linuron	Buctril	bromoxynil
Bladex	cyanazine	Spike	tebuthiuron	Moxy	bromoxynil
Cy-Pro	cyanazine			Tough	pyridate
Sencor	metribuzin	**Uracils**			
Lexone	metribuzin	Hyvar	bromacil	(low doses of the above mimic classical photosynthesis inhibitors - high doses mimic cell membrane disrupters)	
Velpar	hexazinone	Sinbar	terbacil		
Pramitol	prometon				

Premixes containing one of the above:
Basis Gold, Bicep, Bullet, Buctril+Atrazine, Canopy, Conclude, Contour, Cypro AT, Extraxine II, FulTime, Galaxy, Guardsman, Harness Xtra, Headline, Manifest, Marksman, Laddok, Lariat, Rezult, Surpass 100, Turbo, Moxy AT

III. Herbicides causing immediate localized injury with little or no movement

Cell Membrane Disruptors (Contacts)

Trade Name	Common Name	Trade Name	Common Name	Trade Name	Common Name
Bipyridiliums		**Diphenylethers**		**Others**	
Gramoxone Extra	paraquat	Blazer, Status	aciflurofen	Authority	sulfentrazone
Diquat, Reward	diquat	Flexstar, Reflex	fomesafen	Resource	flumiclorac
		Cobra	lactofen	Action	fluthiacet
				Aim	carfentrazone
				Liberty	glufosinate

Premixes containing one of the above:
Authority Broadleaf, Authority One-Pass, Authorty First/Synchrony Co-pack, Canopy XL, Conclude, Galaxy, Manifest, Rezult, Stellar, Storm, Tornado

IV. Herbicides applied to the soil with the potential to injure emerging seedlings

Seedling (or Cell) Growth Inhibitors

Trade Name	Common Name	Trade Name	Common Name	Trade Name	Common Name
Root inhibitors-Dinitroanalines		**Shoot inhibitors-Chloroacetamides**		**Shoot inhibitors-Carbamothioates**	
Balan	benefin	*BAY FOE-5043	thiafluamide	Eradicane	EPTC
Prowl	pendimethalin	Dual	metolachlor	Sutan	butylate
Pentagon	pendimethalin	Frontier	dimethenamid		
Sonalan	ethalfluralin	Harness	acetochlor		
Treflan	trifluralin	Surpass	acetochlor		
Tri-4	trifluralin	TopNotch	acetochlor		
Trific	trifluralin	Lasso	alachlor		
		Micro-Tech	alachlor		
		Partner	alachlor		
		Ramrod	propachlor		

Premixes containing one of the above:
Bicep, Broadstrike+Dual, Broadstrike+Treflan, Bullet, Lariat, Detail, DoublePlay, Guardsman, Harness Xtra, Surpass100, FulTime, Turbo, Commence, Pursuit Plus, Steel, Squadron, Storm, Shotgun, Tri-Scept, Optill, Passport, *Axiom

HERBICIDE SCREEN - CROP & WEED DIAGNOSTIC KEY

1a. Corn Injured or Dead - **2**

1b. Soybeans Injured or Dead - **7**

1c. Both Corn and Soybeans NOT Injured or Dead - **10**

  2a. Only grasses injured or dead - growing point rotten - *Growing Point Disintegrators*

  2b. Grasses and some broadleaves injured or dead - **3**

    3a. Bleached/white appearance of grasses, broadleaves - *Pigment Inhibitors*

    3b. Canola, soybeans o.k.; clubbing of corn, shattercane roots - *Cell-growth Inhibitors*

  2c. Most broadleaves injured or dead - **4**

    4a. Broadleaves and some grasses stunted; soybeans o.k.; corn stunted;
      canola dead or dying - *Amino acid Inhibitors*

    4b. Broadleaves and some grasses burnt, leaves crispy - *Cell-membrane Disruptors*

  7a. Grasses, few broadleaves injured or dead - **8**

    8a. Soybean leaves crinkled, alfalfa o.k. - *Cell-growth Inhibitors*

    8b. Soybeans stunted, yellow; shattercane dying - *Amino acid Inhibitors*

  7b. Most broadleaves and grasses are dead,
      except fall panicum and shattercane - *Photosynthetic Inhibitors*

  7c. Most broadleaves and grasses are dead; shattercane injured - *Photosynthetic Inhibitors*

  7d. Most broadleaves injured or dead; grasses o.k. - **9**

    9a. Broadleaves crispy, burnt - *Photosynthetic Inhibitors*

    9b. Soybeans, broadleaves are bent, twisted - *Growth Regulators*

  10a. Most grasses injured or dead, few broadleaves dead - *Cell-growth Inhibitors*

  10b. Grasses o.k.; broadleaves dead and/or leaves crispy - *Photosynthetic Inhibitors*

D. Control of Spray Drift

It is important for spray drift to be controlled. The soybeans across fence row may be stunted by the 2,4-D you are applying for post emergence control of weeds in your corn crop. Waiting for the wind to subside may not be the economical thing for you to do to avoid spray drift. Drift can be lessened by:

1. lowering the spray boom
2. using larger orifice sized nozzles
3. reducing pressure of herbicide delivery

It is important to note that recalibration of the sprayer will need to be done to achieve proper rates of application.

E. Use of a Wetting Agent

A postemergence herbicide will be more effective if a wetting agent is used. The cuticle of leaves retards movement of materials into the internal structure. The wetting agent is a detergent and will break the waxy cuticle enough to allow movement of the herbicide into the leaf. Water tension is reduced with the use of a wetting agent. This insures a good kill. Without its use most of the herbicide rolls off the plant in the water droplets.

F. Nozzle Types

The flow rate through a nozzle varies with size of nozzle tip and nozzle pressure. Although there are many specialized types of nozzles only a few are commonly used for applying pesticides to field crops. See the description of nozzle types, their suggested uses, and the type of pattern they produce.

G. Herbicide Label Parts

Important information is presented on the herbicide label. You should be familiar with this information. See the following sample label.

Nozzle Types for Use on Field Crops

Type	Suggested Use	Recommended Pressures	Types of Spray Pattern
Reg. Flat-fan	Preemergence and post-emergence herbicides; some insecticides	15-30 psi - never over 40 psi for weed spraying	Fan-like pattern of medium droplets
Even Flat-fan	Preemergence for banding chemicals in row	15-30 psi - do not exceed 40 psi	Even fan like pattern, full dosage over entire width
Flooding flat	Preemergence and post-emergence herbicides where drift is problem, liquid fertilizers	8-25 psi for maximum drift control; below 30 psi	Fan-like pattern with numerous coarse droplets for weed control
Hollow cone	Most insecticides and fungicides	40-80 psi and above; below 40 psi if used for weed control	Circular with light application in center; fine spray droplets
Raindrop	Preemergence herbicides, broadcast or incorporation kits	20-60 psi should not exceed 70 psi	Very large drops in a hollow cone pattern to reduce drift
Whirl chamber	Preemergence herbicides for incorporation kits	5-20 psi high drift potential above 20 psi	Hollow cone pattern with fan angles up to 130 degrees

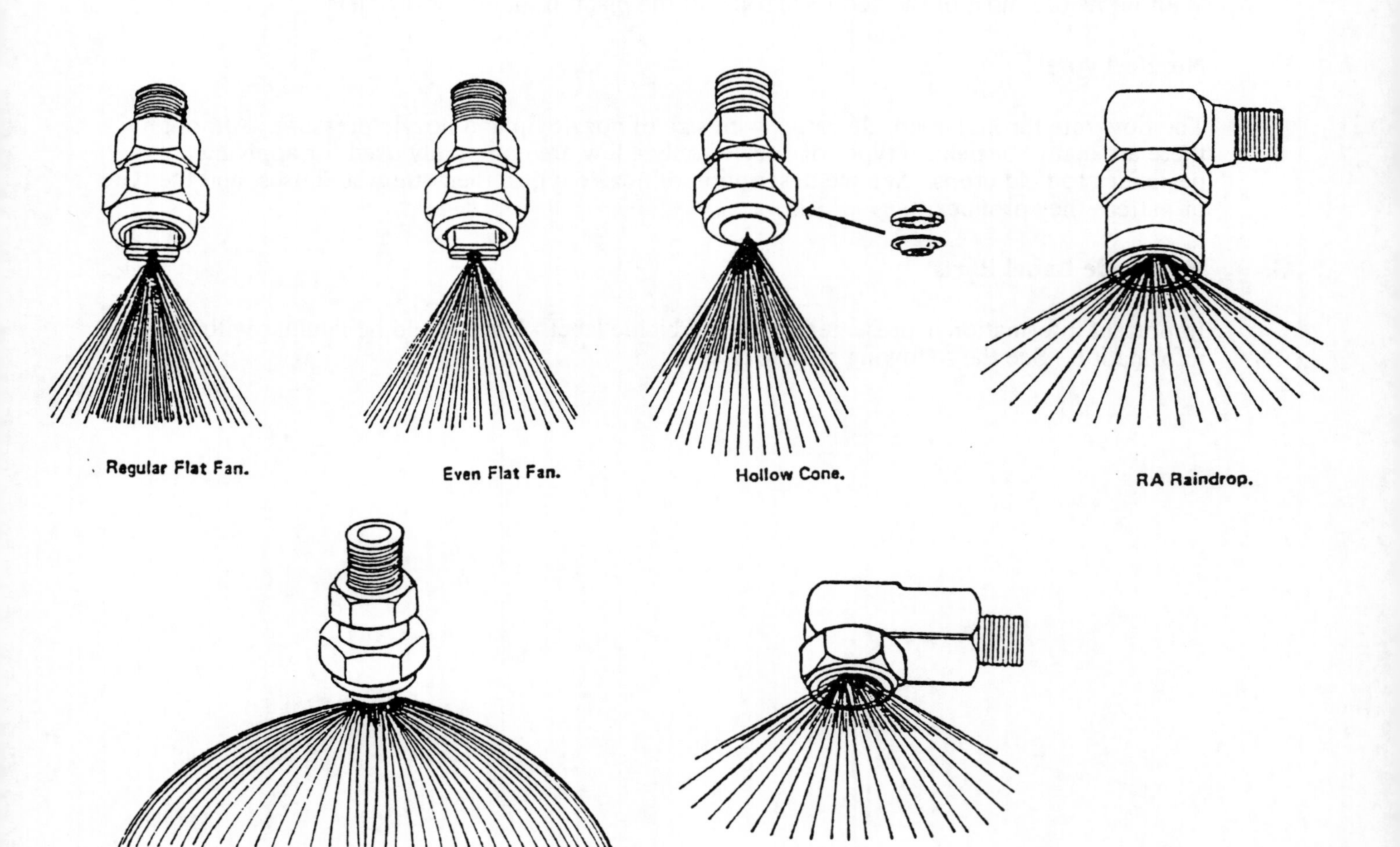

Regular Flat Fan. Even Flat Fan. Hollow Cone. RA Raindrop.

Flooding Flat Fan. Whirl Chamber.

2

AAtrex®

5

4L

SAMPLE LABEL

Herbicide

For season-long weed control in corn, sorghum, and certain other crops

Active Ingredients:
3 → Atrazine: 2-chloro-4-ethylamino-6-isopropylamino-s-triazine .... 40.8% ← 4
Related compounds .... 2.2% ← 6
7 → Inert Ingredients: 57.0%
Total: 100.0%

2½ ← 9
Gallons
U.S. Standard Measure

AAtrex 4L contains 4 lbs. active ingredients per gal.

Shake well before using.

Use entire contents at one time.

Caution ← 8

Keep Out of Reach of Children.

Harmful or fatal if swallowed. Do not get in eyes. Avoid contact with skin, inhalation of vapors or spray mist, and contamination of food and feed.

Do not contaminate domestic or irrigation water supplies or lakes, streams or ponds.

Do not reuse container.
Destroy when empty.

AAtrex® trademark of CIBA-GEIGY for atrazine

EPA Est. 100-LA-1

EPA Reg. No. 100-497

See directions for use inside booklet.

10

Agricultural Division
CIBA-GEIGY Corporation
Greensboro,
North Carolina 27409

CGA 7L38K 090

1 → CIBA–GEIGY

CGA 130-561

HERBICIDE LABEL

1.	TRADEMARK -	name or symbol used by a company on its products, i.e.: ELANCO, DOW
2.	TRADE NAME -	name given to a product for identification and marketing purposes, i.e.: AATREX, DALAPON
3.	COMMON NAME -	name used by the public to designate the product, i.e.: ATRAZINE, DIPHENAMID
4.	CHEMICAL NAME -	name of the chemically active ingredient, i.e.: 2,4-Dichlorophenoxyacetic acid
5.	FORMULATION -	form in which herbicide is prepared (W.P., E.C., Granular)
6.	ACTIVE INGREDIENT -	toxic or carrier, expressed in percent
7.	INERT INGREDIENTS -	nontoxic or carrier, expressed in percent
8.	WARNING -	precautionary statements concerning the safe use of the product
9.	WEIGHT OR VOLUME -	
10.	MANUFACTURER AND ADDRESS -	

NATIONAL CORN HANDBOOK

PEST MANAGEMENT NCH-39

Weeds and Weed Control Systems – What Is A Weed?

C. W. Swann, The University of Georgia

Reviewers

E. L. Knake, University of Illinois at Urbana-Champaign
W. B. Duke, Cornell University
A. R. Martin, University of Nebraska
R. L. Ritter, University of Maryland

A weed is an unwanted plant and is commonly defined as a plant out of place. From this standpoint, a plant is considered a weed if it interferes or hinders efforts to utilize land, water, or animal resources.

Characteristics of Weeds

Weeds encompass a wide range of plants and represent numerous plant families, but despite their botanical diversity, they share many characteristics which adapt them to their role as weeds. Most weeds are adapted to survival under adverse conditions as well as conditions favorable for crop growth. Generally, they are prolific seed producers with some species producing more than 100,000 seeds on an individual plant. Weed seeds are persistent, and many are capable of remaining dormant in the soil for many years despite periodic tillage and cultivation; they can germinate at a later time when conditions are favorable for survival. Weeds are adapted for dispersal—the seed of some species are transported by wind or water and others are moved from place to place by clinging to fur or clothing. Seeds are also dispersed over great distances by contaminating crop seed, equipment, hay, straw, and a wide array of other carriers.

Perennial weeds are especially troublesome since many are capable of reproduction by seed and by vegetative structures such as tubers, bulb and bulblets, rhizomes, and stolons.

LOSSES DUE TO WEEDS

Competition

Weeds reduce crop yields through competition for (1) nutrients, (2) soil moisture, and (3) sunlight, primarily because they have approximately the same growth requirements as crops. In general terms, for every pound of dry matter of weeds produced in a field, one less pound of corn will be produced.

Nutrients

Many weeds tend to accumulate nutrients in higher concentrations than does corn. For example, pigweeds and lambsquarters have been reported to contain two to five times more nitrogen, phosphorus, and potassium than does corn. And where soil nitrogen levels are high, weeds such as pigweeds may accumulate sufficiently high levels of nitrates and become toxic to animals when the weedy forage is consumed as green material, hay, or silage; mortality may result.

Soil Moisture

Water is one of the most critical factors limiting corn yields. Moisture stress for just a few days during tasseling and silking can result in crop failure. Since weeds, such as lambsquarters, require twice as much water to produce a pound of dry matter than is required to produce a pound of corn dry matter, even under irrigated conditions, the heavy demand on soil moisture because of weeds can substantially increase moisture stress and potential yield loss.

Sunlight

Interception of sunlight creating shade on corn crops from tall weeds such as cocklebur, pigweeds, johnsongrass, shattercane, and weeds with a vining growth habit, such as morning glories and field bindweed, is an important factor in reducing corn growth and yield.

Harvest Losses

In addition to yield losses caused by competition for nutrients, heavy rank weed growth causes serious harvest losses due to operating inefficiencies of harvest equipment and to gathering losses at the combine header.

In regions with a long growing season where good weed control may have been maintained during the growing season, late-season weed growth may become a problem if the crop is allowed to stand in the field much beyond the stage of physiological maturity. Under these conditions, the competitive shading effect of the crop canopy is lost, and small shade-suppressed weeds will grow rapidly and become a serious harvest problem.

Loss of Quality

The presence of weed seeds and plant parts in harvested crops will often cause spoilage during drying and storage, and such poor quality may result in dockage at the market. The presence of poisonous weeds, such as showy crotalaria, may cause the buyer to refuse the crop until it has been cleaned. Any of these problems will result in poor quality which will translate into a serious financial loss to the producer.

Weeds as Pest Hosts

Weeds serve as hosts to a number of pests that attack corn. For instance, maize dwarf mosaic and maize chlorotic dwarf viruses may be transmitted from johnsongrass to corn by aphids. Southern root knot, lance, sting nematodes, and perhaps other nematodes are hosted by a wide array of weed species. Troublesome insects such as the southern corn rootworm, bill bug, and sugarcane beetle are known to feed upon and be harbored by many different weeds.

Ecology of Weeds in Corn

Climatic conditions and cultural practices have a great impact on the spectrum or kinds of weed species that occur in corn fields. Weed species vary in their adaptation to climatic conditions across the Corn Belt. For example, quackgrass, which is well adapted to cool conditions, is a serious problem in northern corn production areas while johnsongrass, which is adapted to warmer conditions, is a problem in the southern corn production areas.

Cultural practices influence weed species composition. In general, a reduction in tillage tends to favor perennial species and annual grasses. Continued use of selective herbicides may also result in shifting the weed spectrum from one set of problem weeds to another. For example, continuous use of 2,4-D, dicamba (Banvel) or triazines will often result in a reduction of broadleaf species, such as cocklebur and pigweeds, and increase annual grasses such as foxtails, fall panicum, and Texas panicum.

CONTROLLING WEEDS

Prevention

Preventing weed establishment is probably the least-expensive and most-effective means of weed control available to farmers. Prevention can be accompanied by (1) using weed-free crop seed, (2) avoiding weed-seed scatter with farm equipment and farm products, (3) keeping weeds on the farm from going to seed, and (4) reducing the possibility of contamination by seeds drifting from areas adjacent to the farm. Unfortunately, prevention can not always be achieved with complete success because of the contamination resulting from wind, water, birds, and various other natural factors.

Eradication

Complete weed eradication is "the elimination of all vegetative parts and viable seeds of a weed from a site." Eradication is practical and desirable if the infestation of a weed is confined to a limited area. In an on-farm situation, eradication may be possible if a weed occurs as a "spotty" or limited infestation.

Serious perennial weeds, such as johnsongrass, often occur as initial spotty or limited infestations. Spot treatment with effective herbicides or soil sterilants may eradicate these infestations and prevent further spread of the problem. Usually, more than one year of treatment is required to achieve eradication. Wide-spread heavy infestations of weeds are virtually impossible to eradicate.

Control

Weed control is "the process of reducing weed growth and/or infestation to an acceptable or adequate level." However, a control program usually does not result in the complete elimination of weeds from a field or farm, some weeds escape chemical or cultivation treatments and reproduce to result in a continued presence of the species in the field or on the farm. Weed control requires fewer treatments and is far less expensive than weed eradication.

Effective weed control, in any crop, requires integration of the best available cultural, mechanical, and chemical practices and should be considered an integral part of integrated pest management (IPM). Cultural practices including land preparation, fertility, planting date, seeding depth and precision, plant populations, and numerous other factors can strongly influence the capacity of crops to compete with weeds. Any practice that promotes optimum crop growth and gives the crop a competitive advantage over weeds will improve the results of cultural and chemical weed control practices.

METHODS OF WEED CONTROL

Seed Bed Preparation

Tillage practices are in a state of change. Present trends are toward minimizing tillage practices to cut costs and minimize soil erosion. Despite the method of land preparation or planting used, it is necessary to destroy weedy vegetation at planting time. Destruction of weeds just before or during planting is necessary to permit the crop to emerge and become established with minimum weed competition. In minimum-tillage or no-tillage approaches to planting, herbicides must be used to destroy weeds in the seed bed.

Crop and Weed Competition

In corn, and most other row crops, research has established that early-season weed control is critical in minimizing yield loss due to weeds. Data collected in numerous states show that if weeds are controlled for five to seven weeks after planting corn, little or no yield loss will result from later-emerging weeds.

Suppression of weed growth by the crop is a key factor in obtaining satisfactory weed control. It is particularly important to obtain a closed crop canopy (foliage cover) as early as possible in the growing season. Narrow rows and uniform plant populations will shorten the time required to obtain a closed crop canopy.

Cultivation

Cultivation continues to be an important and effective means of controlling weeds. Early cultivation when weeds are small is much more effective than late cultivation. Particular care should be taken to operate sweeps, rolling cultivators, and other machinery as shallow as possible to avoid root pruning, while still permitting weed destruction. Some soil should be thrown into the row to cover small weeds; however, avoid excessive ridging that may interfere with harvesting or bring weed seed to the soil surface.

Cultivation should be used primarily as a weed-control practice. Where soil tilth is good, there is no need to cultivate unless weeds are present. Documentation shows that on crusted soils or soils subjected to heavy packing rains, slight yield increases are due to cultivation. In these cases, breaking the soil crust apparently improves infiltration of subsequent rainfall.

Herbicides

Numerous herbicides, herbicide combinations, and tank mixtures are currently in use for weed control in corn, but weed species vary greatly in susceptibility to control with herbicides. It is necessary to know, or maintain records, of the weed species present in individual fields to permit selection of the most effective herbicide or herbicide combinations for their control. Recommendations for control of annual grasses, annual broadleaf, and perennial weeds are presented in sections 10.3, 10.4, and 10.5, respectively.

The following discussions are brief description of the characteristics commonly used in corn production.

Acetanilines

Herbicides in this group provide good control of most annual grasses and control, or suppression, of yellow nutsedge. Control of broadleaf weeds is somewhat variable; tank mixtures with triazine herbicides are usually required to achieve satisfactory control of most broadleaf weeds.

The acetaniline herbicides may be applied either as preplanting soil incorporated treatments or as surface treatments after planting. Soil incorporated treatments are somewhat erratic in weed control when used on coarse-textured soils where rainy conditions cause leaching.

Corn is quite tolerant of the acetaniline herbicides; however, under cool, wet conditions some crop injury may occur. Injured corn seedlings may leaf out underground and fail to emerge, or injured emerged seedlings may display malformed, knotted, and twisted top growth.

Alachlor: This herbicide provides good control of most annual grasses and certain broadleaf weeds. Use in combination with a triazine herbicide or follow with postemergence herbicides and/or cultivation to improve broadleaf weed control.

Metolachlor: Metolachlor is generally more consistent in yellow nutsedge control and is slightly less effective in broadleaf control than is alachlor. Corn is less tolerant of metolachlor than alachlor.

Thiocarbamates

The thiocarbamate herbicides, EPTC and butylate, are highly volatile and must be soil incorporated prior to planting to prevent loss of herbicidal activity. These products provide control of most grasses and control, or suppression, of both yellow and purple nutsedge. To improve control of broadleaf weeds, EPTC and butylate are often tank-mixed with triazine herbicides.

Corn is moderately tolerant of both EPTC and butylate. These herbicides are formulated with a chemical safener R25788 to minimize the potential for corn injury. Either herbicide may cause uneven stands or stunted corn growth if used with improper or incomplete soil incorporation, or under cool, wet soil conditions. Injury has been noted in some instances on sandy soils where heavy rainfall leaches the safener more rapidly than the herbicide. Injury symptoms include malformed, knotted leaves with a buggy, whip-like appearance.

Recent research has established that both EPTC and butylate degrade more rapidly in soils where these herbicides have been applied previously than when used on soils where the herbicides have not been previously used. The short duration of EPTC and butylate persistence in soils where the herbicide has been previously applied is known to result in ineffective control of weeds such as wild proso millet, shattercane, and Texas panicum. Research is currently underway on the evaluation of products that show promise for extending persistence of thiocarbamate herbicides.

Triazines

Triazines are used on corn acreage in the United States more than any other herbicide group. Corn is tolerant of the triazines because it can detoxify the herbicide within the plant cell, but the degree of corn tolerance varies according to the specific triazine herbicide, corn variety, and environmental conditions.

The performance and persistence of triazine herbicides are strongly influenced by soil pH. Under acid conditions (pH 5.5 and less) the triazines are rapidly degraded and the duration of weed control is short. Under alkaline conditions (pH 7.0 and above), triazines degrade rather slowly and both the duration of weed control and the probability of carry-over damage to succeeding crops are increased. The triazine herbicides are highly effective in control of most annual broadleaf weeds, but they are quite variable in control of annual grasses.

Atrazine (AAtrex, Atrazine): Corn is highly tolerant of this triazine herbicide. It has gained wide popularity because it is highly effective in the control of most broadleaf weeds. Control of grasses is somewhat variable; Texas panicum, fall panicum, and giant foxtail usually are not adequately controlled, and mid-to-late season control of crabgrass may be poor.

Atrazine may be used in preplanting preemergence and postemergence applications. Addition of crop oil or surfactants increases the effectiveness of postemergence applications on weeds, but it may also increase the potential for crop injury.

Cyanazine (Bladex): This triazine is slightly more effective in control of annual grasses and slightly less effective in control of annual broadleaf weeds than atrazine. Corn is less tolerant of cyanazine than atrazine; therefore, cyanazine should not be applied to corn grown on coarse-textured soils that are low in organic matter. Cyanazine has shorter soil persistence than atrazine and is unlikely to cause carry-over injury to rotation crops.

Cyanazine may be applied as preplanting, preemergence, or postemergence treatments. Due to the potential for corn injury, cyanazine should not be applied with crop oils as a postemergence treatment.

Simazine (Princep): Corn is highly tolerant of simazine, but because of lower water solubility, more rainfall is required to activate simazine than is required for atrazine or cyanazine. Potential carry-over hazards to rotation crops may be slightly greater for simazine than for atrazine because of solubility differences. Simazine is less effective than atrazine for control of certain broadleaf weeds; however, control of fall panicum and crabgrass is generally better.

2,4-D (Numerous Trade Names)

This herbicide may be applied either as preemergence or postembergence treatment. Over-the-top applications can be made until the crop is 5- to 8-inches tall. Use only as a directed spray after corn is 8-inches tall. Corn is most subject to injury when growing rapidly and if soil moisture and temperature conditions are high. Even so, over-the-top applications are more likely to cause injury than directed treatments. Injury symptoms resulting from 2,4-D include: onion-leafing, abnormal brace-root development, stalk brittleness, and occasionally, incomplete pollination.

Preemergence use of 2,4-D has declined in recent years due to erratic weed control and potential for corn injury. Postemergence treatments of 2,4-D are commonly utilized for control of broadleaf weeds that escape control with preemergence herbicides such as atrazine and cyanazine. But where 2,4-D sensitive crops are present, use amine formulations and coarse spray to minimize drift hazards.

Dicamba (Banvel)

This herbicide is similar to 2,4-D in its effects on both corn and weeds. Dicamba may be applied either as a preemergence or postemergence treatment. Due to a high probability of crop damage, dicamba should not be used as a preemergence treatment for corn planted on coarse-textured soils (sands, loamy sands, or sandy loams) or on soils containing less than two percent organic matter. Since dicamba is primarily a broadleaf control herbicide, it is often tank mixed with alachlor for improved control of annual grasses.

Corn is more tolerant of postemergence applications of dicamba than 2,4-D; however, 2,4-D sprays directed to the base of the crop are less likely to cause corn injury than are over-the-top applications. On the other hand, dicamba should not be applied within 15 days of tassel emergence since poor pollination may result.

Spray particle and vapor drift of dicamba may cause injury to adjacent sensitive crops such as soybeans. The use of directed sprays and/or application of dicamba when adjacent soybeans less than 10-inches tall will minimize the potential for serious injury or soybeans.

Miscellaneous

Bentazon (Basagran): This herbicide may be applied over-the-top of the crop and weeds for postemergence control of yellow nutsedge, and certain broadleaf weeds such as cocklebur. A second application within seven to ten days will often be required for yellow nutsedge control. Corn is highly tolerant of bentazon; however, do not apply to corn that is under stress from weather or previous herbicide treatments.

Ametryn (Evik, Linuron (Lorox): Apply any one of these herbicides by using a sprayer equipped with skids, shoes, shields, or similar devices that insures precision adjustment of the spray pattern to avoid spray contact with leaves or whorl of the crop. Apply with a surfactant to improve coverage of sprayed weeds. Ametryn should not be applied to corn within three weeks of tasseling since pollination problems may result.

Glyphosate (Roundup): Glyphosate may be applied before, during, or after planting—but prior to corn emergence for control of emerged annual and perennial weeds in no-till or minimum-tillage cropping systems. Certain perennial weeds are often not controlled with glyphosate when applied at or near corn-planting time since these weeds are usually not at the appropriate growth stage for best results. Perennial weeds, such as bermudagrass and johnsongrass, are controlled more effectively by glyphosate treatments applied in late summer or early fall following corn harvest than where treatments are applied at or near planting. Glyphosate has no soil activity; therefore, it should be tank mixed with a preemergence herbicide or herbicide combination to obtain extended weed control.

Paraquat (Paraquat): Paraquat may be applied before, during, or after planting but prior to corn emergence for control of emerged annual grasses and broadleaf weeds in no-till or minimum-tillage cropping systems. To improve spray coverage, this herbicide should be used with a surfactant.

Paraquat has no soil activity and should be tank-mixed with a preemergence herbicide or herbicide combination to obtain extended weed control. It may also be used as a postemergence directed spray applied in the same manner as discussed earlier for ametryn and linuron.

Bromoxynil (Buctril, Brominal): This herbicide has recently been registered as an over-the-top treatment for control of certain broadleaf weeds. Bromoxynil must be applied when susceptible weeds are 3 inches or less in height. This herbicide has less potential for drift problems than either 2,4-D or dicamba. Bromoxynil causes temporary leaf burn on corn; however, the crop usually outgrows this condition rapidly.

Tridiphane (Tandem): Tridiphane is a new product which, when tank mixed with a triazine herbicide, may be applied over-the-top of corn for control of annual grasses. Tridiphane has virtually no herbicidal activity when applied alone. For best results, tridiphane/triazine tank mixtures must be applied when annual grasses are at the four-leaf stage or smaller.

SUMMARY

When planning a weed-control program, it is necessary to consider all aspects of your farming operation. You must consider soil texture, organic matter, pH and weather conditions, and their effect on mechanical or chemical control treatments. The degree of seedbed preparation will greatly influence mechanical and chemical options available. Weeds vary greatly in their response to herbicides, so select appropriate herbicides on the basis of the anticipated problem weeds and their response to the available treatment. Consideration must be given to expected-rotation crops since certain herbicides such as the triazines have a carry-over potential which restrict crop rotation options.

A weed-control program will be successful only if all other aspects of crop production are well balanced.

Reference to products in this publication is not intended to be an endorsement to the exclusion of others which may be similar. Persons using such products assume responsibility for their use in accordance with current directions of the manufacturer.

A publication of the National Corn Handbook Project

NEW 5/86 (5M)

Cooperative Extension work in Agriculture and Home Economics, state of Indiana, Purdue University and U.S. Department of Agriculture cooperating. H. A. Wadsworth, Director, West Lafayette, IN. Issued in furtherance of the acts of May 8 and June 30, 1914. The Cooperative Extension Service of Purdue University is an affirmative action/equal opportunity institution.

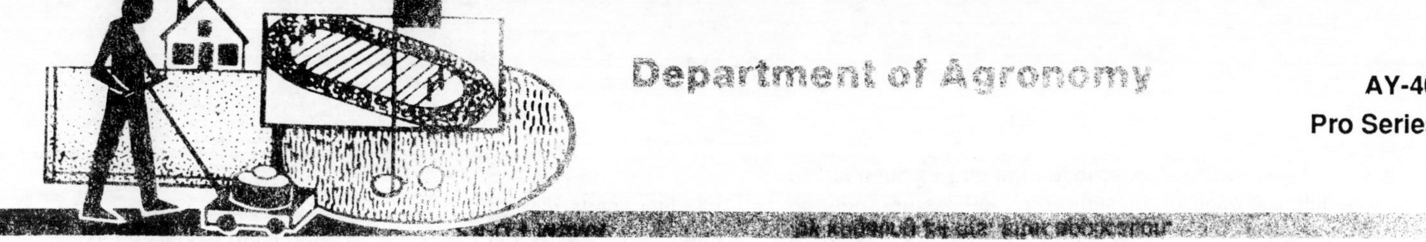

Department of Agronomy

AY-40
Pro Series

Weed Control Recommendations for Turfgrass Areas

by Jeff Lefton, Clark Throssell, and Zachary Reicher, Extension Specialists

Weed	Suggested time of application	Herbicides	Comments
1. Newly seeded turf:			
Crabgrass, foxtail	Can be applied at seeding time or on young seedling turf.	siduron [Tupersan]	Pre-emergent. Needs at least 1/2 inch of water within 3 days after the treatment. Short residual.
2. Established turf:			
Dandelion, buckhorn plantain broadleaf plantain	Fall-mid-September to early November. Spring-mid-April to mid-June.	2,4-D + dicamba (Banvel) 2,4-D + mecoprop (MCPP) 2,4-D + dichlorprop (2,4-DP) [Weedone DPC] 2,4-D + mecoprop + dicamba [Trex-San, Trimec, Three-Way, etc.] 2,4-D + triclopyr (Turflon-D) 2,4-D + dichlorprop + dicamba [Super Trimec] MCPP, MCPA + dicamba [Trimec Encore] chlorflurenol [Break-Thru] +triclopyr + dicamba triclopyrt + clopyralid [Confront] MCPA, MCCP + dichlorprop [Weedestroy]	Post-emergent. Fall application usually the best. Apply when 65-85°F, good soil moisture and sunny skies. Apply to mature turf only. Avoid spray drift on non-target broadleaf plants.

Purdue University Cooperative Extension Service, West Lafayette, Indiana 47907

Weed	Suggested time of application	Herbicides	Comments
Chickweed, henbit, white clover, heal-all, sheep (red) sorrel, curly dock, chicory, yellow rocket, speedwell, ground ivy, etc.	Fall mid-September to early November Spring-April to early May before flowering.	Same as above.	Same as above. Repeated spot treatments following the first application may be necessary for hard to control weeds. Use low volume for the repeat treatment.
Spurge, Oxalis, Knotweed. Purslane, Thristle, etc.	After plants have emerged in the spring or early summer.	2,4-D + dichlorprop (2,4-DP) [Weedone DPC] 2,4-D + trichlopyr [Turflon-D] 2,4-D+ dichlorprop + dicamba [Super Trimec] chlorflurenol [Break Thru] + triclopyr + dicamba	Apply to young plants. Spot spraying at low volume preferred. A repeat treatment may be necessary. Volatilization and chemical drift need to be a concern when making applications near ornamentals, flowers, gardens and fruit trees.
Wild violet	Late spring at the time of flowering (May) or before the frost in the fall.	2,4-D + triclopyr [Turflon-D]	Post-emergent. Apply at low volume. Requires at least 2 years of treatment for control.
Wild onion/garlic	Spring	2,4-D ester	Apply after mowing the turf. Primarily burns back the tops—limited control. Spray during each of 2 successive springs; 2nd spring clean-up essential because plants regrowth from bulbs.
Nutsedge and wild onion/ garlic	Read the label	imazaquin (Image)	Do not use on cool-season grasses; labelled for zoysia-grass and bermudagrass.
Lespedeza	Early summer	2,4-D + triclopyr [Turflon-D]	Low volume preferred.
Yellow nutsedge	When nutsedge is actively growing under good soil moisture conditions	bentazon [Basagran]	Thorough leaf coverage is essential. Apply at 10 to 14 day intervals if necessary. Do not apply more than 3 qts. / acre in one season. Delay mowing for at least 3 days. Repeate treatments may be necessary.
Yellow nutsedge, crabgrass and broadleaf weeds	Young weeds, moderate temperature and good soil moisture	2,4-D + MCPP + diacamba + MSMA [Trimec Plus] 2,4-D + MSMA	Best weed control - 70 to 80°F for KBG; 70 to 90°F bermuda-grass; good soil moisture; reduce discoloration by irrigating 24 hrs. after application.

Weed	Suggested time of application	Herbicides	Comments
Pre-emergent control of crabgrass, foxtail barnyardgrass, goosegrass and many other weeds	In southern Indiana before April 15 and in northern Indiana before May 15. Apply before: 1. Night temp. >65°F. 2. Day temp. consistent between 55 and 65°F. 3. Soil temp. 55-60°F for 7 to 10 days. 4. Moist seedbed	benefin [Balan] benefin + treflan [Team] bensulide [Betasan,Pre-San, Lescosan, Betamec] DCPA[Dacthal]* pendimethalin [Pre-M, Halts Weed grass control] oxadiazon [Ronstar]*/** oxadizon + bensulide** [Goosegrass/Crabgrass Control] dithiopyr [Dimension] prodiamine (Barricade)	*DCPA (Dacthal) and oxidiazon (Ronstar) are labelled as pre-emergent's for sandbur. **Goosegrass control—it germinates 2 to 4 weeks after crabgrass. Control is difficult. Treat 6 weeks after grabgrass germinates.
Post-emergent control of crabgrass, goosegrass, foxtail, and barnyardgrass	Early stage of weed growth preferred. Application rate dependent on the stage of growth up to the 3 tiller stage.	fenoxaprop-ethyl [Acclaim]	Thorough coverage is important. Wait 2 days before and after treatment to mow. Some broadleaf herbicides nullify effectiveness — check the label. Suppressed bermuadagrass.
Post-emergent control of crabgrass, foxtail, barnyardgrass, yellow nutsedge dallisgrass sandbur and oxalis	Apply to young, actively growing weeds during warm weather between 80 and 90°F. Do not apply to drought-stricken turf.	methanearsonate (MSMA, CMA, DSMA, AMA)	Thorough coverage important. Do no water turf for 24 hrs. after the application. Turfgrass may be temporarily discolored. Two or more repeat treatments at 14 day intervals may be needed.
Pre-emergent control of spurge and oxalis	Apply in the spring and again in 6 to 8 weeks.	pendimethalin [Pre-M, Weed grass Control, Halts] DCPA [Dacthal]	Treatments will help reduce the influx of spurge and oxalis. Complete control will require additional strategies.
Pre-emergent broadleaf control of white clover, purslane, knotweed, oxalis, chickweed, henbit and spurge.	Spring or late summer	isoxaben (Gallery)	Combine with pre-emergent crabgrass control product; requires 1/2 inch of moisture after the application; ornamental weed control product call Snapshot.

Weed	Suggested time of application	Herbicides	Comments
(Non-selective)			
Tall fescue clumps, bentgrass patches, nimblewell, bermuda-grass, quackgrass, timothy	Apply when the target grass is actively growing and the soil moisture is good	glyphosate [Roundup, Kleenup] amitrole [Amitrole-T] dalapon	These herbicides will kill all weeds and other plants that are green at the time of spraying. Treated areas must be reseeded or sodded with desirable grasses. Amitrol and dalapon may persist up to 6 weeks. Glyphosate has no residual activity in the soil. However, it is suggested to wait 7 days before renovating the area. A repeat treatment may be necessary for complete control.
(Select)			
Tall fescue clumps	Apply when T.F. and desirable turf actively growing.	chlorsulfuron [Lesco TFC]	Selectively controls T.F. in cool-season grasses (except ryegrass); low volume preferred; irrigate with 1/2 inch of water after treatment; slowly eliminates T.F. over an 8 week period.
Annual bluegrass *(Poa annua)*	Spring & fall Early fall Fall	Growth regulators Pre-emergence herbicides Pre- & post-emergence herbcides	Refer to AY-41 *Control of Poa annua* for control strategies.

*No endorsement of named products by the author or Purdue University is intended, nor is criticism aplied for products that are not mentioned. Follow closely the direction on the label. Many different formulations and combinations of these materials are sold under various trade names, and the quantity of use will vary with the formulation obtained.

RR 1/93 (5M)

Cooperative Extension work in Agriculture and Home Economics, state of Indiana, Purdue University, and U.S. Department of Agriculture cooperating; H. A. Wadsworth, Director, West Lafayette, IN. Issued in furtherance of the acts of May 8 and June 30, 1914. The Cooperative Extension Service of Purdue University is an affirmative action/equal opportunity institution.

Definitions

1. Annual - A plant that completes its life cycle from seed in 1 year.
2. Band application - An application to a continuous restricted area such as in or along a crop row.
3. Biennial - A plant that completes its life cycle in 2 years. The first year it is vegetative and the second it is reproductive.
4. Broadcast application - An application over an entire area.
5. Carrier - The liquid or solid material added to a chemical compound to facilitate its application in the field.
6. Concentration - The amount of active ingredient or acid equivalent in a given volume of liquid or in a given weight of dry material.
7. Contact herbicide - A herbicide that kills primarily by contact with plant tissue rather than as a result of translocation.
8. Emulsifying agent - A surface active material which facilitates the suspension of one liquid in another.
9. Herbaceous plant - A vascular plant that does not develop woody tissue.
10. Perennial - A plant that lives for more than 2 years.
11. Nonselective herbicide - A chemical that is toxic to plants generally without regard to species.
12. Noxious weed - A weed arbitrarily defined by law as being especially undesirable, troublesome, and difficult to control.
13. Postemergence - After emergence of specified weed or crop.
14. Preemergence - Prior to emergence of specified weed or crop.
15. Preplanting - Any time before the crop is planted.
16. Rate - The amount of active ingredient in a herbicide applied to a unit area.
17. Selective herbicide - A chemical that is more toxic to some plant species than to others.
18. Spray drift - The movement of airborne spray particles away from the crop.
19. Translocated herbicide - A herbicide which is moved within the plant.
20. Wetting agent - A compound which, when added to a spray solution, causes it to contact plant surfaces more thoroughly.

QUESTIONS

1. What are the principal methods of weed control?

2. Why is it so important to control weeds?

3. What is a post-emergence herbicide; a pre-emergence herbicide?

4. What is a contact herbicide?

5. What are classifications of herbicides, by mode of action?

6. What is meant by broad-spectrum weed control?

7. How can spray drift be lessened?

8. What is the use of a wetting agent in post emergence herbicides?

INTEGRATED PEST MANAGEMENT

PART I

Objectives

1. Define the concept of I.P.M.
2. List the four fundamental components of I.P.M.
3. List five situations which led to the development of I.P.M.

Integrated pest management, or I.P.M., is a concept of crop production that incorporates effective and economical crop pest control methods and minimizes the undesirable side effects of pesticide reactions. Through I.P.M., potential problems and the need to control the problems are identified. As a result, crop yields and quality are increased (Fig. 1). I.P.M. uses the most economic approach possible; therefore producers maximize their profits.

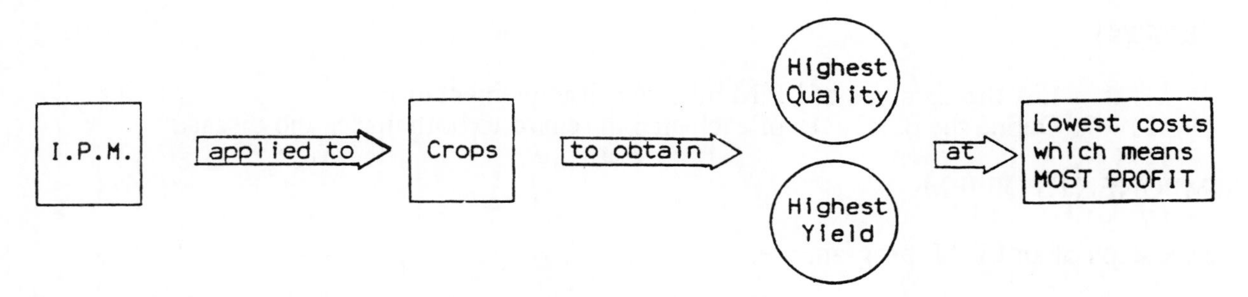

Fig. 1. The concept behind I.P.M.

The key to a successful I.P.M. program is careful monitoring (scouting) of crops for pests, beneficial organisms, and other factors that affect production. The actual scouting procedure involves the first two steps of the total six step process of I.P.M.

I.P.M. deals with more than just insect pests. The term "Integrated" in I.P.M. means that the expertise and management techniques from many agricultural disciplines are used. By utilizing I.P.M., expenses will be held in check while profits will be maximized.

There are four fundamental components which are assumed in any I.P.M. program:

1. Pest management is a part of a total crop production system.
2. Pest control actions can result in undesirable effects.
3. Crop production must be economical.
4. Pest management refers to all pests, including:

   a. insects
   b. weeds
   c. plant pathogens
   d. nematodes
   e. vertebrates.

Why Have I.P.M.?

Since World War II, farmers have relied heavily on new chemicals to eliminate pest populations. Intensive use of these chemicals frequently violated basic ecological principles. Consequently, the following situations evolved and the need for I.P.M. emerged.

1. Pests developed strains resistant to many chemicals.
2. Increasing amounts of pesticides were used to kill more resistant insects, thus the cost increased.
3. Certain pesticides persisted in the environment.
4. Adverse effects on non-target organisms developed.
5. Rachel Carson's book Silent Spring increased public awareness of the increasing amounts of non-biodegradable pesticides released into the environment.

PART II

Objectives

1. List the six steps of I.P.M. in order of implementation.
2. Define the objectives of each step in regard to both insect and disease.

COMPONENTS OF IPM

The six steps of an I.P.M. program are:

1. Identification
2. Sampling
3. Analysis
4. Method Evaluation
5. Implementation
6. Reevaluation

Each step is important and must be fulfilled in order for the total I.P.M. program to be successful. Steps 1 and 2 (Identification and Sampling) are referred to as SCOUTING and are performed in the field simultaneously.

During the scouting procedure, selected and random areas of the field are **sampled** for symptoms of insect and disease damage. The insect or disease causing the damage is then **identified**. In addition the type of crop and its growth stage are identified.

The data collected during scouting is then **analyzed** (Step 3) to determine if any action is necessary. If the damage or potential for damage is severe enough, all possible control **methods** and their potential outcomes are **evaluated** for effectiveness (step 4). The control method selected is correctly **implemented** (step 5). The sixth and final step reevaluates the five previous steps for possible error. **Reevaluation** is essential even when the I.P.M. program appears successful.

PART III

Objectives

For each of the following steps of I.P.M., list or define the terms or characteristics specified:

1. Identification
   - life cycles of insects and diseases
   - damaging stages of insects and diseases
   - types of crop damage.

2. Sampling
   - items to inspect when sampling the pest
   - items to inspect when sampling the crop.

3. Analysis
   - economic injury level, economic threshold, action threshold.

4. Method Evaluation
   - four methods of controlling pests.

5. Implementation
   - three factors to consider when implementing a control method.

6. Reevaluation
   - develop a format for rechecking the entire process.

1. Identification

A. Pest

1. Life Cycles of Insects and Diseases

Two contrasting types of life cycles exhibited by many agronomically important insects are complete and incomplete metamorphosis (Fig. 2).

Complete metamorphosis consists of four distinct stages in the insect's life cycle. These are the egg, larva, pupa, and adult. Incomplete metamorphosis consists of the egg, nymph, and adult stages. The appearance of the insect in the nymph stage is similar to the adult.

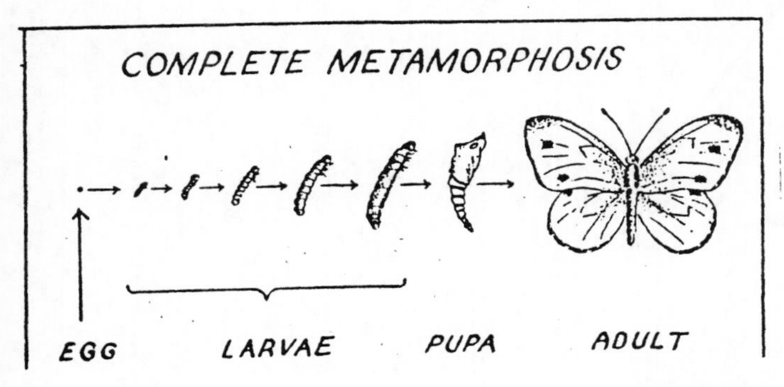

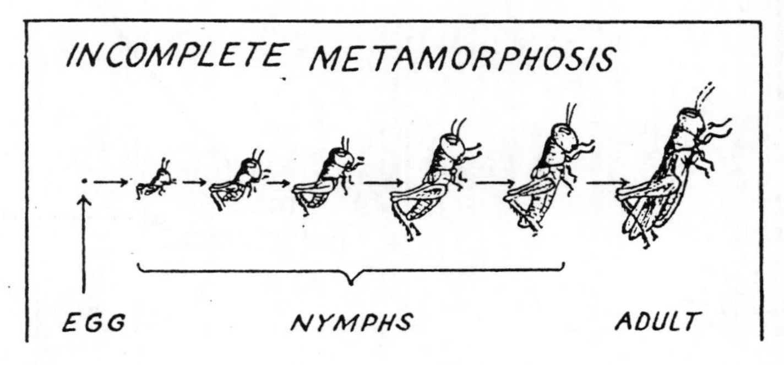

Fig. 2. Complete and Incomplete Metamorphosis

Plant disease symptoms change as the disease progresses, therefore disease cycles are not as clear-cut as are insect cycles. The complete series of symptoms as a disease progresses is referred to as the symptom complex. The symptom complex includes all visible changes from the onset of the disease until the recovery or death of the plant.

Diseases can be classified into two major groups - nonparasitic and parasitic. The following are examples of each.

a. Nonparasitic
   1. Extremes in temperature
   2. Extremes in soil moisture
   3. Wind or hail damage
   4. Pollution damage

b. Parasitic diseases are caused by:
   1. Bacteria
   2. Fungi
   3. Nematodes
   4. Viruses

There are three requirements for the development of a parasitic disease (Fig. 3). If one or more of these requirements is missing, the disease will not develop. The three requirements are:

a. Pathogen - the disease causing organism.
b. Susceptible Host - susceptibility may vary with vigor and stage of growth of the host plant.
c. Proper environment - both the host and the pathogen are influenced, but the pathogen is usually more sensitive than the host.

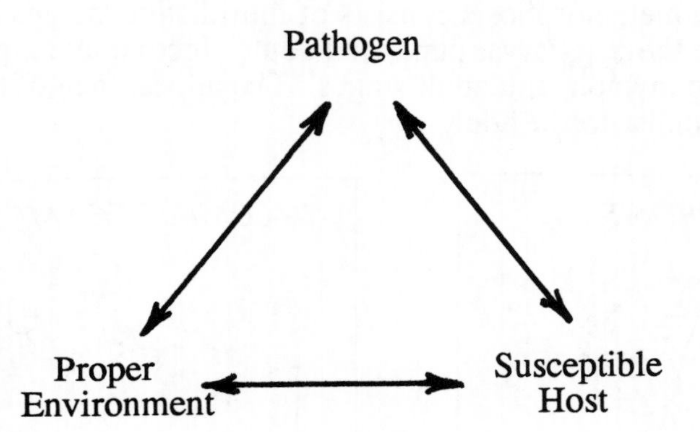

Fig. 3. Interaction between Environment, Pathogen, Host and Disease.

2. Damaging Stages

The stage of insect development which does the most plant destruction determines the necessity for and the type of control methods to be used, and therefore should also be noted. Corn rootworm larvae are very destructive, while adult and nymph grasshoppers are destructive.

B. Crop

Insects harm plants in many different ways. For example, defoliation of soybeans may be either complete or veinicular. Piercing-sucking insects may feed only on soybean pods, while other insects feed only on plant roots. Some insects discolor plant tissue from normal green to different shades of yellow by their piercing sucking mouth parts. When identifying a plant disease, the symptoms may be classified according to:

1. Abnormal coloration of host tissue
2. Wilting of the host plant
3. Death of host tissue
4. Complete defoliation and fruit drop
5. Abnormal growth increase of host plant
6. Stunting of host plant
7. Replacement of host tissue

2. Sampling

A. Pest

Sampling leads to further identification and quantification of the pest attack. More specifically, sampling determines:
1. Number of pest (insect/disease) types in a given area
2. Number within each type in a given area
3. Age of pest (developmental stage)
4. Number of predator and/or parasitic pests

B. Crop

When sampling the crop, inspect for:
1. Number of damaged plants within a given area
2. Type of plant damage
3. Stage of plant growth at which damage occurred

3. Analysis

A. Pest

Now you must decide whether further management action is necessary. The data to be analyzed were obtained during the first two steps, identification and sampling. The following concepts are useful in determining if pest control measures are necessary:

1. Economic Injury Level - the lowest pest population density that will cause economic damage (Fig. 4).
EXAMPLE - The value of crop damage prevented by a pesticide application is $12 per acre and the cost associated with treatment (pesticide material, applicator equipment costs, fuel, labor) is $20 per acre. Therefore, the pesticide application is not economical. In this instance, the pest population has not exceeded the economic injury level.

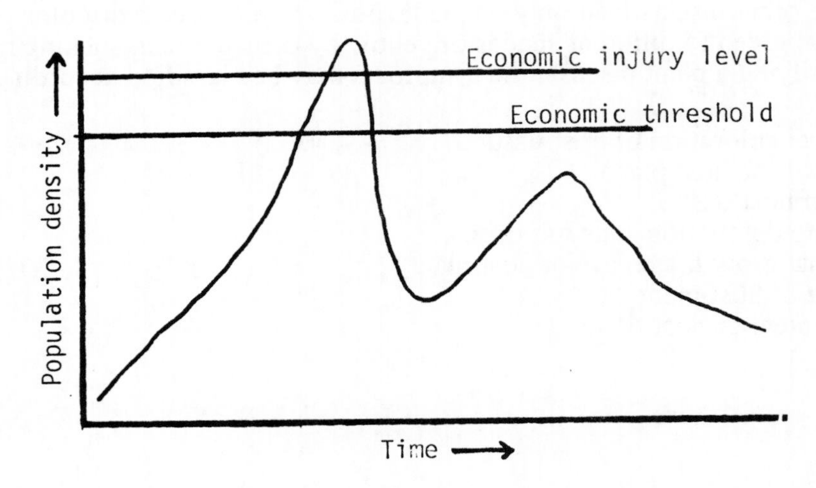

Fig. 4. Insect population density in relation to time.

2. Economic Threshold - the pest population level at which controls are employed to prevent the population from exceeding the economic injury level.

B. Crop

Crop analysis is mainly concerned with plant maturity. The stage at which the plant is damaged is important in planning effective management practices.

4. Method Evaluation

A. Pest

This step involves assessing each possible control method before any control method is implemented. When controlling the actual pest (insect or disease), there are four methods of control. These are:

1. **Biological control** - includes the use of any living organism such as:

a. predator or parasitic insects
b. animals
c. disease organisms to which insects and other diseases are susceptible

2. **Chemical control** - involves applying some compound to which the insect or disease is susceptible. Crop pests can be controlled by chemical means if proper rates, application techniques, and timing are observed. In general, selective chemical control is easier with insects than with weeds or disease organisms.

3. **Cultural control** - consists of using farming practices such as destroying crop debris (which may harbor insects and several disease organisms), using crop rotations, proper fertilizer management, and the use of species or varieties that are resistant to certain insects and diseases.

4. **Mechanical control** - consists of physically killing or removing the pest. Common methods of mechanical pest control include cultivating or mowing. Mechanical methods are more commonly used to control weeds than insects or diseases.

B. Crop

The stages of development of a crop must be taken into account when selecting a control method. At any stage, it is important to know whether the crop can be sprayed, harvested, cultivated, or if it should be left alone under present conditions.

5. Implementation

When implementing any of the four control methods the primary goal is to control the pest with the least amount of damage to the crop. Factors to consider when implementing a control method include:

A. Pest

1. Rate of chemical to be applied.
   The rate must be effective but should not harm the environment.
2. Time to use control method.
   For example, cutworms are nocturnal insects, therefore, if chemical control is the chosen method, an evening application would be most beneficial.

B. Crop

1. Rate of chemical
   The rate must be effective in controlling the pest but should not harm the crop.
2. Timing
   The stage of development of the crop was considered when the possible control methods were evaluated, therefore, the crop should be able to withstand the control method selected.

   For example, if a crop is flowering and the open flowers are susceptible to a chemical spray, an evening application (when the flowers are closed) would be better than a morning application (when the flowers are open.)

In addition to pest and crop factors, environmental conditions may influence the effectiveness of a control method.

For example, certain pesticides need to be activated by rain, and therefore should be applied prior to a chance of rain. Control procedures should be scheduled to optimize their effectiveness. Nevertheless, total farm management may dictate that other farming operations such as alfalfa harvesting may be more economically advantageous than controlling cutworm on five acres of late planted corn.

6. Reevaluation

A. Pest

This final step of an effective I.P.M. program is generally the most overlooked. To determine if your management was effective use check strips or control plots. Small control plots should be somewhat isolated to prevent further insect, disease, or weed migration. After the plot check, reevaluate the five preceding steps and check for possible error (Figure 5).

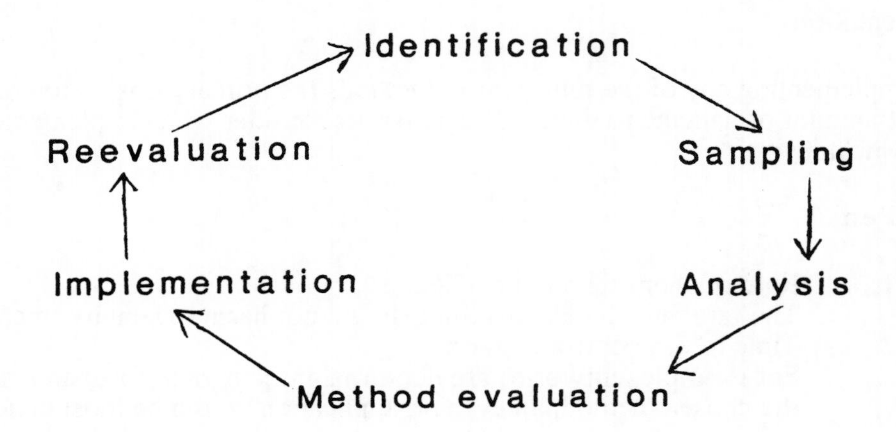

Fig. 5. The I.P.M. process is cyclical in nature.

B. Crop

The main goal of profitable crop production should be MAXIMUM ECONOMIC YIELD with minimal effects on the environment.

When considering only the crop, check other neighboring hybrids or varieties with similar crop pest problems. Changing to a more resistant hybrid or variety may be an alternative. Another alternative may be producing the crop in a crop rotation.

CALIBRATION AND OPERATION OF CROP PLANTING EQUIPMENT

OBJECTIVES

1. Identify components of a row crop planter and a grain drill.
2. Describe how a row crop planter and grain drill operates.
3. Calibrate a row crop planter and grain drill for proper field operation.

Most agronomic crops are seeded with either a row crop planter or a grain drill. To properly seed a crop these machines must perform five functions:

* Open a furrow in the soil
* Accurately meter the desired location in the soil
* Place the seed at the desired location in the soil
* Cover the seed with the proper amount of soil
* Firm the seedbed to assure soil-seed contact

Observe the parts of a row crop planter and grain drill that perform each of these function.

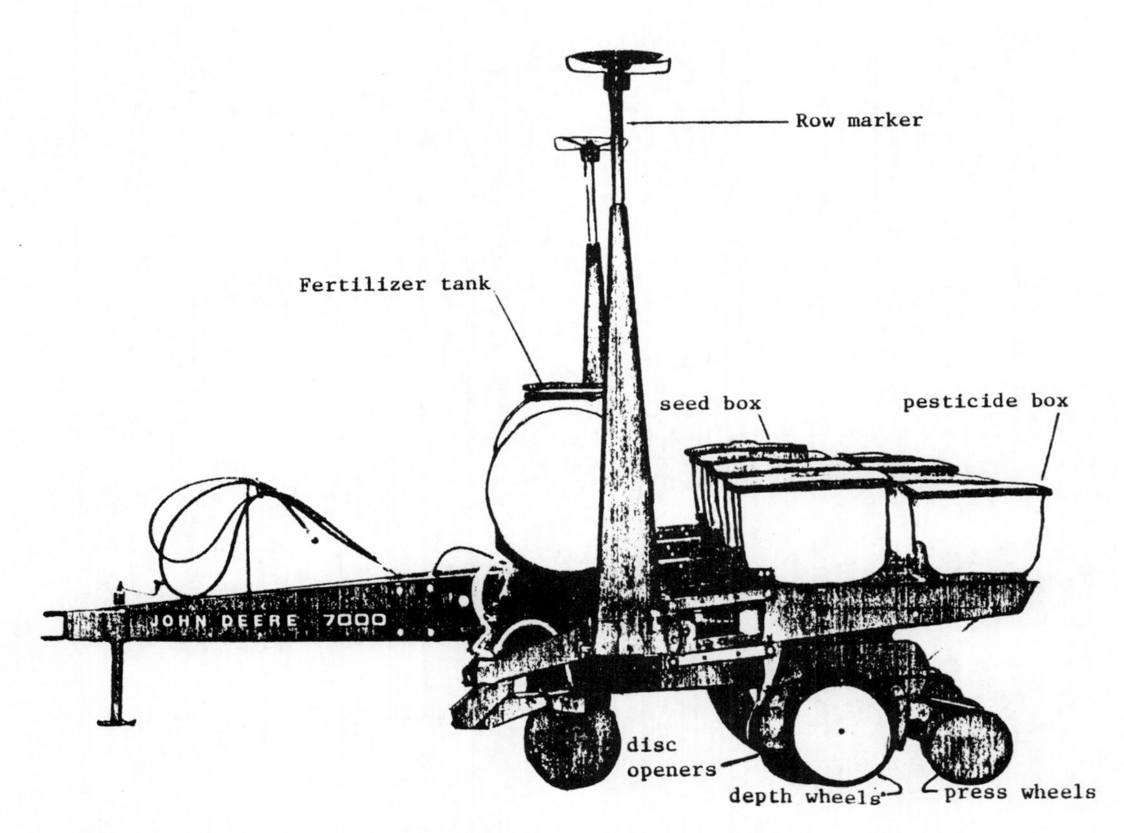

Fig. 1. Some of the components of a typical corn planter.

Fig. 2. A Cyclo Air planter that uses air pressure to meter and carry seeds to the row

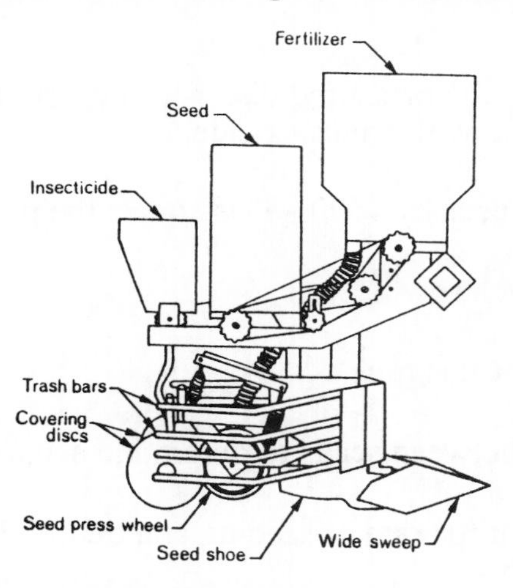

Fig. 3. Major components of a till planter unit

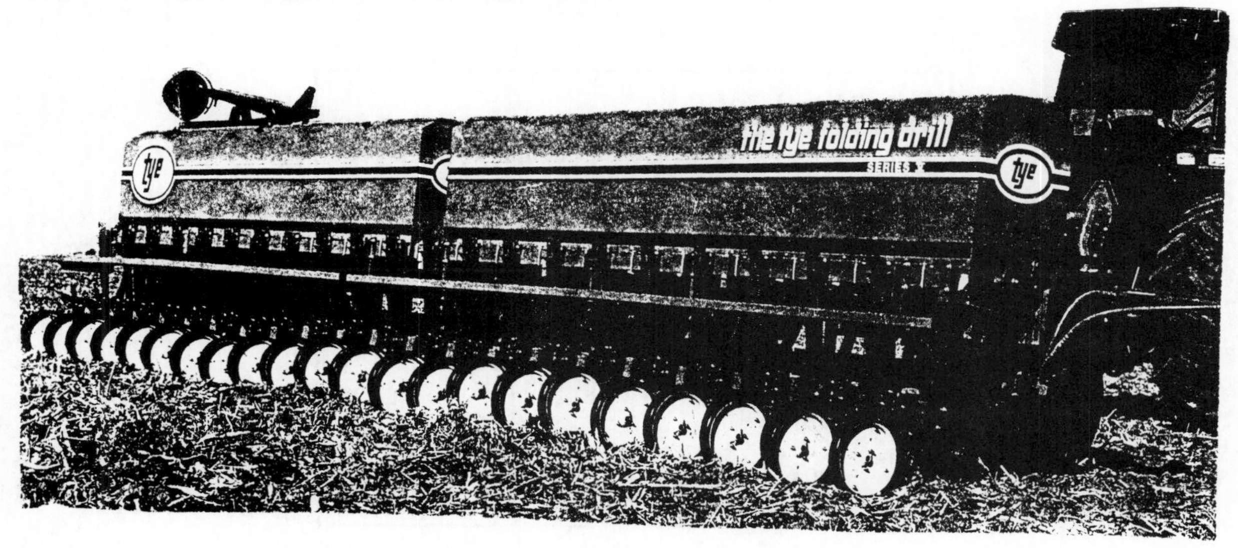

Fig. 4. A drill adapted to conservation tillage systems

I. FIELD CALIBRATION OF ROW CROP PLANTERS

The operator's manual serves as a guide to setting the row-crop planter to plant the desired rate of seeds per acre.

The recommended settings are based on average soil conditions and average grading of seed. The actual rates of seeding can be determined only by field checking. If the rates are too low or too high, adjustments must then be made.

Improper calibration can result in plant populations far below or above desired levels. The following method can be used to calibrate the seeding rate of a row crop planter.

1. Fill the seed hoppers at least half full to simulate average planter weight. Add powdered graphite if recommended by the planter operator's manual.

2. Mark row distance equal to 1/1000 acre. Use Table 1 to determine this distance.

3. Plant the measured distance at the speed you intend to use during planting. To simulate accurate field conditions, start planting before reaching first mark and continue past last mark.

4. Uncover the seeds within the measured distance and count them. Do not use bare hands if the seed is treated with an insecticide.

5. Multiply the number of seeds by 1000. This equals the planting population for one acre.

6. To be accurate, check each row.

7. Measure planting depth at this time.

8. Check average distance between seeds to determine accurate seed placement.

9. Make required adjustment for rate of seeding and depth of planting and recheck by repeating steps 1 through 4.

Table 1 Length of Row for Various Row Widths in Which to Make Plant or Seed Counts for Population Check.

Row width	Length of row to equal 1/1000 acre
in.	ft.-in.
20	26-2
28	18-8
30	17-5
36	14-6
38	13-9
40	13-1
42	12-5

II. Grain-Drill Calibration

Grain-drill calibration is necessary if accurate seeding is to be obtained. The rate table found in the operator's manual or on the hopper lid is based on a standard weight per bushel for various crop seeds. The grain drill meters volume and not weight.

Seed, wheat for example, is sold by the bushel (weight 60 lbs.). Due to differences in varieties, the seed size may vary. One bushel of large-seed wheat will come in a larger bag than a bushel of small-seed wheat.

This example show how the pounds per acre of seed drilled is affected by the size of the seed.

Differences between drilling rates shown in the rate table and actual pounds per acre drilled may be due to several factors:

1. Tire inflation
2. Drive-wheel slippage
3. Differences in the seeds (weight, size, type, variety, moisture content, kind of seed, amount of foreign material in seed)

Therefore, the quantity actually being drilled must be checked.

The following method may be used to calibrate a grain drill.

1. Jack up the drive wheel.
2. Place seed in the hopper and a tarp or canvas under all feeds. (Seed may also be collected in sacks at each seed tube.)
3. Make fee-cup settings on drill for desired quantity per acre as shown on seed chart.
4. Turn drill wheels the required number of turns (some operator's manuals give this information) for 1/20 acre as calculated using the following formula:

$$\text{Rotations to equal 1/20 acre} = \frac{8319.3}{W' \text{ X } D''}$$

W' = Drilling width in feet

D" = Diameter of wheel in inches (rolling diameter)

Example: Assume grain drill is a 16 X 7 and the drive wheel rolling diameter is 28 inches. Desired seeding rate is 60 pounds per acre, therefore:

$$W' = \frac{16 \times 7}{12} = \frac{112}{12} = 9.33 \text{ ft.} \qquad D'' = 29 \text{ inches}$$

$$\text{Rotations for 1/20 acre} = \frac{8319.3}{W' \text{ X } D''} = \frac{8319.3}{9.33 \text{ X } 28}$$

$$= \frac{8319.3}{261.24} = 31.8 \text{ or about 32 rotations}$$

5. Carefully weigh the seed collected and compare that to the weight shown on the seed chart. Divide weight in seed chart by 20 because calibration was on 1/20 acre: 60 % 20 = 3 lbs. for 1/20 acre.

6. Adjust the feed cup setting to compensate for any variation and repeat the test until the desired quantity is obtained.

Plant population at row widths used for soybeans

Row width	Linear feet	Plants per foot of row					
(inches)	per acre	12	9	8	6	4	3
7	74,674	896,088	672,066	597,392	448,044	298,696	224,022
10	52,272	627,264	470,448	418,176	313,632	209,088	156,816
14	37,337	448,044	336,033	298,696	224,022	149,348	112,011
20	26,136	313,632	235,224	209,088	156,816	104,544	78,408
21	24,891	298,692	224,019	199,128	149,346	99,564	74,673
24	21,780	261,360	196,020	174,240	130,680	87,120	65,340
28	18,669	224,028	168,021	149,352	112,014	74,676	56,007
30	17,424	209,088	156,816	139,392	104,544	69,696	52,272
36	14,520	174,240	130,680	116,160	87,120	58,080	43,560
40	13,068	156,816	117,612	104,544	78,408	52,272	39,204

Seed Spacing and Harvest Populations (Assuming 15% Normal Stand Reduction) for Selected Seeding Rates and Row Widths.

Seeding rate	Harvest population with 15% stand loss	Space between seeds when row width is –					
		20 in.	28 in.	30 in.	36 in.	38 in.	40 in.
no.	no.	inches between seeds					
16,000	13,600	19.6	14.0	13.1	10.9	10.3	9.8
18,000	15,300	17.4	12.4	11.6	9.7	9.2	8.7
20,000	17,000	15.7	11.2	10.5	8.7	8.3	7.8
22,000	18,700	14.3	10.2	9.5	7.9	7.5	7.1
24,000	20,400	13.1	9.3	8.7	7.3	6.9	6.5
26,000	22,100	12.1	8.6	8.0	6.7	6.3	6.0
28,000	23,800	11.2	8.0	7.5	6.2	5.9	5.6
30,000	25,500	10.5	7.5	7.0	5.8	5.5	5.2
32,000	27,200	9.8	7.0	6.5	5.4	5.2	4.9
34,000	28,900	9.2	6.5	6.1	5.1	4.9	4.6
36,000	30,600	8.7	6.2	5.8	4.8	4.6	4.4
38,000	32,300	8.2	5.9	5.5	4.6	4.3	4.1
40,000	34,000	7.8	5.6	5.2	4.4	4.1	3.9

PRECISION FARMING

OBJECTIVES

Upon completion of this lab, you should be able to:

1. Describe how global positioning systems (GPS) use satellite signals to determine field location.
2. List causes of variability in a field.
3. Explain how to quantify field variability by using soil maps, grid soil sampling, yield monitors, and remote sensing.
4. Explain how a yield monitor works.
5. Describe how remote sensing detects crop characteristics.
6. List types of equipment that can utilize variable rate technology (VRT).

INTRODUCTION

The continuing replacement of labor with machinery in agriculture has allowed farm operators to be more efficient. Fewer people are farming more acres. But along with increased mechanization has come the tendency to devote less attention to detail within a field.

The adoption of precision farming techniques allows producers to prescribe specific management practices to areas within fields that previously may have gone unnoticed. If inputs are used more effectively, production costs decrease, crop yields increase, and environmental benefits accrue.

DETERMINING EXACT FIELD POSITION

Global positioning systems interpret signals received from Department of Defense satellites to determine exact ground locations. Each satellite sends a complex array of signals at specific times. Since it takes time for a signal to travel, a GPS unit can calculate its distance from each satellite based on *when* it receives specific signals. Signals travel at the speed of light, so a critical component of each satellite is a clock accurate to within a fraction of a second.

Fig. 1 GPS satellite in space

While a GPS unit can tell its distance from a satellite, it has no indication of direction. Thus, it takes three or more satellites to calculate, or triangulate, a ground position. The following exercise illustrates the concept of triangulation:

Mark the location of the point that is 7 miles from point A, 5 miles from point B, and 9 miles from point C. Each square in the grid is one square mile:

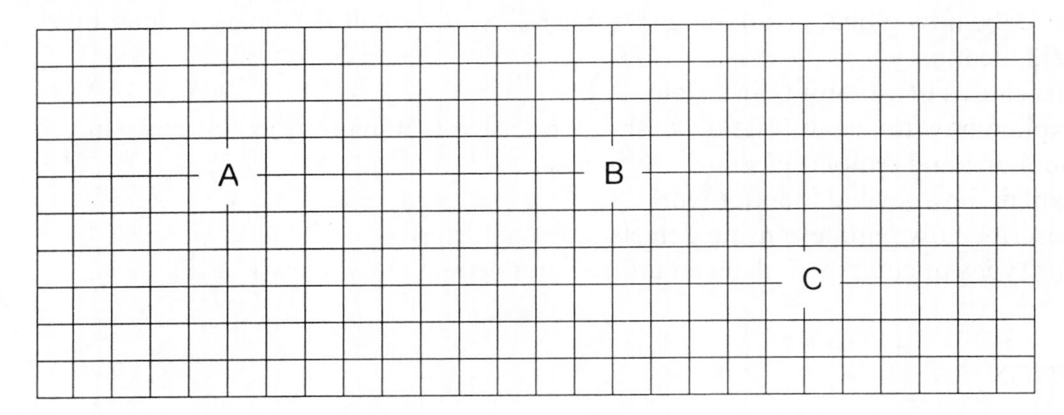

Sources of error in GPS. Satellite signals are highly accurate, but not without error caused by atmospheric interference, irregularity of orbit, and signal bounce off objects. Also, the Department of Defense purposely introduces error into satellite signals so the highest accuracy is not available to unfriendly users. To get high accuracy, farmers practicing precision farming may use a ground-based signal also, often called a "correction signal."

MEASURING VARIABILITY IN FIELDS

Variability is normal in agricultural settings. The "cookie-cutter" appearance of agricultural fields as one drives along a highway is deceptive. Ask any farm operator who has flown in a small aircraft over his fields in early summer how much variation can be seen. Watch readings on a yield monitor vary as a combine goes across a field. Note how soil type and soil color vary across the fields in Fig. 2:

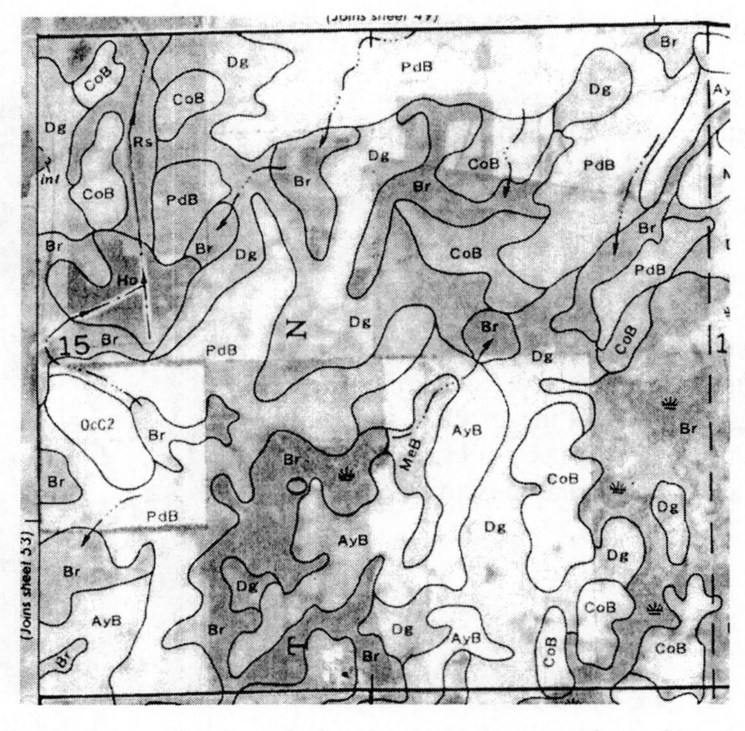

Fig. 2. Map of soil variation in a 640-acre section of land.

Information on variability can come from various sources, such as the NRCS (Natural Resources Conservation Service) map above showing differing soil types. Other sources include grid soil samples, yield monitor information, and remote sensing.

Grid Soil Sampling. Instead of sampling whole fields as a unit, a grid sampling approach breaks up the field into smaller units, commonly two to three acres each. This approach allows the detection of variation of pH, organic matter, and various plant nutrients across a field, although it increases sampling time and expense. Fig. 3 shows soil probes for sampling mounted on a tractor and a pickup truck, and a map of a field divided into a grid.

Fig. 3. Equipment used to grid soil sample a field. (courtesy of Case IH)

Yield Monitors. A typical combine yield monitor for corn, soybeans, or wheat usually consists of these major components:

- grain flow sensor
- grain moisture sensor
- ground speed sensor
- in-cab monitor

The grain flow sensor is a metal plate positioned so it is hit by grain flowing through the combine (see Fig. 4). As yields increase, there is more grain flow, and the plate is deflected more. Field comparisons have shown that grain yield monitors, if calibrated, can be highly accurate.

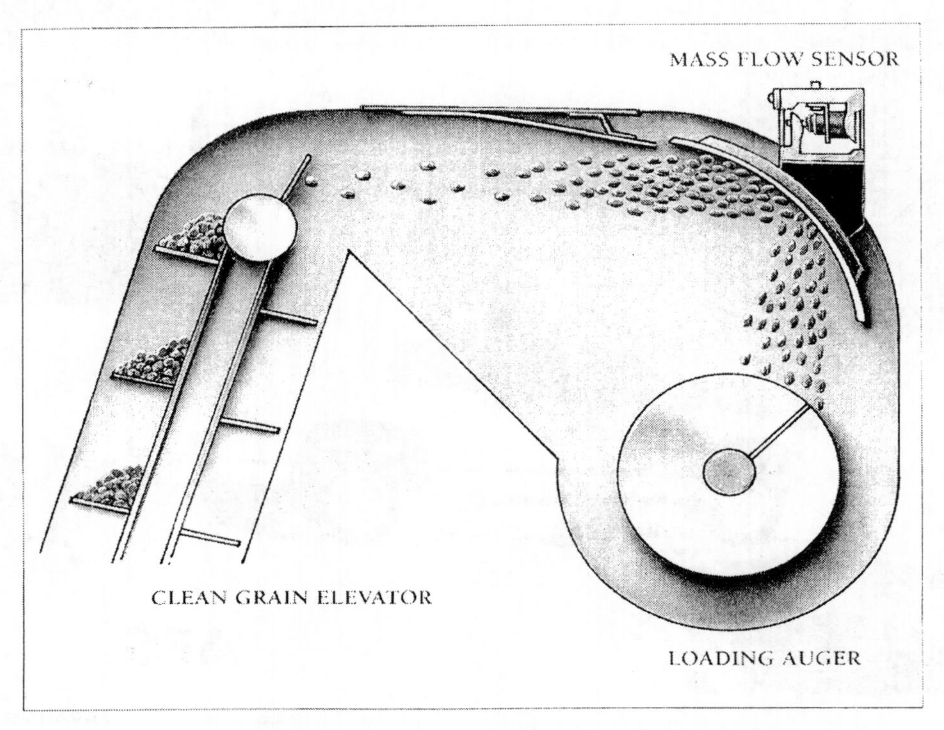

Fig. 4. Mass flow sensor, a component of a yield monitor.
(courtesy Deere and Company)

In addition, by coupling a yield monitor to GPS, the producer can create a database that ties yields to exact locations. From this, yield maps are created (see Fig. 6.)

Remote Sensing. Sunlight that strikes a field is either absorbed by the soil and crops, or reflected back into space. The reflected portion can be measured by sensors mounted just above the crops, in aircraft, or satellite.

The human eye only sees visible light, a portion of the energy reflected. Plants utilize this type of light in the process of photosynthesis. But soils and plants also absorb and reflect energy in infrared, radar, radio, microwave, and other types of energy. The way this energy is reflected varies according to the characteristics of the scene, as shown in Fig. 5.

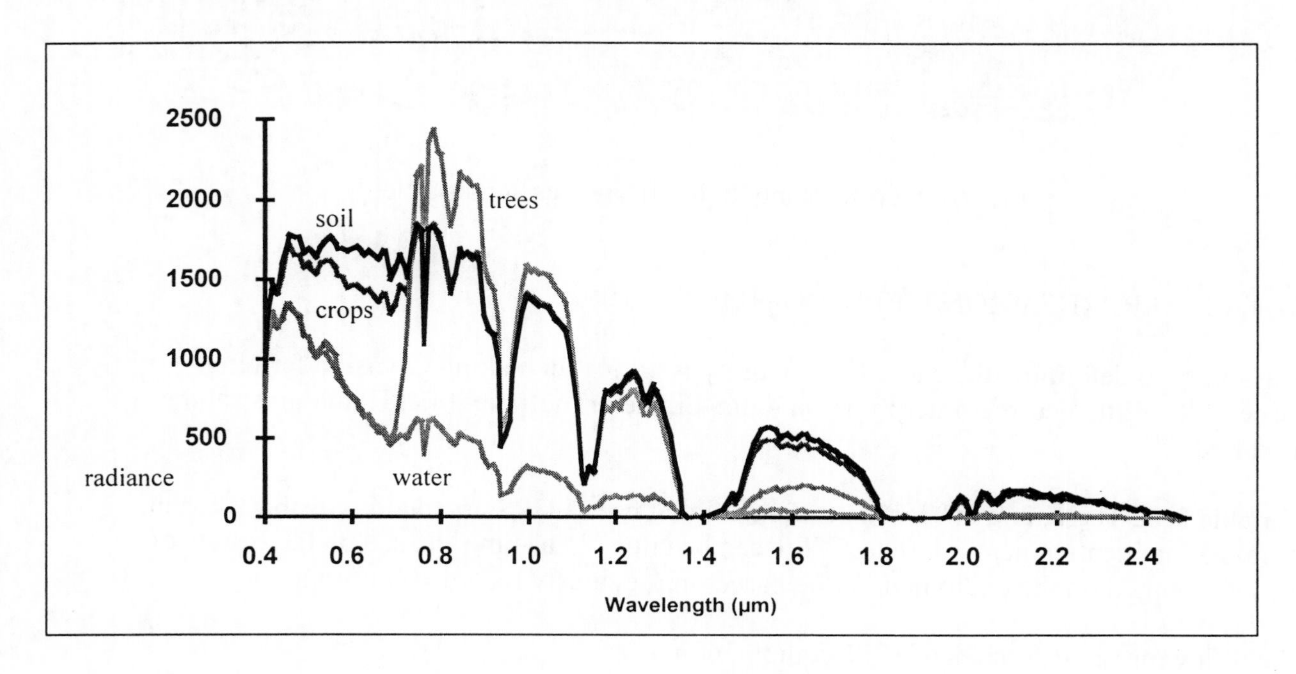

Fig. 5. Radiance vs. wavelength for four different basic materials (courtesy Dave Landgrebe).

The *possibilities* of remote sensing are endless, but it has been difficult to relate what is seen from space to what is actually happening on the ground. Remote sensing has been used successfully to identify crop species on a large scale, to verify acres planted in a nation or state. It can detect differences in crop condition, or "greenness", on a state by state or local basis. As the technology improves, satellite and aircraft sensors are able to see a smaller area on the ground. Thus, the detection of weed infestations, nutrient deficiencies, or other types of crop "stresses" becomes possible.

ASSEMBLING INFORMATION INTO USEFUL FORMATS

The ability to store and process the huge amounts of data generated by precision farming applications has only become practical by computer advancements in the last few years. But generating enormous quantities of data is of little value if it is not in useful form. For most agricultural applications, this takes the form of a map, showing how factors such as yield or soil test information vary across a field. Fig. 6 is a black and white copy of a color yield map of a single field, a common way to portray yield information.

Fig. 6. Map showing yield differences across a field.

DATA INTERPRETATION/FIELD MODIFICATION

The ability to determine the cause of variability is critical in making precision farming successful. With a correct interpretation, unfavorable or inefficient field situations can be modified

Variable Rate Technology (VRT) is equipment that uses a GPS unit and is capable of changing its rate of application across a field. VRT can be either "real-time", based on the results of a sensor mounted on the equipment, or be based on previously recorded information.

The following are examples of VRT equipment in use:

- Variable rate seeding equipment
- Fertilizer spreaders that vary the rate of nutrient application in a field
- Sprayers that dispense herbicide only when weeds are detected

Individual farm operators can practice various components of precision farming. For example, many farm operators use a yield monitor as a beginning step into precision farming. But to realize a benefit from precision farming, some change of management is necessary.

At present, only a small percent of farmers are practicing precision farming. As with any new technology, there is a learning curve and a time lag that precedes the adoption. For example, even though hybrid corn was developed about 1900, it wasn't until the 1950's that most Midwest farmers were using it on their farms. The components of precision farming that will be embraced in the future will depend on their ability to increase profits, reduce expenses, or provide environmental benefits.

GRAIN HARVESTING

Objectives:

Upon completion of this exercise you should be able to:

1. Describe the major components of a combine.

2. Differentiate between pre-harvest and machine harvest losses.

3. Measure harvest losses of either corn or soybeans.

4. List methods of keeping harvest losses below desired loss levels.

One way of increasing harvested yield in the machine shed is to properly adjust your combine. Regardless of the design or brand of combine, the machine will have five major components. All of these components must function properly for efficient harvesting to occur. The five components of a combine are:

1. Gathering component
2. Threshing component
3. Separating component
4. Cleaning component
5. Handling / conveying component

Each component will have various techniques for adjustment to obtain optimum harvest efficiency. Manufacturer recommendations and procedures are outlined in the owner/operator manual that comes with the machine. Any one of these components can cause excessive grain loss if not set properly for the crop and field conditions.

The following information from the University of Missouri-Columbia Extension Division describes different methods used by combine manufacturers to accomplish the five functions listed above.

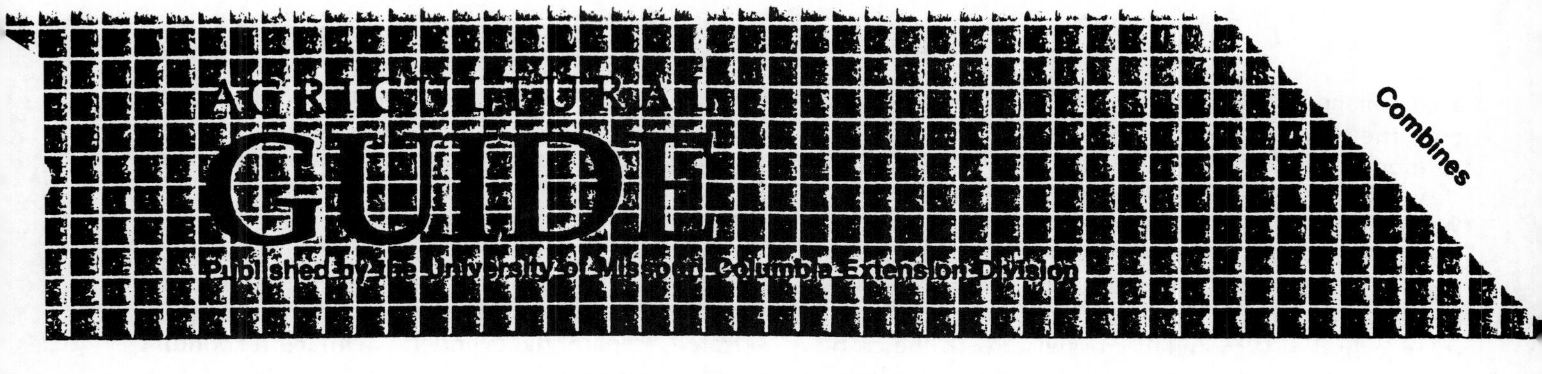

AGRICULTURAL
GUIDE
Published by the University of Missouri-Columbia Extension Division
Combines

Combines

William G. Hires and Charles Ellis
Department of Agricultural Engineering
College of Agriculture

The conventional combine

Figure 1 illustrates the operation of a conventional grain combine. The crop is fed tangentially into a cross-mounted, cylinder-concave assembly. Threshing occurs through the impact of the cylinder bars on the incoming crop. Much of the separation occurs through the open grate concave. Separation of the remaining grain from the straw is accomplished with straw walkers. A cleaning shoe, with chaffer and sieve, is used for scalping and final cleaning.

Other types of combines

New Holland (Figure 2) has two longitudinally mounted, axial, threshing and separating rotors. Threshing occurs in the concaves at the front of the rotors. Separation of the grain from the straw is accomplished along the full length of the rotor. A rear beater-grate assembly performs the final separation. A conventional cleaning shoe is used for scalping and final cleaning. Three different sizes of this combine are currently being produced. They are the TR76, TR86 and TR96.

Case International (Figure 3) has a single, longitudinally-mounted, axial-flow, threshing and separating rotor. Threshing occurs at the front section of the rotor; separation of the grain from the straw is accomplished along the full length of the rotor in both the threshing and separation concaves. A rear beater aids in straw discharge. A conventional cleaning shoe is used for scalping and final cleaning. The company produces five models: the 1620, 1640, 1660, 1680 and 1470 Hillside, as well as a pull type version.

Massey-Ferguson and **White** (Figure 4) also have a single, longitudinally-mounted axial threshing and separating rotor. Threshing occurs at the front section of the rotor. Separation of the grain from the straw is

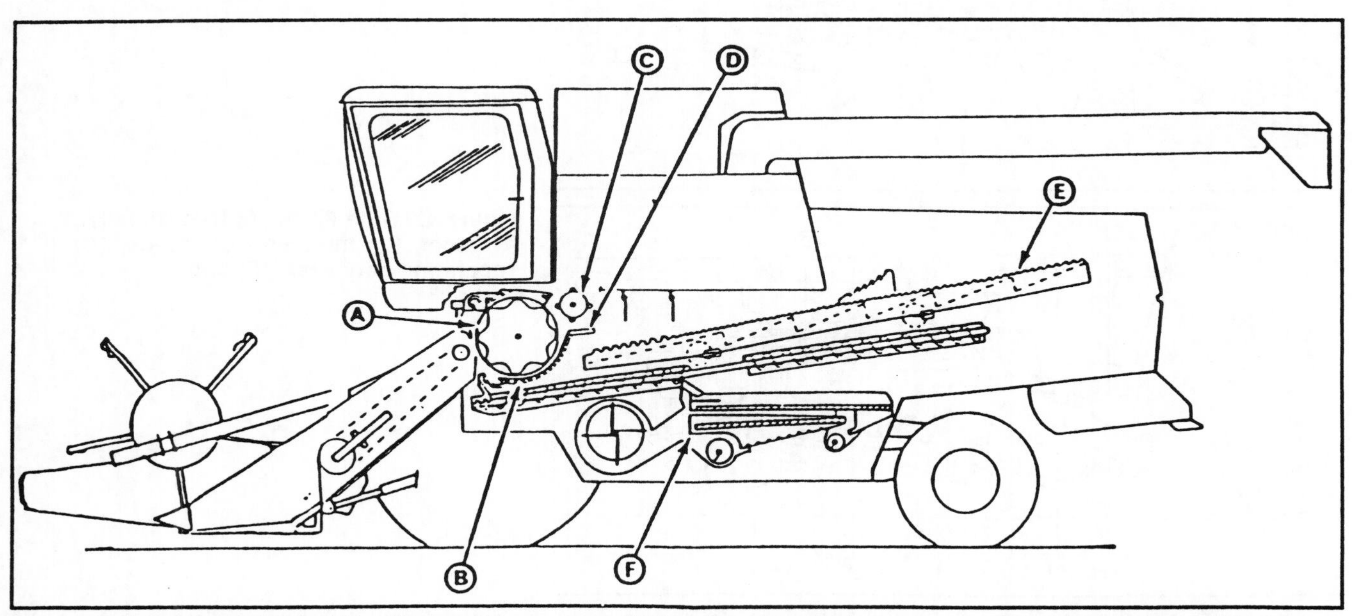

Figure 1. John Deere 8820: (A) cylinder, (B) concave, (C) back beater, (D) beater grate, (E) straw walkers, (F) shoe.

accomplished along the full length of the rotor in both the threshing and separation concaves. A conventional cleaning shoe is used for scalping and final cleaning. Massey-Ferguson is producing the 9720. The White models are the 9320 and 9720.

Deutz-Allis/Gleaner (Figure 5) has a different design than New Holland, Case International, and Massey-Ferguson combines. The threshing and separating rotor (cylinder) is mounted crosswise, with the crop fed tangentially into one end of the rotor. Threshing and separation occurs along the full length of the rotor, as the crop spirals sideways along the rotor. A paddle and impeller assembly discharges the crop from the outlet end of the rotor. A conventional cleaning shoe, combined with accelerator rolls and a high velocity air blast, is used for scalping and final cleaning. Three sizes of this combine are available, the R5, R6, and R7.

Claas of America, Inc. (Figure 6) has three models, the 106, 112CS, and the 116CS, which use a cylinder system. In place of the conventional system of separating and cleaning, Claas has eight synchronized, serrated, separating cylinders, with corresponding concaves, following a six-rasp-bar threshing cylinder. The synchronized, positive-output cylinders carry the crop over the concaves in a thin layer. Grain falls onto the cleaning sieve, and crop residue passes through to the double straw spreaders.

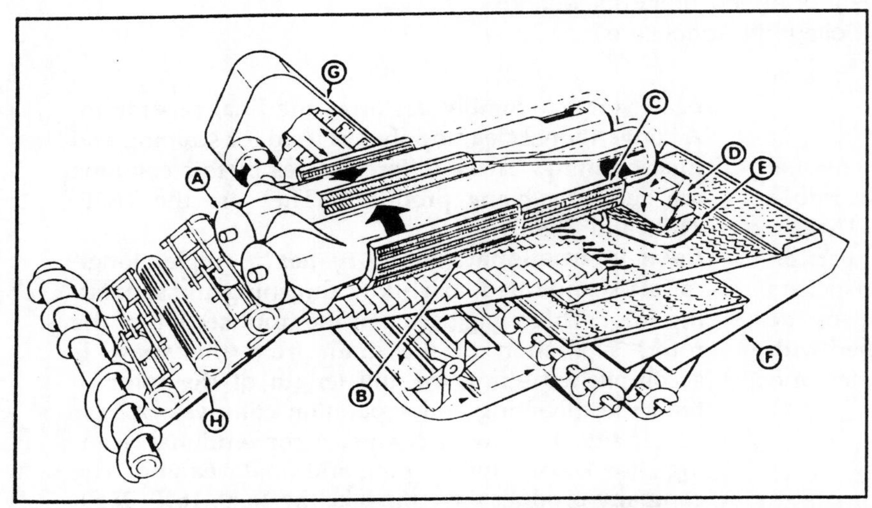

Figure 2. New Holland TR96: (A) rotors, (B) threshing concave, (C) separating concave, (D) back beater, (E) beater grate, (F) shoe, (G) tailings return, (H) stone ejection roller.

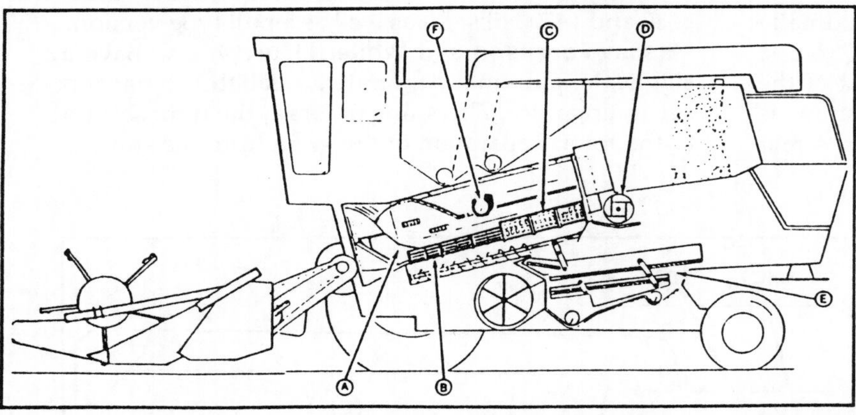

Figure 3. Case International 1660: (A) rotor, (B) threshing concaves, (C) separating concaves, (D) back beater, (E) shoe, (F) tailings return.

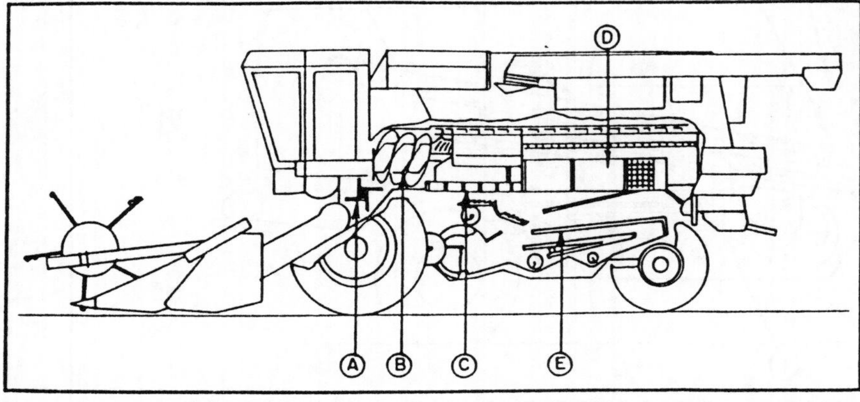

Figure 4. White 9720: (A) feed impeller, (B) rotor, (C) threshing concaves, (D) separating concaves, (E) shoe.

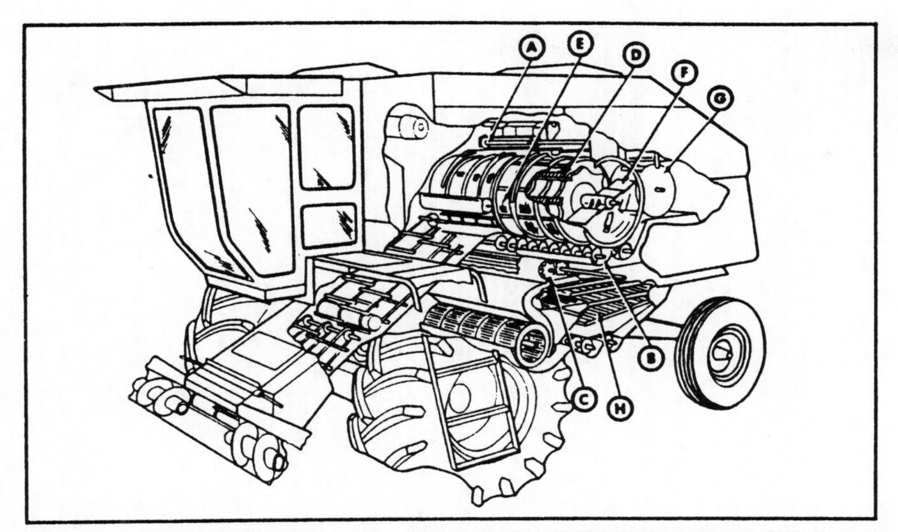

Figure 5. Deutz-Allis R6: (A) cage sweep, (B) distribution augers, (C) accelerator rolls, (D) rotor, (E) concave, (F) paddles, (G) impeller, (H) shoe.

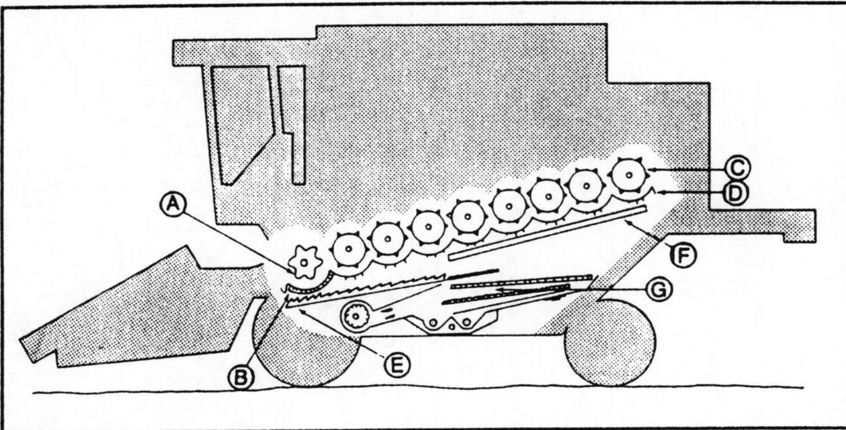

Figure 6. "Cylinder System Combine": (A) cylinder, (B) concave, (C) separator cylinder, (D) separator concave, (E) preparation pan, (F) returns pan and (G) upper and lower frog mouth sieve.

Tips For Reducing Soybean Harvest Loss

Typically, more than 80% of the machine loss occurs at the gathering unit. The following suggestions will help keep these losses to a minimum.

1. Make sure that knife sections, guards, wear plates and hold-down clips are in good condition and properly adjusted.

2. Operate the cutterbar as close to the soil surface as possible at all times.

3. Use a ground speed of 2.8 to 3.0 miles per hour. To determine ground speed, count the number of 3 feet steps taken in 20 seconds while walking beside the combine. Divide this number by 10 to get the ground speed in miles per hour.

4. Set reel speed approximately 25 percent faster than ground speed or for 42 inch diameter reels, use a reel speed of 11 R.P.M.. for each 1 mile per hour ground speed.

5. Reel axle should be 6 to 12 inches ahead of cutter bar. Reel bats should leave beans just as they are cut. Reel depth should be just enough to control the beans.

6. A six bat reel will give more uniform feeding.

Field Trip Data Sheet

Corn

Preharvest loss (#3/4 lb ears) __________ bu/A

Machine ear loss (#3/4 lb ears) __________ bu/A

Machine kernel loss (# per sq ft.) (Average of 3 counts)

Count 1: ______ Count 2: ______ Count 3: ______

Average __________ bu/A

Total corn harvesting loss (sum of above losses) __________ bu/A

Soybeans

(4-5 beans=1 bu.)

Preharvest loss (Average of 3 counts) __________ bu/A

Count 1: ______ Count 2: ______ Count 3: ______

Average __________ bu/A

Stubble loss (Average of 3 counts)

Count 1: ______ Count 2: ______ Count 3: ______

Average __________ bu/A

Average stubble loss __________ bu/A

Machine loss (Average of 3 counts)

Count 1: ______ Count 2: ______ Count 3: ______

Average __________ bu/A

Average machine loss __________ bu/A

Total Soybean harvesting loss (sum of above losses) __________ bu/A

September 1991

Checking Corn Harvest Losses

Samuel D. Parsons
Extension Agricultural Engineer
Purdue University

Ear Corn Loss

Ear corn weighs 70 pounds per bushel at 15.5% moisture content. At slightly higher moistures, it takes more pounds to represent one bushel, say 75 lbs, or 75 lbs in one acre, or 7.5 lbs in 1/10th acre, or 0.75 lbs in 1/100th acre. Or, each ¾-lb ear lost during harvest represents about 1 bushel per acre loss. Preharvest loss can be estimated the same way. Ear loss found after harvesting minus the estimated preharvest loss gives an estimate of machine loss.

Row Width	Row Width	Row Length for 1/100th Acre*
(in)	(ft)	(ft)
30	2.50	174
36	3.00	145
38	3.17	136
40	3.33	131

*1 acre = 43,560 sq. ft.
1/100 acre = 435.6 sq. ft.
Row Length = 435.6/Row Width (ft)

If P is the pounds of ear corn found (ear loss) in 1/100th of an acre,

then, ear loss in bushels per acre is:

$$\frac{P}{0.75} \quad \text{or} \quad \frac{P \times 4}{3}$$

Corn Kernel Loss

Loose kernels on the ground and kernels attached to cobs (or cob pieces) can be counted to estimate loss level. Approximately 2 "normal size" kernels per square foot -- *average* -- represents one bushel per acre. Do ***NOT*** check just one square foot. Rather, check a larger area, then divide the number of kernels counted by the square feet of area checked to get *average* kernels per square foot, **THEN** divide by 2 to get bushels per acre loss.

Loose Kernels Under the Combine. Stop the combine, backup the length of the combine, check on the ground in front of the combine. Loose kernels here were shelled at the header. For better accuracy, check for loss on *all* rows of the header.

Kernels Attached to Cobs or Cob Pieces. This is threshing loss. Check area should be at least 2 feet along the row direction and as wide as the "processor width" -- directly behind the combine -- if a chopper or spreader is *not* used (or is turned-off for checking), **OR** as wide as the "header width" if a chopper or spreader is being used (and is not turned-off for checking). In the latter case, 1 sq. ft. = 1 sq. ft. Calculate bushel per acre loss as noted above. But in the former case (checking directly behind combine), 1 sq. ft. ≠ 1 sq. ft. It represents a *larger* effective harvest area. Compute bushel per acre loss as noted above, then multiply this value by the "processor width" and divide by the "header width" -- using consistent units for the "widths" -- both in feet or both in inches.

Loose Kernels Behind the Combine. These may be header loss, separation loss or cleaning shoe loss. Header loss can be subtracted-out but separation and cleaning shoe loss can not be distinguished (by counting loose kernels on the ground). Check areas should be the same as those discussed under "Kernels Attached to Cobs", being careful to distinguish between a *true* square foot and one that represents a larger effective harvest area.

Checking Soybean Harvest Losses

Each 4-5 "normal size" beans per square foot (average) represents one bushel per acre loss. Count beans in pods still attached to stubble as stubble loss -- plants were cut above these pods. Count beans still in pods that have gone through the combine as threshing loss. Loose beans on the ground *behind* the combine may be due to preharvest shatter (check and subtract out if interested), or to shatter loss at the header (check "under" the combine and subtract out if interested), or to separation and cleaning shoe loss -- which cannot be distinguished by simply counting loose beans on the ground. When checking preharvest loss, stubble loss, and shatter loss at the header, a sq. ft. is a *true* sq. ft. But for checking threshing loss and separation-cleaning shoe loss, a sq. ft. may represent a larger effective harvest area, depending on how the combine is equipped and operated, and where the loss check is made.

ESTIMATING CROP RESIDUE COVER

OBJECTIVES:

1. Describe the significance and the benefits of managing crop residues.

2. Discuss factors that influence crop residue cover.

3. Demonstrate three methods for estimating crop residue cover.

A. Why should you manage crop residues?

Crop residue management is important for a variety of reasons. Savings are realized by the farmer in soil, labor and fuel, along with water by quality improvement. With the continuation of government programs, compliance with the 1985 and 1990 Farm Bills necessitates the control of cropland erosion and the concern for water quality.

Conservation tillage is defined as any tillage planting system that leaves at least 30% of the field surface covered with crop residue after planting has been completed. The easiest, most efficient, and economical method of controlling cropland erosion is through managing crop residues in conservation tillage systems. Therefore, crop residue management is economically, environmentally, and politically sound.

B. Factors that Influence Crop Residue Cover

Several factors influence the crop residue cover that exists in the field. Some of these are rather obvious while others are somewhat hidden in nature.

Type of tillage system

One of the most obvious factors influencing crop residues is the type of tillage system you select. Field operations have varying impacts upon residue cover. A moldboard plow in a conventional tillage system leaves a maximum of 3 to 5 percent cover remaining while a no-till planter leaves 80 - 90 percent cover remaining.

Type of crop

The type of crop plays a major role in residue remaining. Corn leaves about twice the residue that soybeans do. Soybean residue is more brittle than corn residue. Therefore field operations degenerate soybean residue more rapidly than corn residue. Different crops also vary in chemical makeup. Hence some crop species decay quicker than others.

Weather

Weather plays a role in residue decomposition. Adequate moisture, warm temperatures, air and microorganisms are four factors that affect rate of decomposition. Why are most of the residues present in the fall still present the following spring?

Crop management practices

The general health of the crop can influence the amount of residue produced. A crop deficient of nutrients or water will produce fewer stems and leaves. Likewise, insects can decrease the amount of plan dry matter produced. Poor management of pesticides and other field situations such as compaction can stunt plant height resulting in less crop residue.

Winter cover crops

Seeding a winter cover crop that is then sprayed with a contact herbicide the following spring is a method that can be used to increase the amount of residue. Erosion is greatest on uncovered soils. The cover crop will reduce erosion and serve as residue cover.

C. Field Trip Study Guide

Observe the type and amount of crop residue cover in the field. What factors do you think had an effect on the residue cover you see? List these factors.

Influences upon Crop Residue

1.

2.

3.

4.

5.

Methods for estimating crop residue cover

Now that you have observed factors that influence residue cover, it is important to know the amount of crop residue cover left in the field because of its influence on soil erosion potential.

There are three methods in use for estimating crop residue cover. The **line transect** method and the **photo comparison** method are conducted in the field. The **calculation** method can be done without direct field observations. The key to accuracy in all methods is to avoid overestimation and to be consistent. Refer to AY-269, Estimating Corn and Soybean Residue Cover, for more information.

1. Line Transect Method

This method is the easiest and the most reliable way to determine residue cover. This method can be completed by using either a measuring stick or a 50 foot measure.

Using the measuring stick method requires the following four steps:

a. Place a measuring stick on the soil across the rows

b. Measure the total number of inches of residue covered soil under the measuring stick, looking straight down onto it.

c. To calculate residue cover, divide the total you obtained in step (b) by the length of the measuring stick.

Example:

$$\frac{\text{20" (total inches of residue covered soil)}}{\text{36" (if using a yardstick)}} = \text{56\% cover}$$

d. Repeat steps a, b, and c in enough locations to obtain representative values, and obtain an average value for the field.

The fifty foot measuring tape method is somewhat similar. You could also use a 50 foot piece of rope that is knotted or marked at every one foot increment instead of a 50 foot tape measure. The fifty foot measure method uses the following steps:

a. Stretch and stake the tape to 50 feet in length at a 45¯ diagonal across the rows.

b. Count the number of foot marks that intersect or lie directly over a piece of residue. Look straight down onto each foot mark and stand on the same side of the tape when counting.

c. Multiply the number of foot marks by two to obtain the percent of residue cover.

d. Take an average of at least five measurements at sites typical of the entire field.

Using the linear transect method is the most reliable and best for year to year determinations. You should avoid taking measurements in areas that are atypical of the overall field, such as turn row areas and waterways. You should try to figure averages in areas that are similar. To further avoid overestimation while measuring, ask yourself, "If a raindrop falls at this point, would it hit residue or bare soil?" When in doubt don't count it.

2. Photo-Comparison Method

Another method used to estimate residue cover involves carrying photographs of known residue covers into the field and matching a picture to the actual field conditions. This provides a quick estimate, but this method is not as accurate as the line-transect method. Use photos in AY-269 for comparison.

As with any method, use sites that are typical of the field. Look straight down onto the residue. Scanning a field from a road or fence row will cause you to overestimate residue cover, as the exposed soil behind the residue is hidden from view.

3. Calculation Method

This method, unlike the previous two methods, does not require field observations. This method is best used for long range conservation planning. Generalizations are made and then calculated to obtain a residue cover estimate. Use Purdue Agronomy extension publication AY-269, Estimating corn and soybean residue cover, to calculate residue estimates.

In summary, crop residues are beneficial in preventing soil erosion. Even wind erosion is reduced by standing residue and by large quantities of residue. Management of crop residues is relatively simple and easy on the pocketbook. With the current issues at stake in the nineties, residue management makes good in economic dollars and environmental sense!

Written by M.G. Evans; August, 1991.
Adapted from Purdue University Cooperative Extension Service publication AY-269, "Estimating Corn and Soybean Residue Cover," by P.R. Hall, J.V. Mannering and J.R. Wilcox.

DATA SHEET

FIELD I

1. Type of tillage system: ________________

2. Crop: ________________

3. Residue Cover Estimation:

	Line Transect			
	Yardstick	50' Tape	Photo Comparison	Calculation
	________	________	____________	
	________	________	____________	
	________	________	____________	
	________	________	____________	
Average	________	________	____________	________

FIELD II

1. Type of tillage system: ________________

2. Crop: ________________

3. Residue Cover Estimation:

	Line Transect			
	Yardstick	50' Tape	Photo Comparison	Calculation
	________	________	____________	
	________	________	____________	
	________	________	____________	
	________	________	____________	
Average	________	________	____________	________

Agronomy Guide

Purdue University Cooperative Extension Service

SOILS (TILLAGE) AY-269

Estimating Corn and Soybean Residue Cover

Kenneth J. Eck, Purdue Extension Soil Conservation Education Specialist
Peter R. Hill, Purdue Extension Soil Conservation Education Specialist
Jesse R. Wilcox, USDA Soil Conservation Service Agronomist

Crop residue management through conservation tillage is one of the best and most efficient methods of controlling soil erosion. Each year, about 84 million tons of topsoil are eroded from Indiana croplands. Research shows that this can be greatly reduced by maintaining a crop residue cover on the soil surface of at least 30% after all tillage and planting operations. Leaving that amount will cut water-caused erosion to about half of what it would be if the field was clean-tilled; a higher percentage left reduces soil losses even more.

Line-transect method using a knotted rope for estimating percent residue cover.

Conservation tillage and residue management not only save soil, labor, and fuel, but also help improve water quality—all of which are important to the agricultural producer.

Conservation tillage is defined as any tillage-planting system that leaves at least 30% of the field surface covered with crop residue after planting has been completed. For example, two diskings in corn residue that left a 40% cover would be classified as a conservation tillage system, whereas just one disking in soybean residue resulting in less than 20% cover could not be considered conservation tillage because too much of the fragile residue is destroyed.

Conservation tillage is one of the most effective means of cropland erosion control. Uniformly distributed residue shields the entire soil surface from rainfall impact, thus reducing soil particle detachment and eventual erosion. Also, the residue creates small dams which slow the rate of runoff, allowing more time for water to infiltrate into the soil. A slower rate and reduced volume of runoff means less soil removed from the field.

Residue can also protect soil from the erosive forces of wind. To what extent, however, depends on the amount of residue present and its orientation (i.e, whether upright or flat). Standing residue is more effective than flattened residue in reducing wind erosion.

Methods for Estimating Residue Cover

Cropland residue cover estimation is not only useful in planning field operations to maintain erosion control, but is sometimes needed to determine if a particular field qualifies for certain federal, state, or local conservation programs.

Following are three methods for estimating percent of residue cover. The first two are accomplished with field observations; the third, which requires generalizations and calculations, is used primarily for conservation planning purposes.

Produced as an educational resource for Indiana's "T by 2000" erosion/sediment reduction program

West Lafayette, Indiana

Line-Transect Method

The line-transect method is an easy, reliable way to determine residue cover. It involves stretching a 50-foot measuring tape or rope (knotted or otherwise marked at six-inch intervals) diagonally across the crop rows (see Figure 1). Percent of cover is then determined by counting the number of marks that intersect or lie directly over a piece of residue.

The key to accuracy with this method is avoiding over- or under-estimation. To do that, look straight down on each mark and take all readings on the same side of the tape or rope, asking yourself, "If a raindrop falls at this point, would it hit residue or bare soil?" In general, the size of residue should be 3/32-inch in diameter or larger--the average size of a raindrop or roughly the size of a healthy wheat straw. If there is any doubt at all, do not count it.

Take at least five measurements at sites typical of the entire field, and average them to obtain the residue estimate. Do not take measurements in end rows or small areas of the field that have been affected by flooding, drought, compaction, weed or insect infestations, or other factors that may have affected crop residue levels.

Photo-Comparison Method

Residue cover can also be estimated by comparing actual field conditions to photographs of known percentages of cover (see corn and soybean residue cover photos below). This method provides a quick estimate, but is less accurate than the line-transect.

To use the photo-comparison method, go to a site typical of the field, look straight down, and compare what you see with what the photographs show; then estimate your percentage of residue cover based on that comparison. Repeat the procedure at four other typical sites, and average your estimates.

Scanning a field from the road or field boundary is not adequate. You will tend to overestimate percentage of cover because the exposed soil behind the residue is hidden from view.

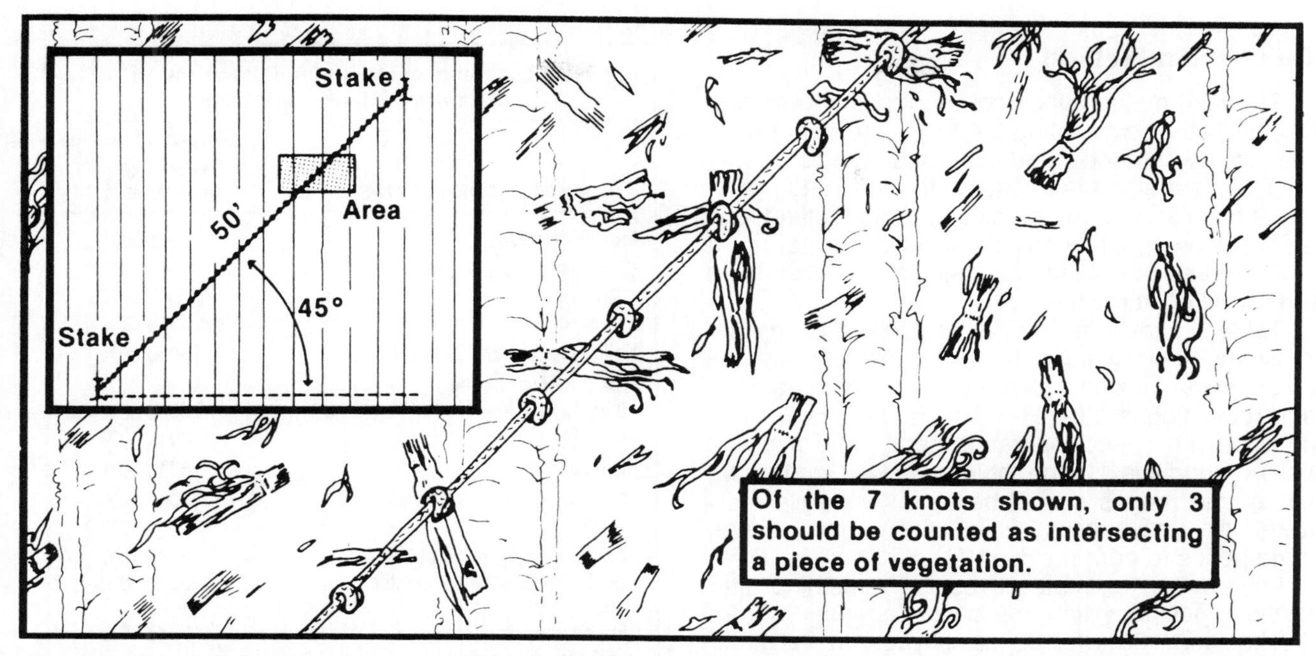

Figure 1. Overview (inset) and close-up of the line-transect method. (Source: Illinois Cooperative Extension Service.)

SOYBEAN RESIDUE COVER

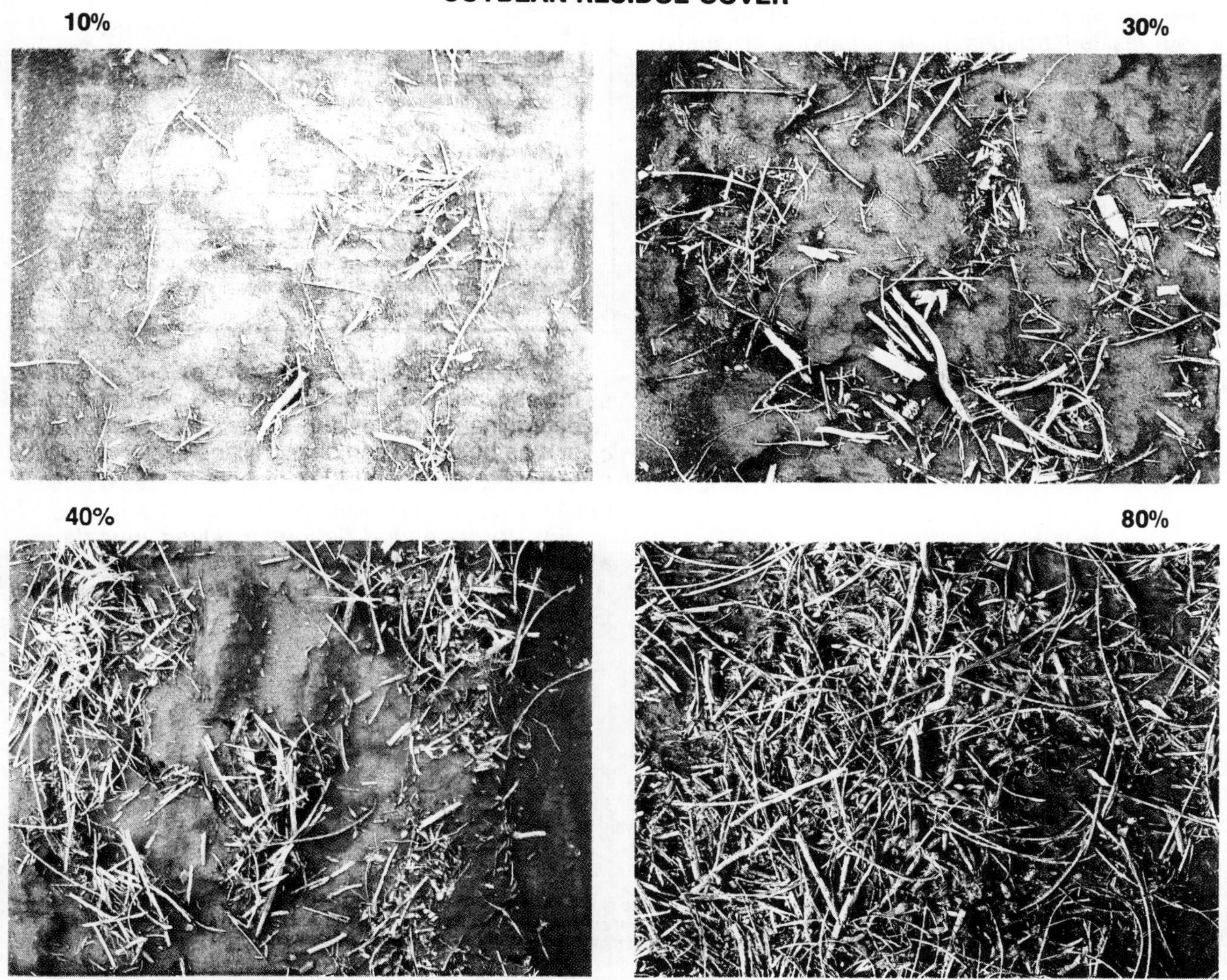

Calculation Method

The third method does not require field observation, but rather calculating the likely percent of residue after weathering and individual tillage operations. This method is adequate in long-range conservation planning for predicting tillage effects on residue cover, although it's less accurate on a year-to-year basis due to variation in weathering and tillage equipment use.

Table 1 shows the ranges in percent of residue remaining after various tillage or planting operations. For a given implement, the actual percentage remaining is a result of several factors, including operating speed, operating depth, and soil and residue condition. In the table, the lower end of the percentage ranges should be used for fragile residues like soybeans, while the upper range corresponds to corn residue.

For a rough estimate of residue remaining after planting, multiply initial crop cover (approx. 95% for 120-bu. corn, 85% for 38-bu. soybeans) by winter weathering loss (if not fall moldboard plowed) and then by the appropriate percentage for each operation that makes up your tillage-planting system.

For example, assume the system used on a field of corn residue includes three operations: (1) spring chisel plowing with straight points, (2) disking 3 inches deep, and (3) planting with narrow ripple no-till coulters. To calculate final residue cover, multiply the following percentages from Table 1:

initial	x	weathering	x	chisel	x	disk	x	plant	=	final
95%		85%		80%		80%		90%		47%

In this example for corn residue, the three operations listed would fit the definition of a conservation tillage system. But they wouldn't for soybean residue, as the following calculations show:

initial	x	weathering	x	chisel	x	disk	x	plant	=	final
85%		75%		50%		40%		85%		11%

Table 1. Influence of Various Field Operations on Surface Residue Cover Remaining.

Tillage and planting implements	Percent of residue remaining after each operation*
Moldboard plow	3 to 5%
Chisel plow (coulter chisel)	
Straight points	50 to 80%
Twisted points	30 to 60%
Knife-type fertilizer applicator	50 to 80%
Disk (tandem or offset)	
3 in. deep	40 to 80%
6 in. deep	30 to 60%
Field cultivator	50 to 80%
One-pass combination tool	30 to 60%
Planter	
Smooth, ripple, or no coulter	90 to 95%
Bubble or fluted/wavy coulter (less than 1" wide)	85 to 90%
Fluted/wavy coulter (1" wide or greater)	80 to 85%
Sweeps, double disks, or horizontal disks (ridge-till)	60 to 80%
Drills–conventional	
Disk openers	90 to 95%
Drills–no-till	
Ripple or no coulter	85 to 90%
Bubble or fluted/wavy coulter (less than l" wide)	80 to 85%
Fluted/wavy coulter (1"wide or greater)	70 to 75%
Winter weathering	75 to 85%

* Use higher values for corn residue and lower values for soybean residue. These numbers apply to primary tillage operations only. Secondary tillage operations may not reduce percent cover by the same amounts on second or third passes.

The calculation method provides only rough estimates since the variables involved prevent accurate determination of residue cover. However, Table 1 can be helpful in comparing tillage systems because they give a general idea of how much residue will remain after specific tillage and planting operations.

For an expanded listing of tillage machinery and remaining residue levels after usage, refer to *AY-280, The Indiana Crop Residue Management Guide for Farm Machinery.*

Adapted from Nebraska Extension publication G86-793, "Estimating Residue Cover," by E. C. Dickey, P. J. Jasa, and D. P. Shelton

REV 1/94 (5M)

Cooperative Extension work in Agriculture and Home Economics, state of Indiana, Purdue University, and U.S Department of Agriculture cooperating; H. A. Wadsworth, Director, West Lafayette, IN. Issued in furtherance of the acts of May 8 and June 30, 1914. The Cooperative Extension Service of Purdue University is an affirmative action/ equal opportunity institution.

Agronomy Guide

Purdue University Cooperative Extension Service

SOILS (CONS. LAND USE) AY-271

Wind Erosion Concerns in Indiana

Gary C. Steinhardt, Extension Agronomist, Purdue University
Charles L. Schoon, and Robert L. Guillaume,
Extension Soil Conservation Education Specialists, Purdue University
Jesse L. Wilcox, U.S. Department of Agriculture Soil Conservation Service

Wind erosion is a significant problem for Indiana soils having a sand, loamy-sand, sandy-loam, or muck-surface texture. Although not as extensive as water erosion, wind erosion can lower crop yields and reduce the productivity of the soil. In addition, wind erosion can damage standing crops by a process called "sand blasting."

Some farming methods increase the risk of wind erosion on susceptible soils, especially intensive cropping systems, including tillage methods that leave less than 30% residue cover. The failure to use cover crops or crop rotations can leave a field subject to long-term damage.

Another problem is increased field size. Removal of fences and similar windbreaks during the last 20 to 30 years to make for more efficient field operations has left many areas less protected. However, the most damaging practice is fall tillage with a moldboard or chisel plow that leaves little surface residue in the field. These wind-erodible soils remain unprotected until crop development late in the following spring.

Problems Caused by Wind Erosion

On-farm Problems

Wind erosion *removes finer particles* from the soil. Clay and organic matter, the finest and lightest particles, hold on to nutrients and are important for holding moisture. Wind erosion can rob a sandy soil of the very components that are most needed for productivity. Researchers have shown that the wind-blown sediments can contain twice as much nitrogen and phosphorus and 20% more potassium than the soil left in place. The fertility can be replaced, but the loss of moisture retention due to removal of finer particles is permanent.

Wind erosion can *reduce crop yields.* Blowing sand can adversely affect young, tender seedlings–especially corn. Abrasive sand, propelled by strong winds, can actually wear away layers of tissue and eventually cut off seedlings. Wind erosion can reduce stands, stunt plant growth and reduce yields. Vegetables and melons are greatly affected in terms of yield, disease susceptibility, and quality.

Wind erosion helps *spread weed seeds* along with the other sediments. Weed seeds can be moved long distances from unprotected areas.

Off-farm Problems

Wind erosion can help *fill roadside and drainage ditches,* which is one of the most costly aspects of both water and wind erosion to the general public. The fewer the windbreaks, or other traps, the more likely wind-blown sediments are to end up in ditches. Ditches that drain organic soils can be filled very rapidly.

Wind erosion can create a *safety hazard* from the reduced visibility on roads and highways caused by airborne sediments. In Indiana, this problem has been particularly noticeable on I-65 north of Lafayette and along U.S. 41 in southwestern Indiana.

Wind Erodible Soils in Indiana

Figure 1 is a map showing the six soil resource areas of Indiana. Table 1 contains the acreage of wind-erodible soils in each area. The majority of the wind-erodible soils are located in Soil Resource Area A, which includes large areas of sandy deposits from glacial outwash as well as from wind-blown sand. Soil Resource Area D also has wind-erodible soils principally along the major rivers.

Table 2 contains a list of Indiana soil series and surface textures that are considered to be wind-erodible. Conservation practices are needed to prevent short- and long-term problems associated with wind erosion on these soils.

The Process of Wind Erosion

Wind erosion is produced by the three types of soil movement illustrated in Figure 2.

Saltation occurs with fine- to medium-size sand particles (0.1 to 0.5 mm). Wind blowing across the ground surface causes these particles to rotate at several hundred revolutions per second, which, in turn, causes them to jump 1-2 ft. into the air. Returning to the ground, they collide with other particles and break down soil aggregates. Saltation accounts for up to 80% of wind-erosion movement. The other

Figure 1. Soil Parent Materials of 13 Regions of Indiana Included in Each.

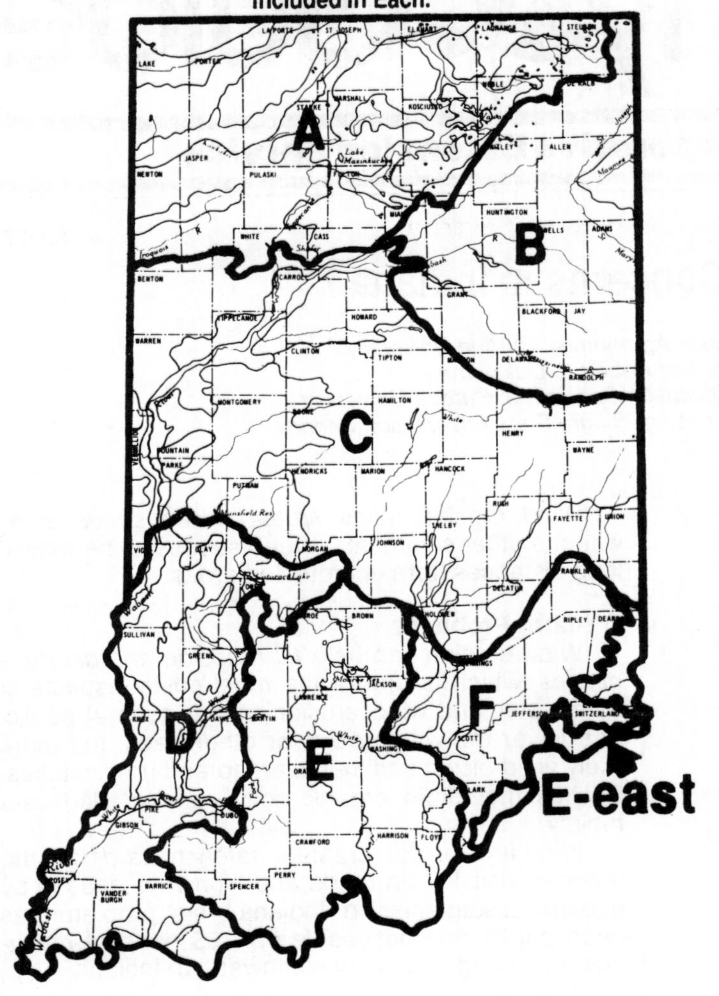

forms of wind-erosion movement do not occur without it.

Suspension affects mineral particles of very fine sand, silt, clay (all less than 0.1 mm) and organic matter. Soil particles that move by saltation strike these finer particles and knock them into the air. The fine particles are light and easily carried by strong wind currents, often for miles before being deposited. Suspension-type movement is very damaging because this lost segment of the soil contains the most nutrients, even though the particles moved by suspension are a small percentage of the wind-eroded sediments.

Surface creep is the way in which the coarse sand-size particles (0.5 to 1.0 mm) are moved by wind. These are the largest particles affected by wind erosion. The factors that cause surface creep are much like those that cause saltation; however, the particles are too heavy to become airborne. They are rolled along the surface instead. This type of wind erosion can also damage vegetation.

Soil avalanching refers to the tendency for the number of soil particles in movement to increase downwind. The degree of damage grows with the increased size of an unprotected area. At the protected edge of a susceptible area, wind erosion is near zero, but when moving downwind, the number of particles of all sizes in movement increases almost like a chain reaction. Soil avalanching can be observed most clearly on sand dunes and to a lesser extent on bare fields. The results of soil avalanching can be observed afterwards by the smoothing of the eroded area and the deposition of sediments in an area that traps the wind-blown materials on the downwind side.

Conditions that Influence Wind Erosion

Cloddiness of surface. Soil clods and aggregates, particularly those resulting from tillage, help a soil resist the damaging effects of wind erosion. Clods or aggregates on the surface are too large to be moved by the wind, thereby protecting the component particles. But clods made up of sand, loamy sand, or sandy loam, which are the most wind-erodible soils, break down much more quickly than do soils of finer texture. Conversely, soils with more clay are held together in much stronger aggregates. Lack of strong aggregates is a major contributing cause to increased wind erosion from coarser soils.

Roughness of surface. An irregular surface from ridge-till or chisel tillage does much to reduce wind damage by absorbing and deflecting the erosive energy of the wind. An irregular surface can slow down and alter the movement of particles involved in saltation and soil creep.

The one disadvantage is that formation of ridges by tillage exposes the top of the ridge increasing the potential for erosion, particularly if the ridge is much higher than 5 in. This is not a great concern if crop residues are left in place as is the case with ridge-till.

Wind velocity. Soil particles do not start to blow until wind velocity reaches 12-13 mi. per hr. 1 ft. above the surface. The effectiveness of the wind to erode soil is related to the wind velocity. When wind velocity doubles, the effectiveness of the wind to erode is not doubled, but rather increased eight times.

Soil moisture. In general, wind erosion increases as soil moisture decreases. Mathematically stated, if soil moisture is reduced by half, then soil movement is increased by four times.

Field length. The efficiency of the wind to erode increases directly with distance from a protected edge as the wind moves across an unprotected field. Soil movement by wind reaches a limit for a given wind velocity with certain soil and roughness conditions. Most fields in the Midwest are too small to see what researchers believe to be the maximum effect of a wind. However, the recommendations for strip cropping provide clues as to the critical length (lim-

Table 1. Indiana Estimates of Wind-Erodible Soils.

Soil resource area	Total area 1000 acres	Wind erodible 1000 acres	%
A	3,957.0	1937.0	49
B	3,264.6	104.9	3
C	7,267.5	101.5	1
D	3,187.9	203.0	6
E	3,935.3	15.0	<1
F	1,535.7	28.2	2
Total	23,148.0	2,389.6	10

ited to 100 ft.) where soil movement by wind becomes important. The 100 ft. limit has been shown to be a critical distance on many soils. In another example, windbreaks are considered to provide effective protection downwind 10 to 12 times the height of the windbreaks. For trees approximately 40 ft. tall this would mean 400 to 500 ft., which appears to be on the upper limits of open fields.

Vegetative/residue cover. Cover on the surface is the most effective way to control wind erosion. There are three aspects of cover that can be used to evaluate effectiveness. They are quantity, kind, and orientation. In general, the more residue, the less soil is eroded by wind. Although a variety of factors must be considered, a 30% cover is probably a good estimate of the amount needed for adequate protection. Crop residues have been shown to have different abilities to limit wind erosion. For example, wheat stubble is, pound for pound, more effective than corn stubble; however, corn stubble is more effective than soybean residue. The orientation (flat or standing) and condition of the residue is important. Small-grain stubble standing is 2.5 times as effective as the same residue when it is flat on the ground.

Knoll effects. The topography of northeastern Indiana includes a number of sandy ridges. The ridges are a combination of beach and dune deposits developed during glaciation. Oak trees are the natural vegetation; and the trees, if present, make these land forms easy to identify. When these trees are removed, however, the soils are very susceptible to erosion, not only because of the sandy surface texture, but also because of slope. The wind velocity increases as it blows across a knoll, and when it does, the compressed air flow at the top of the knoll creates wind erosion that is much greater on the upper third of a slope than on the lower parts. Erosion on the downwind side of the knoll can increase because eroding particles coming from the top of the hill can start the saltation process on the lower parts of the slope.

Figure 2. Wind Erosion Process.

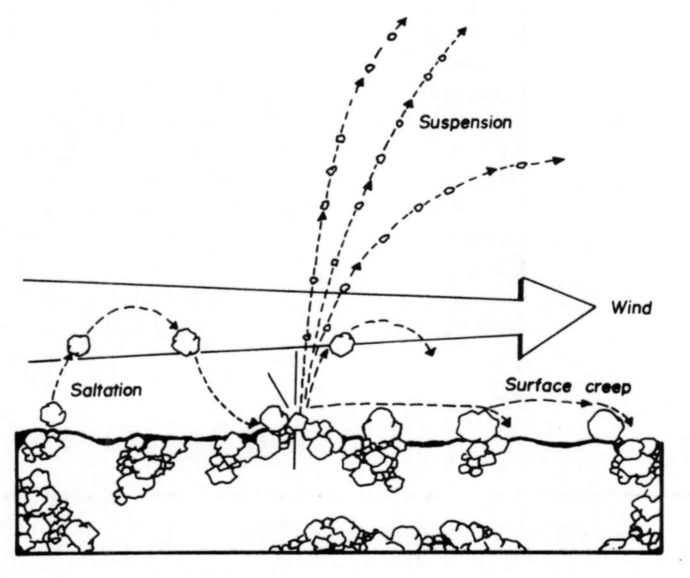

Estimating Wind Erosion on a Field

Researchers have developed a formula to estimate wind erosion on a field. Soil conservationists use this formula for developing a conservation plan. The amount of soil eroded by wind can be calculated using the following equation:

$$E = f(I, K, C, L, V)$$

The average annual soil loss in tons-per-acre by wind erosion (E) from a given field is determined by interaction of these five factors:

I, the soil-erodibility index, is related to the amount of sand-size soil aggregates and further influenced by soil slope.

K, the soil-roughness factor, is determined by the height and spacing of ridges or clods.

C, the climatic-factor, takes into account the anticipated wind velocity and surface-soil moisture.

L, the length-factor, is the unobstructed distance across a field along the direction of the prevailing wind.

V, the vegetation-factor, is related to the present crop, its quantity and orientation.

Components of this wind-erosion equation are specific to a given field and not suitable for general calculations. County soil and water conservation districts have the information needed to complete calculation of wind-caused soil loss on specific fields. This equation is currently being reviewed. It has been very useful on the Great Plains, but some feel that it under estimates wind erosion in the humid Midwest.

Methods to Control Wind Erosion

Wind-erosion control techniques can be grouped into two general categories: practices that cover the soil and practices that slow or disrupt the wind. .

Practices that Cover the Soil

Conservation tillage. This is the practice most easily applied, and most effective, for soils where wind erosion is a problem. There are several types of conservation tillage. One helpful publication, AY-210, *Adaptability of Various Tillage-Planting Systems to Indiana Soils,* discusses each rather thoroughly, and then rates them as to suitability on a given soil series. Before selecting any method for controlling wind erosion, consult AY-210 and your county soil survey to be sure the method selected is the best fit for all the soils on the landscape.

It is best to delay tillage to leave as much residue in place for as long as possible. Row direction should be at right angles to the prevailing wind, usually from the southwest. Unfortunately, this is difficult with the field orientation common to Indiana.

Cover crops. Cover crops of wheat rye or oats planted in the fall can prevent serious wind erosion by covering the soil surface and providing plant roots to hold the soil together. Fortunately, cover crops are well-adapted on the soils that most need them. In the spring, these crops can be plowed under or chemically controlled and then used as a mulch for no-till planting.

Table 2. Indiana Soil Series that may be Wind Erodible.

Soil series	Erodible surface texture*
Abscota	FSL, LFS, S
Ackerman	MU
Ade	FSL, LFS
Adrian	MU
Algansee	FSL, LS, LFS
Alida	FSL, LFS
Alvin	FSL, LFS, SL
Aubbeenaubbee	FSL, SL
Ayr	FSL, LFS, LS
Ayrmount	LFS
Ayrshire	FSL, SL
Barry	FSL
Belleville	LS
Belmore	FSL
Berrien	LFS
Billett	FSL, SL
Bloomfield	FS, LFS, LS, S
Bobtown	LFS
Boots	MU
Bourbon	SL
Boyer	LS, SL
Brady	FSL, LFS, SL
Branch	LS
Brems	FS, S, LFS, LS, FSL
Bronson	LS, SL
Bruno	FSL
Carlisle	MU
Carmi	SL
Casco	SL
Celina	FSL
Ceresco	FSL, SL
Chatterton	FSL, SL, LS
Cheektowaga	FSL
Chelsea	FS, LFS, LS
Cohoctah	FSL, SL
Coloma	FS, LFS, S
Conotton	SL
Conrad	FS, LFS
Corwin	FSL
Craigmile	FSL, SL
Crosby	FSL
Crosier	FSL
Darroch	FSL
Desker	SL
Dickinson	SL
Edwards	MU
Elston	FSL, SL
Foresman	FSL
Fox	FSL, SL
Genesee	FSL, SL
Gilford	FSL, SL
Granby	FSL, LFS, LS
Gravelton	LS
Grovecity	FSL
Hanna	SL
Haymond	LS
Hillsdale	FSL, SL
Homer	FSL, SL
Hononegah	LS, FSL
Hoopeston	FSL
Houghton	MU
Huntington	FSL
Iroquois	FSL
Jasper	FSL
Junius	LS
Kalamazoo	SL
Kentland	FS
Kosciusko	SL
Landes	FSL, SL, LS
Linkville	SL
Linwood	MU
Lyles	FSL, SL
Markton	LS, S
Martinsville	FSL, SL
Martisco	MU
Maumee	FSL, LFS, LS, S
Metamora	FSL
Metea	LFS, LS
Miami	FSL, SL
Montmorenci	FSL
Morley	SL
Morocco	FSL, LFS, LS
Moundhaven	FSL, SL
Mudlavia	SL
Muskego	MU
Mussey	SL
Napoleon	MU
Nesius	FS, LS, LFS
Newton	FSL, LFS, LS
Nineveh	SL
Oakville	FS, LFS, S
Ockley	SL
Octagon	FSL
Onarga	FSL
Ormas	LFS, LS, S

Table 2. Indiana Soil Series that may be Wind Erodible (cont).

Soil series	Erodible surface texture*
Oshtemo	CSL, FSL, LFS, LS, SL
Ouiatenon	FSL, LS, SL
Owosso	FSL, SL
Palms	MU
Papineau	FSL, SL
Parr	FSL
Piankeshaw	SL
Pinevillage	SL
Pinhook	SL
Pipestone	LFS
Plainfield	FS, S, LS, LFS
Princeton	FSL
Prochaska	LS
Rawson	FSL, SL
Rensselaer	FSL, SL
Riddles	FSL, SL
Ridgeville	FSL
Riverdale	LS
Roby	SL
Rockton	FSL
Rodman	CSL, SL
Ruark	SL
Saugatuck	LFS
Seafield	FSL
Sebewa	SL
Selfridge	LFS
Selma	FSL
Seward	LFS
Shipshe	SL
Simonin	LS
Sparta	FS, LS, S, LFS
Spinks	S, FS
Stockland	SL
Stonelick	FSL, LFS, SL
Tawas	MU
Tedrow	FS, LS, LFS
Toto	MU
Tracy	SL
Tyner	LFS, LS
Warners	FSL
Warsaw	SL
Watseka	LFS, LS
Wauseon	FSL
Wawasee	FSL, SL
Wesley	FSL
Wheeling	FSL
Whitaker	FSL, SL
Willette	MU
Wirt	FSL, VFSL
Zaborsky	FS
Zadog	LS
Zipp	SL

*CSL - Coarse Sandy Loam
FS - Fine Sand
FSL - Fine Sandy Loam
LFS - Loamy Fine Sand
LS - Loamy Sand
MU - Muck
S - Sand
SL - Sandy Loam
VFSL - Very Fine Sandy Loam

Practices that Slow the Wind

Strip cropping. Just as strip cropping disrupts water runoff, it also disrupts wind-current flow. For wind-erosion control, use long, narrow strips, oriented at right angles to the prevailing wind. Crops of varying heights are planted in each narrow strip. Both row orientation and crop-height difference disrupt wind currents. One limitation of this method is that small grains and hay may not be tall enough to completely protect the soil during critical periods.

A variation of the above technique can be used after corn harvest if some fall plowing is desired. For each 40 rows of cornstalks plowed under, leave eight rows unplowed. This usually disrupts wind currents enough to reduce wind erosion under most Indiana conditions. Again, corn rows need to be at right angles to the prevailing wind.

Windbreaks. In extensive areas of wind-erodible soils, tree or grass windbreaks at right angles to the prevailing wind are useful. Although it removes some land from production, the long-term productivity is saved on the protected land, and the yield benefit of enhanced moisture collection through increased snow collection should offset this loss.

In general, the effectiveness of a windbreak extends in distance ten to twelve times the height of the windbreak, depending somewhat on climate, soils, and the other practices used. Spacing should take into account equipment size. If irrigation is present, windbreak designs and plantings are available which do not interfere with established patterns. Windbreaks are most effective when used over fairly large areas and when used in combination with other wind erosion-control practices. Windbreaks can be designed to serve a variety of purposes in addition to erosion control, such as wildlife habitats and livestock protection.

Snow has a tendency to collect around windbreaks, and may delay some spring tillage near the windbreak, but part of this area may actually have a higher yield because of moisture collection. Lower yields are likely to occur very close to the windbreak because of trees competing with crops for moisture and nutrients.

Emergency tillage. The primary means of controlling wind erosion are conservation tillage, cover crops, strip cropping, and windbreaks. If because of intense wind, very dry conditions or some other reason these should fail, emergency tillage may be needed. The purpose of emergency tillage is to produce a cloddy surface and thus break-up wind patterns near the surface.

A heavy-duty chisel plow is the implement best suited. The chisel points may be a variety of types, but the important thing is that they leave a rough surface. Speed of operation is important. If you drive too slow, not enough clods are raised. If you drive too fast, the soil is pulverized. A speed of 3.5 to 4 mi. per hr. is usually about right. Begin on the windward side of the field and chisel occasional strips every 100 ft. or so. The entire field may need to be chisel plowed.

The sandier the soil, the more difficult it is to stop the erosion of soil by emergency tillage. Thus the importance of improving the soil over the long-term by cover crops, rotational sod crops and good residue management cannot be over emphasized.

Wind Erosion Control Planning

Precise estimates of wind erosion are difficult. It is clear from the Indiana soil map that wind erosion is a significant problem in the northwestern part of Indiana. This publication has discussed some of the principles of controlling wind erosion, but the ultimate decision rests with the farmer.

An important first step is to check the county soil survey report and compare the soils with those listed in Table 2. If there are a number of soils subject to serious wind erosion in a field, then a plan should be developed. Fortunately, wind erosion can usually be controlled by minor changes in current practices. If more extensive changes are needed, your local soil and water conservation district has the expertise to develop such programs.

RR 8/90 (5M)

Cooperative Extension work in Agriculture and Home Economics, state of Indiana, Purdue University, and U.S. Department of Agriculture cooperating; H. A. Wadsworth, Director, West Lafayette, IN. Issued in furtherance of the acts of May 8 and June 30, 1914. The Cooperative Extension Service of Purdue University is an affirmative action/equal opportunity institution.

Purdue University Cooperative Extension Service

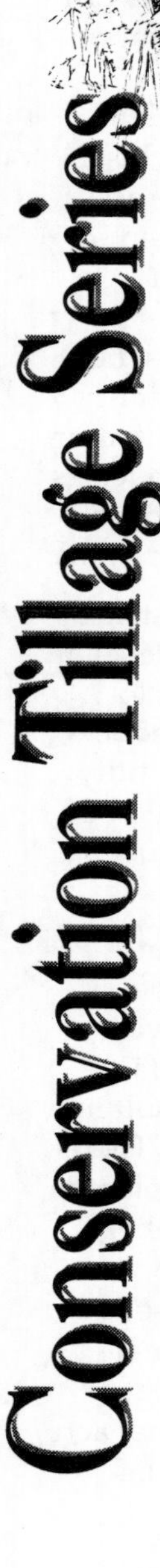

Managing Crop Residue with Farm Equipment

Peter R. Hill, Purdue Extension Soil Conservation Specialist
Kenneth J. Eck, Purdue Extension Soil Conservation Specialist
Jesse R. Wilcox, USDA Soil Conservation Service Agronomist

Over 40 percent of Indiana's 13.8 million acres of cropland is eroding at a rate faster than natural processes can replace it. Research shows that this erosion can be greatly reduced by maintaining a crop residue cover of at least 30 percent on the soil surface after all tillage and planting operations are completed. This type of system is known as a conservation tillage system.

Conservation tillage is one of the most effective means of cropland erosion control. Uniformly distributed residue shields the soil surface from rainfall impact, thus reducing the tearing and washing away of soil particles. The residue also creates small dams which slow the rate of runoff, allowing more time for water to infiltrate the soil. A slower rate and reduced volume of runoff means less soil removed from the field.

Residue can also protect soil from the erosive forces of wind. To what extent, however, depends on the amount of residue present and whether it is upright or flat. Standing residue is more effective than flattened residue in reducing wind erosion.

Several tillage systems, including chisel plow, disk, ridge-till, and no-till systems, can leave 30 percent residue cover or more after planting. However, the number of field operations must be limited. This number has a greater impact on residue cover than the type of implement used. For example, when using a chisel or disk system in high-yielding corn residue, two tillage operations will generally leave about a 30 percent cover. In high-yielding soybean residue, however, no-till is the only system that will consistently leave 30 percent or greater cover.

This guide is intended to be a planning tool only. An ideal residue management program is presented beginning at harvest and proceeding through winter into the spring tillage and planting operations. Ranges are given with respect to how much residue cover remains after single operations of selected tillage machines. Remember, however, that these are general guidelines and actual percentages may vary. **For the most accurate estimation of crop residue levels, actual field measurements are recommended.**

For more information on crop residue management, contact your local USDA Soil Conservation Service or Purdue University Cooperative Extension Service office.

Designing A Crop Residue Management Program

Residue After Harvest

The ideal residue management program for leaving as much residue as possible on the surface after planting begins at harvest. Combines should be adjusted to spread the residue uniformly over as much of the harvested swath as possible. This is usually not a problem for combines that handle 4-row corn heads or 15' or narrower grain tables. However, larger corn heads and grain tables make it difficult to spread the residue evenly over the entire width of the harvested swath. Therefore, chopper attachments (if present) should be adjusted to spread full-width and the addition of a chaff spreader attached to the rear axle should be considered. Chaff spreaders are most effective for spreading wheat and soybean residue because a larger percentage of the harvested residue is handled by the combine's cleaning shoe.

Some brands of combines offer a spreader attachment in place of the chopper. While the spreader distributes the residue more uniformly than the chopper, more cover can actually be obtained with the chopper as the residue is chopped into smaller pieces before spreading. The spreader attachment, by design, spreads whole pieces of residue (soybean stems, wheat straw, whole corn stalks) and consequently does not cover as much of the surface. One drawback, however, is that small pieces of residue decompose quickly and are subject to movement by wind and water.

Residue cover following corn harvest is usually in the range of 85-90%. Low yields (e.g. <100 bu. corn and <30 bu. soybeans), however, may result in significantly lower levels of residue cover. Therefore, with residue management in mind, producers should be aware of residue cover levels after harvest. This will allow for planning of fall and spring tillage operations that will leave the desired levels of residue cover. **Refer to AY-269 for information on methods for estimating corn and soybean residue cover.**

Over-winter Residue Loss

Over the winter months crop residues are decomposed by microorganisms. Warm, moist conditions favor high rates of decomposition. While the months of January and February are quite cold, a thin blanket of snow can actually insulate the surface enough to allow decomposition to take place. For Indiana, over-wintering residue cover losses can approach 40% but typically fall in the range of 15-25%. Field operations conducted prior to winter months can further reduce remaining residue levels.

Residue that has been disturbed or buried by fall tillage or knife-type fertilizer applications is more susceptible to over-wintering and decomposition than undisturbed residue. Partially decomposed residue is easily broken and buried during spring tillage, further reducing its erosion control potential. Producers should take over-wintering losses into account when planning tillage operations.

Tillage and Residue Loss

Ultimately, no-till systems leave the highest levels of residue cover. However, less than 30% of Indiana's cropland is no-tilled. Therefore, a wide variety of primary and secondary tillage implements are used on the remaining cropland. Table 1 summarizes the effects of tillage operations on residue cover. Note that there are two categories for crop residue, non-fragile and fragile. Non-fragile residues mainly include corn and small grains while fragile residues include soybeans, canola, and fall-seeded cover crops.

The numbers in Table 1 are provided for planning purposes, but whenever possible, producers should estimate residue cover after each pass with an implement to ensure that the desired level of residue cover is maintained.

Residue Management Strategies

(1) The number and intensity of tillage operations should be limited. In general, the number of passes can be as important as the type of tillage operation selected. Residue cover is also sensitive to depth and speed of equipment operation and to row spacing. When selecting values from the ranges in Table 1, consider the following general rules of thumb:

- Shallower operating depths can leave up to 15% more residue on the surface.
- Slower operating speeds can leave as much as 20% more residue on the surface.
- Straighter disk blade alignments and straighter chisel plow points may leave as much as 20% more residue than curved or twisted counterparts.
- Under some conditions, field cultivators and other finishing tools with field cultivator gangs may return as much as 20% of the residue incorporated by previous operations.

(2) Ultimately, no-till systems will provide the highest level of residue cover. However, to prevent potential yield reductions, compaction, soil fertility, and other problems should be eliminated before beginning a no-till system .

(3) Nitrogen management techniques may need to be changed. With higher levels of crop residue present, surface-applied nitrogen may result in high volatilization rates. Therefore, nitrogen should be placed beneath the crop residue by either knifing or injection methods. Additionally, soils may be colder and wetter at planting and starter nitrogen rates of 15 to 30 pounds per acre should be considered when planting corn.

(4) Planters and drills may require modifications (e.g. row cleaners or coulters) to ensure proper seed and fertilizer placement. The type and positioning of coulters and row cleaning and fertilizer attachments will affect residue levels, however, and the least aggressive units available for a given operation should be used.

Table 1. Influence of Various Field Operations on Surface Residue Cover Remaining.

Tillage and Planting Implements	Percent of residue cover remaining after each operation [1] Non-Fragile	Fragile
Moldboard Plow	0-10	0-5
Machines Which Fracture Soil		
Paratill / Paraplow	80-90	75-80
V-ripper / 12-14" deep w/ 20" spacing	70-90	60-80
Chisel Plows		
Sweeps	70-85	50-60
Straight or spike points	60-80	40-60
Twisted points (3 or 4")	50-70	30-40
Combination Chisel Plows		
Coulter Chisel Plow with:		
Sweeps	60-80	40-50
Straight or spike points	50-70	30-40
Twisted points (3 or 4")	40-60	20-30
Disk Chisel Plow with:		
Sweeps	60-70	30-50
Straight or spike points	50-60	30-40
Twisted points (3 or 4")	30-50	20-30
Disk or Disk Harrows		
Tandem or Offset		
10" or greater blade spacing	25-50	10-25
9" or greater blade spacing	30-60	20-40
7-9" blade spacing	40-70	25-40
After harvest as primary tillage	70-80	40-50
Field Cultivators (including leveling devices)		
As primary tillage:		
Sweeps 12-20"	60-80	55-75
Sweeps or shovels 6-12"	35-75	50-70
Duckfoot points	35-60	30-55
As secondary tillage:		
Sweeps 12-20"	80-90	60-75
Sweeps or shovels 6-12"'	70-80	50-60
Duckfoot points	60-70	35-50
Finishing Tools		
Combination finishing tools with:		
Disks, shanks, and leveling attachment	50-70	30-50
Spring teeth and rolling baskets	70-90	50-70
Harrows:		
Springtooth (coil tine)	60-80	50-70
Spike tooth	70-90	60-80
Flex-tine tooth	75-90	70-85
Roller harrow (cultipacker)	60-80	50-70
Packer roller	90-95	90-95
Row Cultivators (30" and wider)		
Single sweep per row	75-90	55-70
Multiple sweeps per row	75-85	55-65
Finger wheel cultivator	65-75	50-60
Rolling disk cultivator	45-55	40-50
Ridge-till cultivator	20-40	5-25
Unclassified Machines		
Anhydrous applicator	75-85	45-70
Anhydrous applicator with closing discs	60-75	30-50
Subsurface (injected) manure applicator	60-80	40-60
Rotary hoe	85-90	80-90
Drills		
Conventional w/ double-disc openers	85-95	75-85
No-till with following coulters		
Ripple or no coulter	85-95	70-85
Bubble or fluted/wavy (<1" wide)	80-85	65-85
Fluted/wavy (1" wide or greater)	75-80	60-80
Planters		
Conventional:		
Staggered double-disc openers	90-95	85-95
Non-staggered double-disc openers	85-95	75-85
No-till:		
Smooth, ripple, or no coulter	85-90	75-90
Bubble or fluted/wavy (<1" wide)	75-90	70-80
Fluted/wavy (1" wide or greater)	65-85	55-80
Strip-till:		
2 or 3 fluted/wavy coulters	60-80	50-75
Row cleaning devices (5-10" bare strip)	60-80	50-60
Ridge-till (sweeps/double-discs/horizontal)	60-80	40-60
Climatic Effects		
Over-winter weathering:		
Following summer harvest (wheat/oats)	70-90	65-85
Following fall harvest	80-95	70-80

1 Crop residues are generally classified as either non-fragile or fragile. Following is an abbreviated listing of crops common to Indiana that are classified into these categories:

Non-Fragile: Corn, Wheat, Rye, Oats, Alfalfa or legume hay, Cotton, Tobacco
Fragile: Soybeans, Canola, Rapeseed, Fall-seeded cover crops, Vegetables

(5) Cover crops such as rye, wheat, or hairy vetch should be considered, as they provide additional cover, particularly in low-residue crops such as soybeans or corn silage. In addition, cover crops can suppress weed growth, decrease additional nitrogen requirements, and aid in field moisture management.

Sample Residue Calculations

Following are two examples of how to use the numbers in Table 1. Remember that these numbers are provided for planning purposes only and that the percent residue cover remaining after tillage can vary due to operating speed, operating depth, and soil moisture conditions.

Example #1

A farmer had 150 bushels per acre corn yield last year and wants to chisel plow with 4" twisted points in the fall. In the spring, he will disk twice (tandem, 7-9" blade spacing) and field cultivate once (6" shovels). The new crop will be planted with a conventional planter with staggered double-disc openers. The winter months were considered mild (maximum decomposition).

From Table 1, the following factors can be found for each operation. *Remember, there is no set rule for deciding which number to choose that lies within the listed ranges. The highest number in the range may represent "optimal" conditions (e.g. above average yields that result in high levels of residue cover) while the lowest number may represent "poor" conditions. A conservative general rule of thumb would be to pick the number that lies in the middle of the range.*

Field Operation	Percent Residue Cover Remaining (Non-fragile, from Table 1)
After Harvest (high yield)	95%
Chisel Plow with 4" twisted points	60%
Over-winter (mild winter)	80%
Disk once	55%
Disk once (the second time)	55%
Field Cultivate once	75%
Plant	95%

Simply multiplying the factors together will give the percent residue cover after planting. For this example, the percent residue cover is equal to:

95% x 60% x 80% x 55% x 55% x 75% x 95% = 9.8 or 10% residue cover

This system would not qualify as a conservation tillage system since less than 30% residue cover is maintained after planting.

If spikes (or 2" straight points) are used instead of 4" twisted points, the percent residue cover would equal:

95% x 70% x 80% x 55% x 55% x 75% x 95% = 11.4 or 11% residue cover

Switching points did not significantly increase residue cover after planting since three secondary operations were still used. Switching points can make a difference, however, in systems where secondary operations are limited to one or two passes with less aggressive tools (e.g. field cultivate only once).

Example #2

A farmer had 45 bushels per acre soybeans last year and wants to no-till corn in the spring. He will apply anhydrous ammonia (with closing discs) in the spring and will plant with a no-till planter that has 1" wavy coulters. The winter was cold with little snowfall (minimum decomposition). The factors from Table 1 are as follows:

Field Operation	Percent Residue Cover Remaining (Non-fragile, from Table 1)
After Harvest (high yield)	85%
Over-winter (cold winter)	80%
Apply anhydrous ammonia	50%
Plant	80%

Percent residue cover after planting (calculated in the same fashion as example #1) would then equal 27% which is near the definition of a conservation tillage system. *Producers should remember that soybean residue is very fragile and that even some no-till systems can leave low levels of residue cover after planting.*

This method provides only rough estimates since the variables involved prevent accurate determination of residue cover. However, Table 1 can be helpful in comparing tillage and planting operations. Producers should always consider estimating residue cover after each pass with an implement to ensure that crop residue management objectives are being met.

Adapted from the United States Department of Agriculture - Soil Conservation Service and the Equipment Manufacturers Institute, *Estimates of Residue Cover Remaining After Single Operation of Selected Tillage Machines*, February 1992.

11/94 (2M)

Cooperative Extension work in Agriculture and Home Economics, state of Indiana, Purdue University, and U.S. Department of Agriculture cooperating; H. A. Wadsworth, Director, West Lafayette, IN. Issued in furtherance of the acts of May 8 and June 30, 1914. The Purdue University Cooperative Extension Service is an equal opportunity/equal access institution.

IDENTIFICATION

ACCEPTABLE SPELLING AND CAPITALIZATION FOR CROP AND WEED IDENTIFICATION

Forage Crops

Red clover
White clover
Alfalfa
Hairy vetch
Tall fescue
Ryegrass
Kentucky bluegrass
Smooth bromegrass
Orchardgrass
Sudangrass

Restricted Noxious Weeds

Bitter wintercress
Buckhorn
Corncockle
Curly dock
Dodder
Field peppergrass
Giant foxtail
Horsenettle
Oxeye daisy
Pennycress
Wild Mustard

Field Crops

Soft red winter wheat
Hard red winter wheat
White wheat
Durum wheat
Rye
Barley
Common oats
Grain sorghum

Common Weeds

Velvetleaf
Common smartweed
Ivyleaf morning glory
Redroot pigweed
Jimsonweed
Giant ragweed
Downy bromegrass
Wild carrot
Common lambsquarter
Yellow nutsedge

Prohibited Noxious Weeds

Canada thistle
Field bindweed
Johnsongrass
Hoary cress
Perennial sowthistle
Quackgrass
Wild garlic

Poisonous Weeds

Cocklebur
Castorbean
Jimsonweed
Common milkweed
Poison ivy
Common pikeweed
Sweet clover
Spotted water hemlock
White snakeroot
Wild cherry
Black nightshade

CROP AND WEED IDENTIFICATION

Objectives:

1. To be able to identify by common name, plant and seed specimens of the prohibited noxious weeds, restricted noxious weeds and selected common weeds, forages, and field crops.

2. To be able to discuss characteristics of the plants, such as longevity, methods of propagation, control, and uses.

Terminology

1. alternate leaf arrangement - leaves located singly at a node, with adjacent leaves arising at opposite sides of the stem.

2. annual - plants which complete their life cycle in one year.

3. auricle - ear-like appendage at junction of sheath and blade on a grass leaf.

4. awn - an appendage that extends from the tip or back of lemmas and glumes

5. axil - angle formed between a leaf and a stem.

6. biennial - plants which complete their life cycle in two years. The plant forms a rosette the first year, and flowers during the second.

7. blade - the flat, expanded portion of a leaf.

8. bract - modified leaf, usually subtending a flower or another leaf.

9. bulb - underground bud attached to a central stem.

10. bulblet - a little bulb or bulb-like body, often produced above ground on stems or in inflorescences.

11. caryopsis - one-seeded fruit of grasses; a grain or kernel.

12. crown - the point at, or just below, the soil surface where the stem and root join and much branching occurs.

13. culm - the flowering stem of a grass or sedge.

14. dehiscent - refers to a fruit (pod) that splits open at maturity.

15. floret - the grass flower (or caryopsis) enclosed in a lemma asnd palea.

16. glabrous - smooth, without pubescence.

17. glume - scale-like bract, outermost structure of a grass spikelet.

18. grass - a monocotyledonous plant having round or flattened stems; leaves borne in two ranks; and flowers borne in spikelets.

19. head - dense inflorescence of sessile flowers on a much shortened axis.

20. habitat - place or site where a plant naturally grows.

21. indehiscent - refers to a fruit that does not split open at maturity.

22. inflorescence - flowering portion of a plant composed of one to many flowers.

23. legume - a dicotyledonous plant bearing a characteristic fruit that is single celled and usually dehiscent.

24. lemma - the outer and usually hard bract enclosing a grass flower.

25. ligule - membranous or hairy outgrowth at the top of the sheath in the grasses.

26. lobed - partially divided.

27. longevity - life cycle of a plant, summer annual, winter annual, biennial or perennial.

28. node - swollen portion of a stem from which leaves, branches, and flowers arise.

29. opposite leaf arrangement - leaves are borne in pairs on opposite sides of a node.

30. palea - the inner and usually papery bract enclosing a grass flower.

31. panicle - an inflorescence with compound branching; the flowers are twice removed from the central axis of the inflorescence.

32. pedicels - the stalk of a single flower in an inflorescence.

33. peduncle - the uppermost internode of the main stem which supports the inflorescence.

34. perennial - plants which live longer than two years.

35. petiole - the stalk of a leaf; absent in grasses.

36. propagation - method of plant reproduction, by seed, by root or by stem.

37. pubescence - hairiness.

38. raceme - inflorescence type in which the flowers are borne on pedicels on a rachis.

39. rachis - central axis of an inflorescene.

41. rhizome - an underground, lateral stem.

42. rosette - a basal cluster of leaves produced on a very short stem.

43. sedge - a monocotyledonous plant having triangular shaped, solid stem; leaves borne in three ranks and flowers borne in clusters.

44. sessile - without a stalk, as in some leaves and flowers.

45. sheath - part of a grass leaf which encloses the stem

46. spike - inflorescence in which flowers are borne sessile on the rachis.

47. spikelet - structure in grass inflorescence which is composed of glumes and florets.

48. stipule - leaf-like appendages that may be present at the base of a leaf.

49. stolon - a horizontal, above-gound stem.

50. suture - a seam or crease on a caryopsis.

51. tuber - a thickened, fleshy, underground modified stem; functions in food storage and propagation.

FIELD CROP AND WEED DESCRIPTION

Forage Legumes

1. Red clover - <u>Trifolium</u> <u>pratense</u>

   a. Plant - short-lived perennial acting as a biennial; taproot system; erect, branching habit from a crown, stems pubescent; leaves palmately compound, pubescent, with white variegation, stipules prominent, with distinct green to reddish veining, reddish flowers in head-type inflorescences at end of branches.

   b. Seed- lemon-yellow to violet, with prominent nose nearly half as long as seed, giving appearance of boxing glove; a sample of seed will have a purplish cast.

   c. Uses -

2. White clover - <u>Trifolium</u> <u>repens</u>

   a. Plant - shallow root system; prostrate, stoloniferous stems rooting at the nodes; palmately compound leaf supported by erect petiole, variegations on leaflets; white dense flower head on erect peduncle.

   b. Seed - small, canary yellow to brown, heart-shaped seeds.

   c. Longevity -

   d. Uses -

3. Alfalfa - <u>Medicago</u> <u>sativa</u>

   a. Plant - deep taproot and well-developed crown; much branched stems ascending to 3 feet tall; 3 leaflets in a pinnately compound leaf, pubescent, with serrations on outer 1/3 of leaflet; flowers yellow or purple, raceme inflorescence.

   b. Seed - borne one to several in a spiral pod; variable shaped, kidney to mitt shaped; yellow to green to light brown, turning dull reddish-brown with age.

   c. Longevity -

   d. Uses -

4. Hairy vetch - Vicia villosa

   a. Plant - annual; stems weak, pubescent, twining; leaves pubescent, pinnate with 12-20 leaflets, terminating in a tendril; flowers lavender, several to a raceme; dehiscent pod, with 3 to 8 seeds.

   b. Seed - spherical, dull black about 1/8 inch in diameter.

   c. uses -

Forages Grasses

5. Tall fescue - Festuca arundinacea

   a. Plant - cool season bunchgrass; spreads by seeds only; stem erect, 2 to 3 feet tall; leaf with glabrous sheath, blade flat, with rough upper surface and margin inflorescence an open panicle, with few to many florets per spikelet.

   b. Seed - 1/4 to 3/8 inch long, retains lemma and palea in threshing, rachilla slender and found with cupped knob.

   c. Longevity -

6. Ryegrass - Lolium spp.

   a. Plant - cool season bunchgrass; spreads by seeds only; stem erect, 2 to 3 feet tall, leaf glossy on upper surface, blade flat, and narrow, small auricles, glabrous sheath; short membranous ligule; inflorescence is a slender spike with spikelets attached edgeways to rachis, one glume absent except on terminal spikelet, few to many florets per spikelet.

   b. Seed - somewhat flattened, about 1/4 inch long, does not thresh free, lemma awned, rachilla wedge-shaped, lies close to lemma.

   c. Uses -

7. Kentucky bluegrass - Poa pratensis

   a. Plant - cool season grass; rhizomes; stems erect; leaf blade smooth, flat, or folded, ends in a boat-shaped tip; panicle type inflorescence 4-5 branches at lower node, spikelets crowded at ends of branches.

   b. Seed - less than 1/8 inch long, does not thresh free, persistent rachilla.

   c. Uses -

8. Smooth bromegrass - Bromus inermis

   a. Plant - perennial, cool season; spreads by seeds and rhizomes; leaves smooth, with closed sheaths; inflorescence an open, erect panicle with elongated spikelets, no awns.

   b. Seed - about 3/8 inch long, flattened, dark brown caryopsis shows through lemma and thin palea.

   c. Uses -

9. Orchardgrass - Dactylis glomerata

   a. Plant - perennial, cool-season, bunch grass; flattened leaf sheaths long membranous ligule; panicle-type inflorescence, with spikelets clumped at ends of few stiff branches; spikelets 2 to 5 flowered.

   b. Seed - does not thresh free; awn-pointed, rachilla present; lemma appears as if tip has been twisted quarter turn.

   c. Habitat -

10. Sudangrass - Sorghum sudanense

   a. Plant - annual; 6 to 8 feet tall; leafy stems, inflorescence medium to large, open panicle type.

   b. Seed - brown caryopsis, usually enclosed in golden-tan to purplish colored glumes; pedicels are not knobbed.

   c. Uses -

Field Crops

11. Common wheat - Triticum aestivum

    a. Plant - spike-type inflorescence; awned or awnless, often 3 or more fertile florets per spikelet.

    b. Seed - threshes free of lemma and palea.

        1) Soft red winter wheat - barrel-shaped caryopsis, widest near middle; round, wrinkled back; open suture; usually soft and chalky; large germ.

            Uses -

            Area grown -

        2) Hard red winter wheat - slender, vitreous kernel, widest near germ end; small germ, rounded suture.

            Uses -

            Area grown -

        3) White wheat - white and starchy, barrel-shaped, widest near middle; round wrinkled back; open suture.

            Uses -

            Area grown -

12. Durum wheat - Triticum durum

    a. Plant - inflorescence awned, spike; long coarse awns extending from tip of lemma.

    b. Seed - long, with pinched, pointed germ; caryopsis hard, translucent, amber colored; usually no brush; open suture with high ridge down back.

    c. Uses -

    d. Area grown -

13. Rye - Secale cereale

    a. Plant - spike-type inflorescence; long, slender, often drooping; short awns; kernels partly exposed from glumes.

    b. Seed - size varies from small and slender to large and plump; color from light brown to green to dark brown.

    c. Uses -

    d. Area grown -

14. Barley - Hordeum vulgare

    a. Plant - spike-type inflorescence; usually awned; 3 spikes per rachis node.

    b. Seed - lemma and palea attached, awn broken in threshing:

        1) six-row - all spikelets fertile, 2/3 of kernels have curved suture, 1/3 straight.

    c. Uses -

15. Common oats - Avena sativa

    a. Plant - open panicle-type inflorescence; spikelets with large, papery glumes

    b. Seed - plump; yellow, gray, or red, depending on variety; lemma and palea do not thresh free; awns, if present, attached to back of lemma.

    c. Uses -

    d. Area grown -

16. Grain sorghum - Sorghum bicolor

    a. Plant - inflorescence is a panicle-like raceme; semi-compact; dark glumes.

    b. Seed - rounded 3/32 inch in diameter, usually a rust-red or reddish brown in color.

    c. Uses -

Prohibited Noxious Weeds

17. Canada thistle - Cirsium arvense

    a. Plant - perennial; roots extend several feet deep and several feet horizontally; stems 2 to 5 feet tall, becoming increasingly hairy as they mature; leaves dark green on upperside, grayish underside, usually with crinkled edges and spiny, somewhat lobed margins; inflorescence 3/4 inch or less in diameter, composed of several lavender flowers and surrounded by bracts without spiny tips.

    b. Seed - light brown, slightly tapered, about 3/16 inch long, with a ridge around the blossom end, almost banana-shaped.

    c. Habitat -

    d. Propagation -

    e. Control -

18. Field bindweed - Convolvulus arvensis

    a. Plant - extensive root system; stems smooth, slender, 2 to 7 feet long, vine like; leaves ovate with spreading basal lobes, somewhat bell-shpaed; flowers white or pink, funnel-shaped, about 1 inch across, usually borne singly in the axils of leaves. Flower stalk has 2 bracts 1/2 to 2 inches below the flower.

    b. Seed - brownish-gray, rough, about 1/8 inch in diameter, with 1 rounded and 2 flattened sides.

    c. Longevity -

    d. Propagation -

    e. Habitat -

19. Johnsongrass -Sorghum halepense

a. Plant- fibrous root system; rhizomes stout, creeping, with purple spots; stems erect, 1 1/2 to over 6 feet tall; leaves alternate, simple, 6 to 20 inches long 1/2 to 1 1/2 inches wide; open panicle type inflorescence.

b. Seed - nearly 1/8 inch long, oval, yellow to reddish-brown, with fine lines on the surface; two knobbed rachilla are on each floret.

c. Longevity -

d. Propagation -

e. Habitat -

20. Hoary cress - Cardaria

a. Plant - roots deep, slender, extending horizontally and vertically as much as 10 feet; stems 1 to 1 1/2 feet tall, branching at top; leaves on lower stem spatulate, tapering to a slender base, upper leaves sessile. Leaf margins wavy with shallow indentation. Leaves covered with whitish pubescence. Flowers white, 4 petaled, small; borne in flat-topped clusters; seed pod 2-parted, heart shaped.

b. Seed - oval, rough about 1/16 inch long, reddish-brown.

c. Longevity -

d. Propagation -

e. Problem in Indiana?

21. Perennial sow thistle - Sonchus arvensis

a. Plant - roots extend several feet horizontally and vertically; produce shoots from buds on roots; stems smooth, 3 to 7 feet tall, erect with a whitish bloom on the surface. Milky juice exudes when stem is broken; leaves are 4 to 8 inches long, alternate, irregularly toothed, lobed, with spiny edges; inflorescence up to 1 1/2 inches across, composed of several deep yellow flowers.

b. Seed - brownish, slender, about 1/8 inch long, slightly flattened, 5 to 7 longitudinal ribs running the length of the seed, a tuft of hair on one end that is easily broken off.

c. Longevity -

d. Propagation -

e. Habitat -

22. Quackgrass - Agropyron repens

a. Plant - fibrous root system, very vigorous; stems 1/2 to 3 feet tall; with smooth culms and 3 to 6 nodes; leaves with auricles, short ligule, hairy lower sheaths, upper sheaths nearly glabrous; spike composed of several spikelets, spikelets with 3 to 7 short-awned florets.

b. Seed - does not thresh clean, florets usually thresh in small clusters rather than singly; rachilla oppressed to palea.

c. Longevity -

d. Habitat -

e. Uses -

23. Wild garlic - Allium vineale

a. Plant - roots small, at base of underground bulb, stems 1 to 3 feet tall, smooth, waxy; leaves slender, hollow, nearly round, attached to lower half of stem; aerial bulblets form in a cluster at top of stem are oval and smooth with shiny covering; flowers greenish-white, small, on short stems above aerial bulblets.

b. Seed - black, flat on one side, about 1/8 inch long; formed only occasionally.

c. Longevity -

d. Propagation -

e. Habitat -

Restricted Noxious Weeds

24. Bitter wintercress - Barbarea vulgaris

a. Plant - winter annual; spreads by seeds and taproot; numerous stems growing from a crown, upright, 1 to 2 feet tall, branching near top; basal leaves with a large terminal lobe; are 2 to 8 inches long and form a dense rosette; stem leaves become progressively shorter and less deeply lobed; flowers are bright lemmon yellow, 4-petaled, borne in spike-like racemes on the end of each branch; pods about 1 inch long, on the end of each branch; pods about 1 inch long, 3/32 inch in diameter, nearly square in cross-section.

b. Seed - dull, grayish-brown, about 3/32 inch in diameter, ripening in May and early June.

c. Habitat -

25. Buckhorn - Plantago lanceolota

a. Plant - perennial; spreads by seed; stems erect, leafless 4 to 12 inches long, terminating with flower spike; leaves at ground level in a basal rosette, hairy, 2 to 10 inches long, 1/4 to 1 inch wide, with 3 to 5 prominent veins running lengthwise, flowers numerous, petals inconspicuous.

b. Seed - brown, shiny, smooth, canoe-shpaed, with indentation in middle of one side, sticky when damp.

c. Habitat -

26. Corn cockle - Agrostemma githago

a. Plant - winter annual; spreads by seed; stems rough, hairy, erect, 2 to 3 feet tall, swollen at nodes, branching slightly; leaves slender and hairy; flowers large, with narrow, green sepals longer than purple colored petals; seed pod holding several seeds.

b. Seed - 1/8 inch in diameter, black, appears folded at one end, covered with several rows of needle-like projections.

c. Where found in Indiana -

27. Curly dock - Rumex crispus

a. Plant - reproduces by seed; large, yellow, somewhat branched taproot, first year plant forms dense rosette, later sends up erect stems, stems 3 or more feet tall branched at top, glabrous; leaves alternate, simple lanceolate, wavy along margins, venation prominent on underside; flowers in clusters on ends of branches, without petals, small greenish, becoming red at maturity.

b. Seed - shiny, brown, triangular, about 1/32 inch in diameter.

c. Longevity -

d. Habitat -

28. Dodder - Cuscuta spp.

a. Plant - parasitic; annual; spreads by seeds; stems thread-like, without chlorophyol, vinelike; leaves reduced to scales; small, white flowers in dense clusters; 2-4 seeds borne in indehiscent capsule.

b. Seed - white to yellow to brown, rough, one side rounded, opposite side with small depression, about 1/32 inch in diameter.

c. Habitat -

29. Field peppergrass - <u>Lepidium</u> <u>campestre</u>

a. Plant - winter annual (south) or annual (north); stems hoary-pubescent or rarely hairless, 6 to 24 inches tall, very leafy; leaves alternate on stem, pubescent, arrow-shaped, vases clasping the stem, flowers inconspicuous, white or greenish, with 4 petals, borne in rather dense racemes at top of plant; seed pods boat shaped, containing 2 seeds.

b. Seed - dark brown, rough coated, pointed at tip.

c. Propagation -

30. Giant foxtail - <u>Setaria</u> <u>faberii</u>

a. Plant - stems 3 to 7 feet tall, weak, lodge easily; leaves covered with short hairs on upper surface; dense panicle, 3 to 8 inches long, drooping near base, 3-6 bristles extend from base of each spikelet.

b. Seed - mostly greenish, about 1/16 inch long.

c. Longevity -

d. Propagation -

e. Habitat -

31. Horsenettle - <u>Solanum</u> <u>carolinense</u>

a. Plant - reproduces from seeds and rootstocks; stems simple or branched, hairy, spiny, 1 to 4 feet tall; leaves alternate, oblong, lobed, sharp spines on petiole, midrib, and veins, flowers white or bluish, about 1 inch across, 5-lobed, borne in clusters; berries yellow, juicy, smooth at first, wrinkled late in season, about 1/2 inch in diameter, containing many seeds.

b. Seed - 1/16 inch in diameter, round flattened, yellowish.

c. Longevity -

d. Habitat -

32. Oxeye daisy - Chrysanthemum leucanthemum

a. Plant - spreads by rhizomes and seeds; root system fibrous; stems smooth, 1 to 3 feet high; leaves althernate, simple lower leaves lobed, upper leaves notched; flower heads 1 to 2 inches in diameter, occurring singly at ends of stems; ray flowers with white petals. Disk flowers yellow.

b. Seed - elongated, one side straight, other side convex, 1/16 inch long, black with 8 to 10 white ridges.

c. Longevity -

d. Habitat

33. Pennycress - Thlaspi arvense

a. Plant - annual or winter annual; stems erect, branched at top, glabrous; leaves alternate, simple, 1 to 2 inches long, lower leaves petioled, upper leaves sessile; few branches bearing white flowers and seed pods lens-shaped, broadly winged, dehiscent, containing 2 to 8 seeds.

b. Seed - flattened, oval, dark reddish-brown to black, 10-14 concentric circles on either side.

c. Propagation -

34. Wild mustard - Brassica kaber

a. Plant - reproduces by seed; taproot; stem erect with spreading branches sparsely pubescent; leaves alternate, pubescent, somewhat oval with irregular lobing, usually deeply lobed at base; many inflorescences, composed of yellow 4 petaled flowers; seed pods long and slender, constricted between seeds.

b. Seed - reddish brown to black, smooth, round, resembles bird-shot.

c. Longevity -

d. Habitat -

Common Weeds

35. Velvetleaf - Abutilon theophrasti

a. Plant - annual; spreads by seed; taproot; stem smooth, often 6 to 8 feet tall, covered with short, velvety hairs; leaves large, heart-shaped, alternate, petioled, with a soft, velvety surface; flowers about 3/16 inch in diameter, 5-petaled, borne on short stalks attached to leaf axils on upper part of stem; seed pod cup-shaped, about 1 inch in diameter, with ring of prickles about upper edge.

b. Seed - grayish-brown, flattened, notched, 1/8 inch in diameter.

c. Habitat -

36. Common smartweed - Polygonum pensylvanicum

a. Plant - spreads by seed only; stems smooth, swollen at nodes, branching 1 to 4 feet tall; leaves smooth, pointed alternate 2 to 6 inches long, with stipule extending around stem and base of petiole; flowers bright pink or rose colored, 5 parted, arranged in a short spike.

b. Seed - shiny, black, smooth, flattened, nearly circular.

c. Longevity -

d. Habitat -

37. Ivyleaf morningglory - Ipomoea hederacea

a. Plant - annual; stems hairy, vinelike, spreading on the ground; leaves ivy-shaped, 3-lobed, alternate, hairy; flowers funnel-shpaed, white, purple, or pale blue, borne singly on long stalks, seed pods egg-shaped, containing 4 to 6 seeds.

b. Seed - about 1/2 inch in diameter, dark brown to black, with 1 round side and 2 flattened sides.

c. Propagation -

d. Habitat -

38. Redroot pigweed - _Amaranthus_ _retroflexus_

a. Plant - spreads by seeds; shallow, red taproot; stems erect, 6 feet or more tall, very rough, branching freely if not crowded; leaves dull green, with long petioles; flowers green, small in dense panicles at ends of stems and branches, each flower surrounded by 3 spiny bracts.

b. Seed - shiny, black, lens-shaped.

c. Longevity -

39. Wild carrot - _Daucus_ _carota_

a. Plant - biennial; first year produces rosette of finely divided leaves and fleshy taproot, second year blooms and dies; stem (second year) erect, 1 to 4 feet tall, hairy, branched at top; leaves alternate, finely divided, hairy, with distinct carrot-like odor; flowers with 5 white petals, small, borne in umbels at the ends of branches; pedicels on outside edge of umbel curve in sharply as they mature.

b. Seed - light brown, up to 1/8 inch long, one side flattened, the other side rounded and showing 4 long-bristled ridges, with smaller ones between.

c. Propagation -

d. Habitat -

40. Jimson Weed - _Datura_ _stramonium_

a. Plant - spreads by seeds; root thick, shallow, extensively branched; stems smooth, thick, erect, branching widely, 2 to 4 feet tall; leaves alternate, large smooth, with irregularly toothed edges and a distinctive rank-smelling odor; flowers large, funnel-shaped, white to pinkish, 2 to 5 inches long, borne singly on short stalks in axils of leaves; seed pod about 1 inch in diameter, egg shaped, covered with short, sharp spines.

b. Seed - dark brown to black, kidney-shaped, flattened, surface irregular and pitted.

c. Longevity -

d. Habitat -

41. Giant ragweed - Ambrosia trifida

a. Plant - spreads by seed; stems coarse, rough, reaching a height of 12 to 18 feet on fertile, moist soils; leaves opposite, large slightly hairy, 3-lobed; male flowers abundant in clusters on tips of branches and stems, female flowers sparse, without petals in axils of upper leaves.

b. Seed - 1/4 to 1/2 inch long, with a woody hull bearing blunt ridges that end in several short, thick spines at the tip.

c. Longevity -

d. Habitat -

42. Downy bromegrass - Bromus tectorum

a. Plant - winter annual or annual; stems erect on spreading, slender, 6 to 24 inches tall; leaves light green covered with long soft hairs, panicle rather dense, soft, very drooping, often purplish, flowering April to May.

b. Seed - long and narrow, bearing awns 1/2 to 3/4 inch long, retains lemma and palea.

c. Propagation -

d. Maturity date -

43. Common lambsquarters - Chenopodium album

a. Plant - reproduces by seeds; short taproot; stems 3 to 8 feet tall, grooved, often with red or light green striations, branches much if not crowded; leaves alternate, 1 to 3 inches long, smooth, usually grayish underside; flowers small, green without petals, borne at ends of branches and in leaf axils.

b. Seed - shiny black, disk-shaped, 1/16 inch in diameter, usually with a gray hull.

c. Longevity -

d. Habitat -

44. Yellow nutsedge - Cyperus esculentus

a. Plant - reproduces by seed and tubers on roots; fibrous root system, often bearing, small, nut-like tubers at tips; stems erect, triangular; leaves 3-ranked, narrow; flowers yellowish, small, arranged in narrow spikelets on umbel-like inflorescence.

b. Seed - yellowish-brown, 3-angled, about 1/16 inch long, with blunt ends.

c. Longevity -

d. Habitat -

SOME POISONOUS PLANTS COMMON TO INDIANA

I. Cocklebur - Xanthium spp.

A. Plant: annual; taproot woody, stout; stem erect, bushy, 2-4 feet tall; leaves alternate, simple, triangular in outline, toothed or lobed rough with long petioles; flowers small in clusters in upper leaf axils; at maturity oval burs are woody, hard, covered with hooked prickles and ending in two curved spines.

B. Seed:

C. Habitat: Cultivated fields, abandoned land, poor pastures, roadsides

D. Poisonous part:

E. Poison affects: usually pigs, also cattle, sheep, chickens, horses

F. Symptoms: intestinal upset, convulsions, impaired liver function.

II. Castorbean - Ricinus communis

A. Plant: perennial; taproot thick with strong lateral roots; stem erect, smooth, branched, partly hollow; leaves glossy green, alternate, palmate, lobed with notched edges, on large petioles; flowers purplish, lance-shaped raceme; fruits are rounded, smooth capsules with three projecting sides and covered with tough spines.

B. Seed: broad, oval, compressed, variegated color--resembles an engorged tick

C. Habitat:

D. Poisonous parts:

E. Poison affects: horses, children

F. Symptoms: intestinal upset, cardiovascular impairment

G. Commercial Uses:

III. Jimsonweed - Datura stramonium

A. Plant: annual; shallow, branched, thick roots; stems smooth, erect, branched widely in upperpart, 2-4 feet tall; leaves alternate, large coarse, smooth, oval with toothed edges and rank odor; flowers large, funnel-shaped, white to pink, on short stalks in branch axils; seed pod about 1 inch in diameter, egg-shaped, covered with short, sharp spines.

B. Seed: dark brown to black, kidney-shaped, flattened, surface irregular and pitted.

C. Habitat: cultivated crops in fertile soils and old feedlots

D. Poisonous part:

E. Poison affects:

F. Symptoms: great thirst, pupil enlargement, delirium, convulsions

IV. Common Milkweed - Asclepias syriaca

A. Plant: perennial; long, spreading root stocks; stems stout and erect 2-5 feet tall, covered with short downy hair, with milky juice leaves opposite, oblong, rounded, 4-8 inches long with prominent veins, upper surface smooth, lower surface covered with short white hairs; flowers sweet-smelling, pink to white, in large, many flowered ball-like clusters at stem tips and upper leaf axils; seed pod gray, hairy, covered with short spines.

B. Seed: brown, flat, oval, with tuft of white silky hairs at the tip

C. Habitat: cultivated fields, pastures, open woods, roadsides

D. Poisonous parts: leaves, stems

E. Poison affects:

F. Symptoms:

V. Poison Ivy - <u>Rhus</u> <u>radicans</u>

A. Plant: woody perennial; a low shrub or climbing vine; leaves consist of 3 shiny leaflets 2-4 inches long, pointed at the tip, leaflet edges smooth or irregularly toothed; flowers small, green, 5-petalled, borne in a 1-3 inch long head. This is a variable species!

B. Seed: borne in small, white, round hard berries

C. Habitat: open woods, fence rows, thickets, orchards, wasteland

D. Poisonous parts:

E. Poison affects:

F. Symptoms: skin blistering, redness, itching

VI. Common Pikeweed (pokeweed, pokeberry) <u>Phytolacca</u> <u>americana</u>

A. Plant: perennial; large taproot; stems stout, erect 3-9 feet high, smooth, branching above, often reddish, dying to the ground each winter; leaves alternate, large but smaller at the top of the plant, short to long petioles; flowers small, white, in long, narrow, unbranched racemes in ends of stems and upper branches; fruit a dark purple, many-seeded berry with red juice.

B. Seed: small, flattened, round, shiny black, 1/8 inch in diameter

C. Habitat:

D. Poisonous parts:

E. Poison affects: all species, usually pigs

F. Symptoms: burning sensation in the mouth, intestinal cramps, vomiting, diarrhea, visual disturbance, convulsions.

VII. Sweetclover - Melilotus spp.

A. Plant: biennial; tap-rooted; stems branched; leaves contain 3 leaflets, leaflet margins toothed all the way around; flowers white or yellow in long, loose racemes; pods loosely attached, containing 1-2 seeds

B. Seed: small, kedney-shaped, green to yellow

C. Habitat: roadsides, field margins, waste places

D. Poisonous parts: leaves, stems

E. Poison affects: cattle, horses, sheep, goats

F. Symptoms:

VIII. Spotted water hemlock - Cicuta maculata

A. Plant: perennial roots tuberous with distinct odor; stems smooth, 3-5 feet tall, branched only at the top, streaked with purple spots; leaves compound 8-12 inches long, alternate, smooth with toothed edges, often spotted, base of petioles clasping stem; flowers very small, 5 white petals in compound umbels.

B. Seed: flat on one side and rounded on the other, ridged length wise with light and dark lines, distinct odor

C. Habitat: swamps and low lands, in water or at water's edge

D. Poisonous parts: entire plant, especially the roots

E. Poison affects:

F. Symptoms:

IX. White snakeroot - Eupatorium rugosum

A. Plant: perennial; roots fibrous and branched; stem 1-3 feet tall, smooth, branched at the top; leaves opposite, elliptical, thin, smooth, with toothed edges and slender petioles; flower heads small, white disk flowers only.

B. Seed: black angular, 1/8" long, with a tuft of white hairs

C. Habitat:

D. Poisonous parts: dairy products of animals that have eaten white snakeroot are poisonous

E. Poison affects:

F. Symptoms: trembling, convulsions, terminal coma

X. Wild cherry - Prunus spp.

A. Plant: perennial, small shrub or tree; leaves alternate, simple; flowers elongated, flat-topped clusters or solitary in leaf axils

B. Seed: a single pit

C. Habitat: pasture, fence rows, wooded areas, waste areas

D. Poisonous parts:

E. Poisonous affects:

F. Symptoms: labored breathing, staggering dizziness, coma, death

XI. Black nightshade - Solanum nigrum

A. Plant: annual; stem erect or spreading, becoming widely branched, 1-2 feet tall; leaves ovate, 1-3 inches long, alternate, edges wavy; flowers white, 5-lobed, 1/4 inch across, in small clusters; berries green turning black at maturity, smooth, 3/8 inch in diameter, containing numerous seeds.

B. Seeds: flattened, 1/16 inch across, dull, pitted, yellow to dark brown

C. Habitat:

D. Poisonous parts: unripe berries

E. Poison affects:

F. Symptoms: loss of feeling, pupil dilation, coma, cramps, convulsions